Lecture Notes in Nanoscale Science and Technology

Volume 43

Lecture Notes in Nanoscale Science and Technology (LNNST) aims to report latest developments in nanoscale science and technology research and teaching – quickly, informally and at a high level. Through publication, LNNST commits to serve the open communication of scientific and technological advances in the creation and use of objects at the nanometer scale, crossing the boundaries of physics, materials science, biology, chemistry, and engineering. Certainly, while historically the mysteries in each of the sciences have been very different, they have all required a relentless step-by-step pursuit to uncover the answer to a challenging scientific question, but recently many of the answers have brought questions that lie at the boundaries between the life sciences and the physical sciences and between what is fundamental and what is application. This is no accident since recent research in the physical and life sciences have each independently cut a path to the edge of their disciplines. As both paths intersect one may ask if transport of material in a cell is biology or is it physics? This intersection of curiosity makes us realize that nanoscience and technology crosses many if not all disciplines. It is this market that the proposed series of lecture notes targets.

Shrikaant Kulkarni
Editor

Next-Generation Nanomaterials for Sustainable Engineering

Springer

Editor
Shrikaant Kulkarni
Faculty of Engineering and Technology
Sanjivani University
Kopargaon, Maharashtra, India

ISSN 2195-2159 ISSN 2195-2167 (electronic)
Lecture Notes in Nanoscale Science and Technology
ISBN 978-3-032-05718-1 ISBN 978-3-032-05719-8 (eBook)
https://doi.org/10.1007/978-3-032-05719-8

This Springer imprint is published by the registered company Springer Nature Switzerland AG
The registered company address is: Gewerbestrasse 11, 6330 Cham, Switzerland

Preface

The imperative for sustainable development has become a defining challenge of our time. As we grapple with the environmental and societal consequences of conventional engineering practices, the quest for innovative and eco-conscious solutions has intensified. At the forefront of this transformative endeavor lies the field of nanotechnology, offering a powerful toolkit for creating materials and systems that are not only high performing but also environmentally benign and resource-efficient.

This book, *Next-Generation Nanomaterials for Sustainable Engineering*, delves into the exciting realm of advanced nanomaterials and their pivotal role in shaping a more sustainable future. Building upon the foundational principles of nanotechnology, we explore the cutting edge of material design, focusing on nanomaterials engineered with sustainability as a core objective. This includes materials derived from renewable resources, those enabling energy efficiency and conservation, platforms for pollution remediation, and components designed for enhanced durability and recyclability.

The journey toward sustainable engineering necessitates a paradigm shift in how we conceive, synthesize, and utilize materials. Next-generation nanomaterials offer a unique opportunity to achieve this shift. Their high surface area to volume ratio, tunable properties, and ability to interact at the nanoscale open doors to unprecedented levels of control over material performance and environmental impact. From lightweight composites that reduce fuel consumption to highly efficient solar energy harvesters and smart sensors for environmental monitoring, the potential applications are vast and transformative.

This book aims to provide a comprehensive overview of the latest advancements in this rapidly evolving field. We will explore the synthesis, characterization, and application of a diverse range of next-generation nanomaterials, highlighting their specific contributions to sustainable engineering practices across various sectors. We will also address the critical aspects of lifecycle assessment, environmental impact, and the responsible development of these technologies to ensure their long-term sustainability.

Intended for researchers, engineers, policymakers, and anyone interested in building a more sustainable future, this book seeks to inspire innovation and foster collaboration at the intersection of nanotechnology and sustainability. By harnessing the power of the ultrasmall, we can engineer solutions to some of the world's most pressing challenges and pave the way for a greener, more resilient tomorrow.

Kopargaon, India Shrikaant Kulkarni

Contents

Abbreviations

1D	One dimensional
2D	Two dimensions
2D	Two dimensional
3D	Three dimensions
3D	Three dimensional
AFM	Atomic force microscopy
AI	Artificial intelligence
ALD	Atomic layer deposition
ANN	Artificial neural network
ASTM	American Society for Testing and Materials
AuNPs	Gold nanoparticles
BC	Bacterial cellulose
CAT	Catalase
CMOS	Complementary metal-oxide-semiconductor
CNCs	Cellulose nanocrystals
CNFs	Cellulose nanofibers
CNT	Carbon nanotubes
CNW	Cellulose nanowhiskers
CO2	Carbon dioxide
CQD	Carbon quantum dot
CT	Computed tomography
CVD	Chemical vapor deposition
DNA	Deoxyribonucleic acid
EE	Electrochemical exfoliation
EESDs	Electrochemical energy storage devices
EPR	Enhanced permeability and retention
EQE	External quantum efficiency
EU	European Union
FDA	Food and Drug Administration
FeNPs	Iron nanoparticles
FLG	Few-layer graphene

FSP	Flame spray pyrolysis
GAs	Graphene aerogels
GBMs	Graphene-based materials
GBNs	Graphene-based nanomaterials
GDP	Gross domestic product
GF	Gauge factor
G-FETs	Graphene field effect transistors
GNPs	Graphene nanoplatelets
GO	Graphene oxide
HCl	Hydrochloric acid
HTS	High-throughput screening
IoE	Internet of everything
IP	Intellectual property
ISO	International Organization for Standardization
KOH	Potassium hydroxide
LED	Light-emitting diode
LIG	Laser-induced graphene
LPE	Liquid phase exfoliation
LSPR	Localized surface plasmon resonance
LTP	Long-term plasticity
MFC	Microfibrillated cellulose
ML	Machine learning
MNPs	Magnetic nanoparticles
MOF	Metal-organic framework
MPa	Megapascal
MRI	Magnetic resonance imaging
MRSA	Methicillin-resistant *Staphylococcus aureus*
nAg	Silver nanoparticle
NaOH	Sodium hydroxide
NCC	Nanocrystalline cellulose
NFC	Nano-fibrillated cellulose
NIR	Near-infrared
NNI	National Nanotechnology Initiative
NP	Nano-particle
P. hypolateritia	*Poria hypolateritia*
PCE	Power conversion efficiency
PECVD	Plasma-enhanced CVD
PLQY	Photoluminescence quantum yield
PMMA	Polymethyl methacrylate
POX	Peroxidase
PPE	Personal protective equipment
PPO	Polyphenol oxidase
QD	Quantum dot
QLED	Quantum light-emitting diode
QM	Quantum mechanics

R&D	Research and development
REACH	Registration, Evaluation, Authorisation and Restriction of Chemicals
rGO	Reduced graphene oxide
ROS	Reactive oxygen species
RRAM	Resistive random-access memory
SERS	Surface-enhanced Raman spectroscopy
SiC	Silicon carbide
SME	Small- and medium-sized businesses
SOD	Superoxide dismutase
STM	Scanning tunneling microscopy
STP	Short-term plasticity
TCR	Temperature coefficient of resistance
TEMPO	2,2,6,6-Tetramethylpiperidin-1-oxyl
TFT	Thin-film transistor
TMC	Transition metal chalcogenides
TMDC	Transition metal dichalcogenides
TS	Technical specification
UPS	Uninterruptible power supplies
UV	Ultraviolet
VOCs	Volatile organic compounds
ZIF	Zeolitic imidazolate framework
Zn	Zinc

Part I
Setting the Stage for Sustainable Nanomaterials

Chapter 1
Scale Up Production of Nanomaterials

Kalai Rajan Poomalaisundar, Nithishkumar Sengottaiyan, Pooja Sivakumar, Jeevanya Sri Murthi, Adarsh Kesava Ramanujam, and Ruby Celsia Arul Selvaraj

1.1 Introduction

The molecular structure of energetic materials stores chemical energy, which can quickly release in response to mechanical, thermal, or chemical stimuli. An ideal energetic material would have minimal sensitivity in addition to excellent performance and thermal stability. The explosives, pyrotechnics, and rocket/gun propellant industries rely heavily on energetic materials. Production of an energetic formulation for real-world uses requires additional components beyond the energetic substance itself. Competitive economies today are putting increasing pressure on the European manufacturing industry. Europe faces competition from established economies, particularly in the high-tech sector, where countries like China are gaining ground. The manufacturing sector accounts for one-fourth of the EU's GDP and provides millions of jobs, even if the unemployment rate is steadily rising. The European Union's industrial sector, which includes world-renowned brands in cars and heavy machinery, continues to dominate on a global scale, even as other economies show signs of improvement (Saleh, 2022; Van der Heijden, 2018). The exponential growth in nanoscience materials synthesis and theoretical capabilities, allied with open access efforts, reduced the challenge in accessing target data necessary to implement machine learning approaches to reduce labor intense synthesis

K. R. Poomalaisundar · N. Sengottaiyan · P. Sivakumar · J. S. Murthi
A. Kesava Ramanujam
Department of Food Technology, K S Rangasamy College of Technology, Namakkal, Tamil Nadu, India

R. C. A. Selvaraj (✉)
Department of Food Technology, K S Rangasamy College of Technology, Namakkal, Tamil Nadu, India

Department of Biotechnology, Prathyusha Engineering College, Tiruvalllur, India

S. Kulkarni (ed.), *Next-Generation Nanomaterials for Sustainable Engineering*, Lecture Notes in Nanoscale Science and Technology 43,
https://doi.org/10.1007/978-3-032-05719-8_1

protocol optimization. However, the development of Naas concepts remains constrained by two significant challenges: lack of low-cost and flexible reactor design and deficiency in reliable and fast unsupervised optimization algorithms. Nanomaterials have become indispensable in modern technology due to their outstanding properties and extensive variety of applications. These materials provide exceptional opportunities for innovation and advancement in several fields such as electronics, health, energy, and environmental remediation. The efficient and economical expansion of nanomaterial manufacturing has been a significant obstacle to their broad acceptance. Nanomaterials, due to their distinctive characteristics and extensive range of possible uses, have become a fundamental component of contemporary technology. These materials have exceptional prospects for innovation and progress in several fields such as electronics, medicine, energy, and environmental cleanup. However, a significant barrier to the widespread use of nanomaterials is the efficient and cost-effective growth of their production.

A major challenge is the lack of efficient and scalable technologies for synthesizing nanomaterials with the appropriate characteristics and quality. When it comes to mass production, traditional synthesis methods typically require too many steps and too much time (Charitidis et al., 2014). Avoiding agglomeration and ensuring uniform distribution of nanomaterials during processing is a challenging task.

To guarantee the performance and dependability of nanomaterials, precise characterization and quality control are crucial. Nanomaterials provide distinct challenges in characterization due to their tiny size and unusual characteristics. Keeping product quality and consistency in mind, it is essential to develop sensitive and effective characterization methodologies.

Another big issue is whether or not producing nanomaterials is economically viable. Commercialization of nanomaterials may be hindered by the hefty expenses linked to synthesis, characterization, and processing. Sustainable growth also necessitates giving serious thought to the environmental effects of producing and disposing of nanomaterials.

It will need a collaborative effort by scientists, engineers, materials scientists, and politicians to overcome these obstacles. Innovative synthesis methods, better characterization techniques, and optimized production processes can only be achieved via collaborative efforts (Gruar et al., 2013). Furthermore, the advancement of nanomaterial production cannot be achieved without expenditures in research and development.

Finally, there is a great potential and a major problem with increasing the manufacturing of nanomaterials. Nanomaterials have the potential to transform many different businesses while also making a positive impact on the environment if we can figure out how to make them economically viable and overcome the challenges of synthesis and characterization. Researchers, businesspeople, and politicians will all benefit from the insights provided by this chapter, which delves more into the unique problems and possible solutions associated with increasing the production of nanomaterials.

1.2 Challenges, Strategies, and Innovations for Industrial-Scale Nanomaterial Manufacturing

There are novel approaches and developing technology that try to overcome the specific obstacles that industrial-scale nanomaterial manufacturing entails. Figure 1.1 shows industrial-scale nanomaterial manufacturing (Li, 2010; Sebastian et al., 2014). Problems, Solutions, and New Developments in Industrial-Scale Nanomaterial Production are Discussed Here:

1.2.1 Challenges

- Maintaining a constant size, shape, and surface chemistry when producing nanomaterials on a large scale is a formidable problem in terms of material consistency and quality control. Minute changes can affect the functionality of nanomaterials in fields as diverse as electronics, health, and energy storage.
- Access to Affordable Energy, Materials, and Equipment: Nanoscale manufacturing processes frequently demand costly energy, materials, and equipment. Finding ways to mass-produce high-quality nanomaterials without significantly raising manufacturing costs is the real difficulty.
- Nanoparticles pose risks to both humans and the environment due to their microscopic size. Industrial operations must prioritize managing exposure hazards during production, handling, and disposal.
- Laboratory Process Scalability: It can be challenging to scale up technologies that need a high level of accuracy, such as chemical vapor deposition (CVD), atomic layer deposition (ALD), or other lab-scale procedures (González et al., 2021; Holkar et al., 2018). A huge challenge is scaling up from small-scale laboratory operations to large-scale industrial processes.

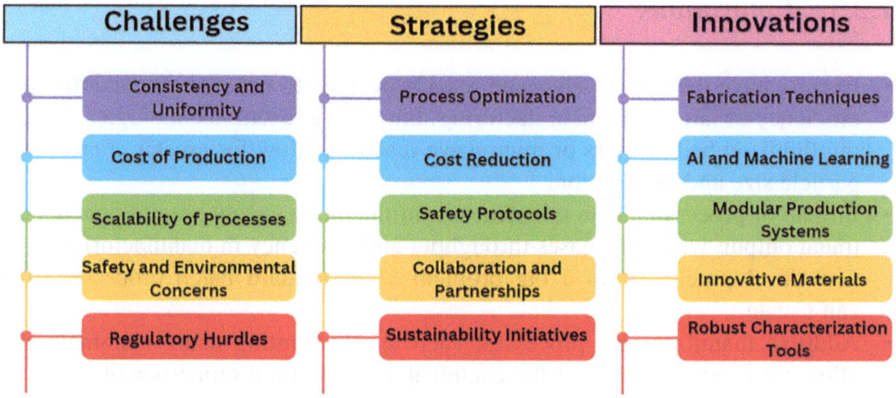

Fig. 1.1 The Industrial-scale nanomaterial manufacturing operation

- Existing System Integration: Nanomaterial integration into existing supply chains or production infrastructures necessitates substantial changes, which in turn increases the complexity and expense of scale-up initiatives.

1.2.2 *Strategies*

- Optimizing the process allows for increased yield without sacrificing quality by honing production techniques such as continuous flow processes and high-throughput synthesis. Optimizing characteristics such as response time, temperature, and pressure can achieve scalability without compromising consistency.
- Machine learning (ML) and artificial intelligence (AI) are used in production monitoring and control to help with quality control by improving reaction conditions in real time, predicting when equipment maintenance is needed, and lowering the chance of human error.
- Reducing costs and addressing environmental issues are two goals of sustainable manufacturing. One goal is to develop greener, less resource-intensive processes. Discovering solvent-free or water-based synthesis techniques is one example. To do this, the use of energy-efficient technology and reusable materials is critical (Chang et al., 2008; Saldanha et al., 2017).
- Partnerships and Collaborations: When academic institutions, businesses, and governments work together, everyone benefits from the pooled knowledge, resources, and technology. Public-private partnerships may scale up faster by dividing up the costs and rewards.
- By designing modular systems that can expand or contract in response to demand, manufacturers can increase output in small increments. This method requires less initial outlay and more adaptability.

1.2.3 *Innovations*

- Experts have developed advanced reactor designs to enhance the viability and efficiency of large-scale manufacturing. Innovative reactor technologies, including fluidized bed reactors or microwave reactors, allow for greater control over particle size and distribution.
- Instead of processing data in batches, continuous flow technologies enable continual output, which increases throughput and efficiency in manufacturing. This new development solves a big problem with standard batch processes: they can't scale.
- Additive manufacturing processes, such as 3D printing using nanomaterials, allow for exact control over the chemical and structural properties of materials

(Patwardhan et al., 2018). This method has the potential to revolutionize the production of bespoke nanomaterials on a large scale.

- Nanomaterial hybrid systems, which combine nanomaterials with other materials (such as polymers or metals) during the manufacturing process, enable the production of multifunctional composites. It is generally easier to incorporate these hybrid systems into existing industrial applications, and they typically have improved features.
- AI-Enabled Material Discovery and Process Design: ML models can foretell the most efficient ways to manufacture nanomaterials and pinpoint the ideal circumstances for increasing production volume. Researchers are also finding novel materials that are easier to mass-produce with the assistance of AI-driven simulations.

Scaling up nanomaterial production from the lab to industrial levels requires overcoming substantial challenges related to cost, safety, and process optimization. However, through innovative reactor designs, process automation, and partnerships between industry and academia, it is possible to achieve efficient and sustainable nanomaterial manufacturing (GadelHak et al., 2024; Naz et al., 2020). Advances in technology and a strategic approach to scaling up are essential for bringing the vast potential of nanomaterials into widespread commercial use.

1.3 Bridging Lab to Industry: Techniques for Efficient Nanomaterial Scale-Up

It is not an easy task to scale nanomaterial production from lab to industrial levels while keeping material quality, economic feasibility, and environmental sustainability in mind. To realize the full potential of nanomaterials for use in electronics, energy storage, and biology, this change is essential (Makgwane & Ray, 2014). Figure 1.2 shows the bridging lab to industry operation. In this article, we will go over some of the most important methods that are letting nanomaterial production scale up efficiently:

1.3.1 Continuous Flow Synthesis

Continuous flow synthesis, as opposed to more conventional batch techniques, may greatly scale up the manufacturing of nanomaterials (de Zárate et al., 2020). This technology improves manufacturing efficiency and consistency by allowing for continuous raw material input and output of finished products.

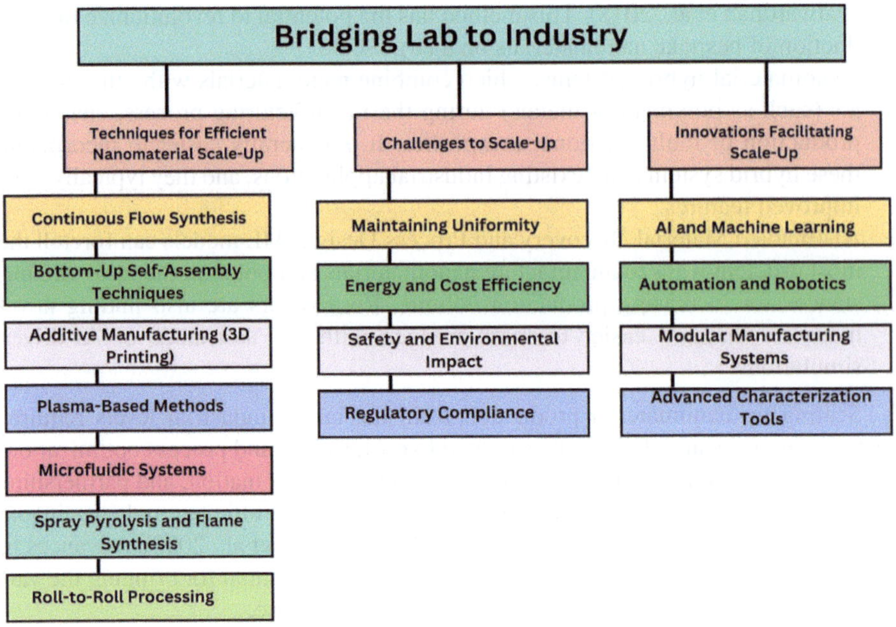

Fig. 1.2 The bridging lab to industry operation

1.3.1.1 Advantages

We have reduced wait times between batches, leading to higher throughput. We can regulate reaction parameters (like temperature and pressure) more precisely to achieve more uniform particle size and shape. Quickly adjust to high-volume industrial manufacturing (Aitken et al., 2006; McCourt et al., 2022). One example is the mass manufacture of carbon nanotubes (CNTs), which may have their growth and alignment carefully regulated over extended production runs using continuous flow reactors.

1.3.2 Microreactor Technology

Microreactors provide a further scalable option for nanomaterial production by allowing for precise regulation of reaction conditions (Kioumourtzoglou et al., 2024). Rapid mass and heat transmission is essential for creating homogeneous nanomaterials, and these miniature reactors' high surface area-to-volume ratios make this possible.

1.3.2.1 Advantages

We ensure consistent and easy-to-replicate product quality. Miniaturization of reactive materials improves safety. The ability to perform reactions in environments that present challenges at larger scales, like those with high pressure and temperature, is crucial. Microreactor technology has proven effective in producing very pure metal nanoparticles for use in electronics and catalysis (Bhanvase et al., 2021; Tulinski & Jurczyk, 2017).

1.3.3 Spray Pyrolysis

Nanomaterials including oxides, metals, and ceramics may be efficiently mass-produced via spray pyrolysis. Spraying a precursor solution into a hot environment triggers chemical reactions and quick evaporation, leading to the production of nanomaterials.

1.3.3.1 Advantages

It is highly scalable, making it possible to produce a large number of nanomaterials. By adjusting the precursor concentration and spray conditions, one may achieve excellent control over particle size and content (Mabrouk et al., 2021). It can be applied to a wide variety of materials. Spray pyrolysis is a common method for creating nanostructured coatings for sensors, batteries, and solar cells.

1.3.4 Sol-Gel Processing

One common method for making ceramic nanoparticles and metal oxides is sol-gel processing. Sol-gel processing transforms a liquid precursor into a gel-like network, subsequently drying and subjecting it to calcination to solidify the material (Abd-Elsalam et al., 2022; Ma et al., 2015). The production of controlled-porosity high-purity nanomaterials is made possible by this scalable technique.

1.3.4.1 Advantages

Its versatility allows for the creation of nanomaterials in various forms such as nanoparticles, thin films, and fibers. Strict control of particle size, porosity, and surface area is possible. Compared to other procedures that use high temperatures, this one uses less energy (Gacem & Abd-Elsalam, 2022).Manufacturing silica nanoparticles on a large scale for use in optics, biomedicine, and catalysis is one example of how sol-gel processes are used.

1.3.5 Plasma Synthesis

Chemical vapor deposition (CVD) and plasma-enhanced CVD (PECVD) are two examples of plasma-based techniques often used in industrial nanomaterial fabrication (Raliya et al., 2019; Ramesh & Ramesh, 2009). With these methods, one may precisely regulate the film's thickness, shape, and composition by depositing nanomaterials on surfaces while subjected to plasma.

1.3.5.1 Advantages

The facility has the capacity to manufacture uniformly characterized nanoparticles of excellent quality. Large-area manufacturing requires scalability to produce thin sheets for electronics and coatings. We are making the most of resources while minimizing waste. Graphene, silicon nanowires, and thin films used in sensors and transistors are all examples of materials made using plasma-based processes.

1.3.6 Advanced Automation and Robotics

To produce nanomaterials on an industrial scale with the level of accuracy and consistency needed, automation and robots are indispensable. Robotic systems, cutting-edge sensor technology, and AI-driven process management enable automated handling, synthesis, and quality control.

1.3.6.1 Advantages

Less human intervention reduces the potential for human error and contamination. We can achieve more effective production with lower operating expenses. Always keeping an eye on manufacturing conditions and finding ways to improve them. Robotic systems, for example, automate the etching and deposition procedures required to create nanoscale details on integrated circuits in the semiconductor manufacturing process (Khan, 2020).

1.3.7 *Large-Scale Electrospinning*

From biomedical scaffolds to filtration, electrospinning is a go-to method for creating nanofibers with a high surface area and porosity. Producers may meet the increasing demand for nanofibers by increasing the production volume of these innovative materials (Malekzadeh & Swihart, 2021).

1.3.7.1 Advantages

The product offers excellent scalability and high throughput for mass manufacturing. It has the ability to control the surface's functionality, orientation, and fiber diameter. It is useful for many different types of composites, ceramics, and polymers (Lines, 2008).Nanofiber membranes for water and air purification systems are one industrial use of electrospinning.

1.3.8 *High-Throughput Screening and Machine Learning*

Efficiently scaling up nanomaterial manufacturing necessitates rapid process optimization. Machine learning algorithms and high-throughput screening methods can expedite the process of identifying optimal conditions for mass manufacturing (Pitkethly, 2004).

1.3.8.1 Advantages

We may identify optimal yield, quality, and cost-effective synthesis conditions more quickly. We have the ability to predict outcomes for untested scenarios through the analysis of extensive experimental datasets. Process scalability and repeatability were enhanced (Behera & Chatterjee, 2022; Pitkethly, 2004). Optimizing metal-organic framework (MOF) synthesis for gas storage and catalysis using machine learning models is one use case.

To guarantee uniformity, efficiency, and sustainability, new methods are needed for the industrial scale-up of nanomaterial manufacturing. Possible solutions to these problems are emerging in areas such as continuous flow synthesis, plasma techniques, microreactor technologies, and others (von Bomhard et al., 2019). Industries may make nanomaterials more accessible for commercial applications by combining automation, improved materials discovery tools, and modular production systems. This will bridge the gap between small-scale research and large-scale manufacture (Cristaldi et al., 2018).

1.4 Overcoming Barriers in Scaling Nanomaterial Production for Commercial Applications

Scaling up nanomaterial production for commercial applications requires addressing a number of significant challenges. These include technical issues, cost inefficiencies, safety concerns, and regulatory complexities. Fortunately, strategic approaches and technological advancements can overcome these challenges (Liu et al., 2010; Yang et al., 2021). Figure 1.3 shows the commercial application. In this discussion, we will focus on the key barriers and provide effective strategies for moving from lab-scale research to full-scale industrial production.

1.4.1 Technical Challenges in Process Scalability

Recreating industrial laboratory procedures without compromising material quality is one of the main obstacles to increasing the production of nanomaterials. Unfortunately, laboratory approaches like mechanical exfoliation or wet chemical synthesis often fail to meet commercial goals due to their slowness or resource requirements (Kelesidis & Pratsinis, 2021).

1.4.1.1 Possible Solutions

Improving manufacturing processes: People are increasingly using methods such as large-scale electrospinning, scalable CVD, and continuous flow synthesis (Sebastian, 2022). Using these technologies, one can achieve continuous, high-throughput

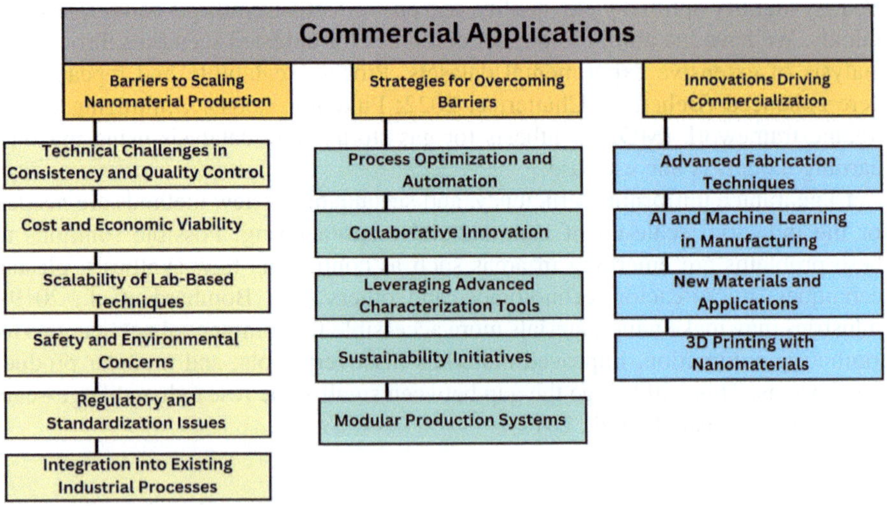

Fig. 1.3 The commercial application

manufacturing while meticulously controlling the size, shape, and composition of nanomaterials.

1.4.2 Process Intensification

Manufacturers can build sophisticated reactor designs and optimize reaction parameters (temperature, pressure, and precursor concentration) to maximize reaction efficiency and reduce material waste (Souto et al., 2020). This enables the production of nanomaterials at a larger scale.

1.4.3 Cost Inefficiencies and Economic Feasibility

Another major obstacle is the high production cost. Producing nanoparticles at a price competitive with conventional materials is challenging due to the high cost of raw ingredients, high-energy consumption, and the requirement for specialized equipment.

One solution is to take advantage of economies of scale, which allow for the spreading of fixed expenditures like equipment and facility costs over a larger amount of output as production volumes rise. As a result, the production costs of nanomaterials can decrease, making them more financially feasible.

1.4.4 Alternative Raw Materials

Scientists are working on less expensive substitutes for the precursors used to make nanomaterials. One way to ensure sustainable manufacturing processes while reducing input costs is by employing earth-abundant resources or biomass waste.

1.4.5 Energy-Efficient Procedures

One way to save energy expenses is to use energy-efficient manufacturing procedures is one way to save money. These include renewable energy sources or low-temperature synthesis. Cutting down on energy use is another benefit of novel approaches such as plasma-based synthesis, which eliminates the requirement for settings with high temperatures.

1.4.6 Consistency in Material Quality and Process Control

Commercial applications require consistent performance; hence, it is crucial to achieve uniformity in nanomaterial attributes such as particle size, surface area, and shape. Nevertheless, it is rather challenging to preserve these qualities across huge manufacturing quantities.

Automated systems, utilizing machine learning algorithms, can monitor production conditions and implement real-time modifications to ensure consistent quality. In order to keep the result consistent, these AI-powered systems can monitor changes in pressure, temperature, or material composition and adjust the process settings accordingly (Lloyd et al., 2011; Thirugnanasambandan & Gopinath, 2023).

Establishing standardized nanomaterial synthesis and processing procedures helps ensure high-quality materials at different facilities. Certification and regulatory systems also aid in establishing uniformity for commercial marketplaces.

1.4.7 Safety and Environmental Concerns

Because of their extremely microscopic size, nanomaterials present novel dangers to both humans and ecosystems. There are valid worries about the safety of producing nanoparticles on a wide scale due to their toxicity and the difficulties associated with their management and disposal.

1.4.7.1 Solutions

Nanomaterial Risk Assessment: It is essential to do thorough risk assessments for the creation, application, and disposal of nanoparticles. We must investigate the effects of nanomaterial exposure on humans, animals, and the natural world. When producers are aware of the dangers their employees face, they may take precautions and provide PPE to keep them safe.

The development of closed-loop methods for synthesizing nanomaterials aids in the reduction of nanoparticle emissions into the atmosphere. By lowering the possibility of contamination and increasing sustainability, these systems collect, recycle, and reuse waste resources.

Nanomaterials that are safe and compatible with living organisms are the subject of active research. To lessen the impact on the environment, we can find safer ways to make nanomaterials on a wide scale, including biodegradable ones or by employing green synthesis techniques, such as plant-based ones.

1.4.8 Integration Into Existing Industrial Processes

Incorporating nanomaterials into goods may be a challenging and time-consuming task for firms since they typically need new production techniques or adjustments to current ones.

As a solution, one option is to combine conventional materials with nanomaterials to form hybrid systems. Another is to use composites. To cut down on the requirement for additional machinery or procedures, nanomaterial composites can be engineered to work with what's already in the production pipeline.

The use of modular systems in manufacturing allows for the gradual addition of nanomaterial production capabilities, as opposed to the complete revamp of production lines (Baron et al., 2007; Jose et al., 2021). Companies may increase their production of nanomaterials gradually without interrupting current activities, and the initial investment expenses are reduced using this method.

Businesses in sectors that potentially use nanomaterials (such as electronics, medicines, and energy) can work together with nanotechnology firms to close the gap. These collaborations can hasten the incorporation of nanomaterials into commonplace industrial uses by exchanging knowledge, assets, and infrastructure.

1.4.9 Regulatory and Market Barriers

Depending on their intended use, nanomaterials are subject to different sets of regulations (e.g., those pertaining to medicine, the environment, and consumer products). Commercialization faces obstacles due to the complexity of complying with these rules and a lack of knowledge about the long-term consequences of exposure to nanomaterials.

1.4.9.1 Problem Solving

Collaborating with Governmental Agencies: To make sure that new nanomaterials are safe and up to code, producers should contact regulatory bodies early and often to learn about the landscape of regulations. Businesses are better prepared to enter new markets if they proactively comply with regulations.

The development of industry-wide standards for nanomaterial qualities, production techniques, and safety regulations can facilitate commercialization by creating common market expectations. Standardized nanotechnology goods will instill greater trust in customers.

Overcoming market hurdles necessitates raising public awareness and educating the public about nanomaterials' pros and cons. Efforts to educate the public about the benefits of nanotechnology in product performance enhancement (such as improved medicine delivery or longer-lasting batteries) have the potential to increase demand and market acceptability (Arole & Munde, 2014).

Scaling up nanomaterial production for commercial applications involves addressing a range of barriers related to technical feasibility, cost, safety, and regulatory compliance. Through innovations in production methods, cost-reduction strategies, enhanced safety protocols, and collaboration with regulatory bodies, companies can overcome these challenges. With a systematic and strategic approach, the commercial potential of nanomaterials can be fully realized across industries such as electronics, energy, healthcare, and environmental protection.

1.5 Advancing Nanomaterials from Research to Mass Production

Many environmental, economic, legal, and technological considerations are involved in the long and winding road from the lab to the factory floor when it comes to nanoparticles. Figure 1.4 shows the advancing nanomaterials from research to mass production operation. A well-planned strategy for scaling up that keeps material performance constant while dealing with commercial and industrial realities is necessary for the success of this undertaking (Gupta et al., 2023). The transition of nanomaterials from the laboratory to industrial-scale manufacturing is outlined in this handbook.

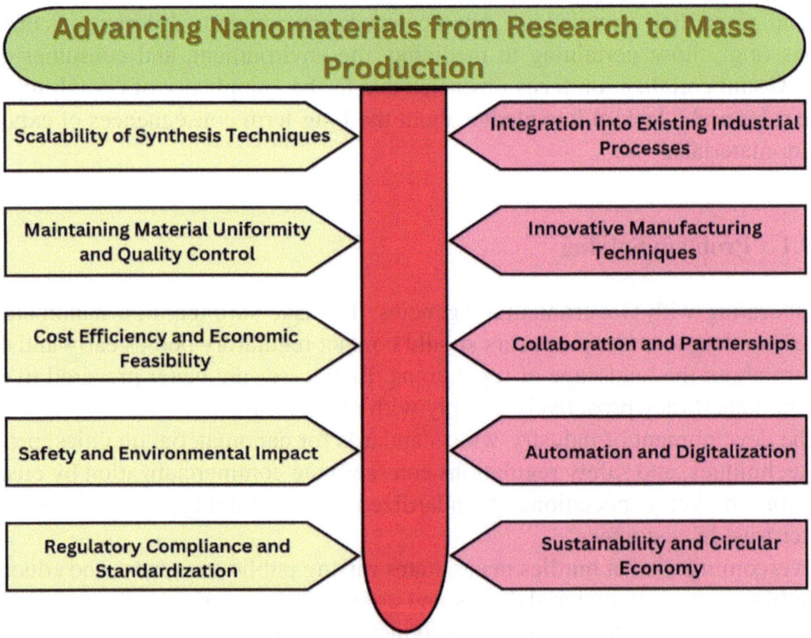

Fig. 1.4 The advancing nanomaterials from research to mass production operation

1.5.1 Scalability of Synthesis Methods

Replicating laboratory synthesis techniques on an industrial scale without compromising material quality or performance is a key concern for increasing nanomaterial manufacturing.

Lab-scale approaches often criticize batch-based methods for their inefficiency at industrial scales, while continuous production methods receive more favorable reviews. You may improve throughput and guarantee constant product quality by switching to automated synthesis or continuous flow technologies.

Custom reactor designs are frequently necessary for scaling up to accommodate big quantities. When producing on a large scale, technologies like continuous flow reactors, microwave reactors, and fluidized bed reactors allow for more precise control over particle size, distribution, and morphology.

State-of-the-art techniques like computational modeling and machine learning can determine the ideal synthesis parameters, such as temperature, pressure, and precursor concentration, to optimize the process for scalable production (Sasidharan et al., 2019).

1.5.2 Material Consistency and Quality Control

For nanomaterials to work reliably in many industrial contexts, it is essential that their attributes (such as size, surface area, and chemical composition) remain constant.

Production lines can integrate in-line monitoring equipment, such as spectroscopy and laser diffraction, to ensure product homogeneity and facilitate fast modifications to synthesis conditions, providing real-time data.

Establishing industry-wide standards for the properties of nanomaterials and their manufacturing processes is crucial to ensure the interchangeability of materials from different manufacturing batches or suppliers (Stieberova et al., 2019).

Procedures for Ensuring Product Quality: To track the physical, chemical, and functional characteristics of nanomaterials, rigorous testing and characterization should occur at every stage of manufacturing.

1.5.3 Cost-Effective Production

When trying to scale up manufacturing of nanomaterials, cost is a major factor, especially when using costly materials or energy-intensive procedures. Achieving economic viability requires a significant reduction in production costs.

Sourcing Raw Materials: Finding sustainable and cost-effective precursors to make nanomaterials can cut manufacturing costs in half. Renewable or bio-based materials can be a fantastic substitute for conventional, costly chemical precursors in some situations.

Energy Efficiency: One obstacle to cost-effective production is the use of energy-intensive techniques like high-temperature synthesis is one obstacle to cost-effective production. Cutting down on energy usage and manufacturing costs is possible with innovations such as solvent-free production processes, microwave heating, or low-energy plasma synthesis.

By spreading fixed expenses over larger quantities, economies of scale allow for an increase in manufacturing volumes and a reduction in per-unit costs. For efficient handling of high quantities, this necessitates investments in infrastructure and process improvement (Zhu et al., 2021).

1.5.4 Environmental and Safety Considerations

Because of their tiny size and distinct characteristics, nanomaterials present hazards to both the environment and human health. To achieve long-term success, it is critical to ensure safe production processes while reducing the environmental impact of producing nanomaterials. Reduce the environmental impact of nanomaterial manufacturing by developing eco-friendly synthesis methods. These approaches might include the use of nontoxic solvents or waste-free procedures. The use of green chemistry techniques in industrial production is on the rise.

It is crucial to reduce and recycle the amount of unreacted precursors, solvents, and by-products produced during the manufacturing of nanomaterials. Reducing production costs and waste is possible with closed-loop manufacturing systems that recycle and reuse resources.

Protecting Employees from Harmful Nanoparticles: In order to ensure workers' safety, facilities that produce nanomaterials must establish and adhere to strict safety measures. Ensuring safety requires ventilation systems, PPE, and technologies to detect nanoparticles in the air.

1.5.5 Regulatory Compliance

There are several regulatory frameworks that nanomaterials, especially those used in healthcare, electronics, and consumer products, must adhere to when they migrate from laboratory to industrial-scale manufacturing. To ensure consumer safety and market clearance, it is critical to meet regulatory standards.

Engaging with regulatory authorities, Businesses can benefit from knowing what safety, environmental, and performance requirements are necessary for market clearance by interacting with regulatory authorities early on in the scale-up process. Regulatory factors might differ depending on the area and the specific use case, adding complexity to the procedure.

Before granting certification, nanomaterials often undergo extensive toxicological studies and environmental impact assessments. Achieving commercial success requires certification and adherence to international standards, such as REACH and ISO.

Building customer confidence necessitates accurate product labeling and transparent information about the nanoparticles used in goods. Transparency on safety measures and environmental effects can also enhance market acceptability.

1.5.6 Integration into Existing Manufacturing Systems

Incorporating nanomaterials into preexisting industrial systems can be difficult because they often require specialized equipment or modified manufacturing procedures. Nevertheless, the commercialization of nanomaterials-enabled goods depends on their effective integration. Composites and Hybrid Materials: By fusing nanomaterials with more conventional materials, we may build hybrid systems that are more amenable to integration into current production procedures. Standard machinery can easily create nanocomposites that enhance the properties of metals, polymers, or ceramics (Dąbrowska et al., 2018).

With the use of modular production lines, companies can increase their production of nanomaterials on a slow but steady basis, eliminating the need to completely revamp their industrial infrastructure. This method enhances flexibility and reduces initial capital expenditures.

Innovative Collaboration: Working together with companies in related industries, including the automotive, electronics, or textile industries, can make it easier to incorporate nanomaterials into current manufacturing processes. These collaborations can bridge the gap between innovative ideas and mass manufacturing.

1.5.7 Market Demand and Commercial Viability

For commercial success, it is vital to ensure that production of nanomaterials is in line with market demand. Understanding customer preferences and market trends is crucial for determining the optimal applications of nanomaterials.

Finding Useful Uses: Pay attention to applications where nanomaterials have a significant impact, such as better energy storage, better medication delivery, or better coatings. Recognizing and pursuing these valuable applications enhances the possibility of achieving financial success.

Companies, especially those in slow-adopting industries, should assess the nanomaterial market. Important elements in penetrating a market include gathering consumer feedback, learning about their wants, and showing how your product is superior to the competition.

Analyzing the Cost Benefit: While nanomaterials can enhance performance, their price tag needs to be considered. If you want to know if producing nanomaterials for a certain application is worth the expenditure, a thorough cost-benefit analysis is a beneficial place to start.

1.5.8 Intellectual Property and Innovation

Since nanomaterials frequently incorporate innovative synthesis methods, materials, or uses, safeguarding intellectual property (IP) becomes paramount as the field progresses from research to mass production.

It is imperative that businesses take the necessary steps to safeguard their inventions in nanomaterial production, processing, and applications by securing patents. Having a strong portfolio of intellectual property gives you an edge in the market and may be a great asset when looking for partners or licenses.

Accelerating the transition from research to mass manufacturing may be achieved by technological licensing and collaboration as well as through collaborating with research institutes. By working together, businesses are able to have access to state-of-the-art resources and knowledge.

Companies in the nanomaterials industry must maintain their dedication to constant innovation in order to keep up with the dynamic nature of the market. To maintain a competitive edge in a dynamic industry, it is crucial to invest in research and development (R&D) to create novel materials, enhance manufacturing methods, or investigate potential new uses.

For nanomaterials to move from the lab to the factory in large quantities, they must overcome technological, economic, environmental, and regulatory hurdles. Businesses must manufacture nanoparticles in a way that complies with regulations, is cost-effective, consistent, safe, and scalable. Nanomaterials can revolutionize many industries and open up new business possibilities in many different fields if we use them strategically and innovate.

1.6 Efficient Pathways and Technologies for Large-Scale Nanomaterial Fabrication

Adopting efficient methods and modern technology is crucial for scaling up nanomaterial creation from lab to industrial-scale production. Nanomaterials are in high demand across many different industries, including electronics, energy, healthcare, and materials research. Figure 1.5 shows the efficient pathways and technologies for large-scale nanomaterials fabrication. The challenge for producers is to produce these materials at a lower cost without sacrificing quality or environmental safety (Niu et al., 2016). To efficiently fabricate nanomaterials on a wide scale, this handbook describes important approaches and technologies.

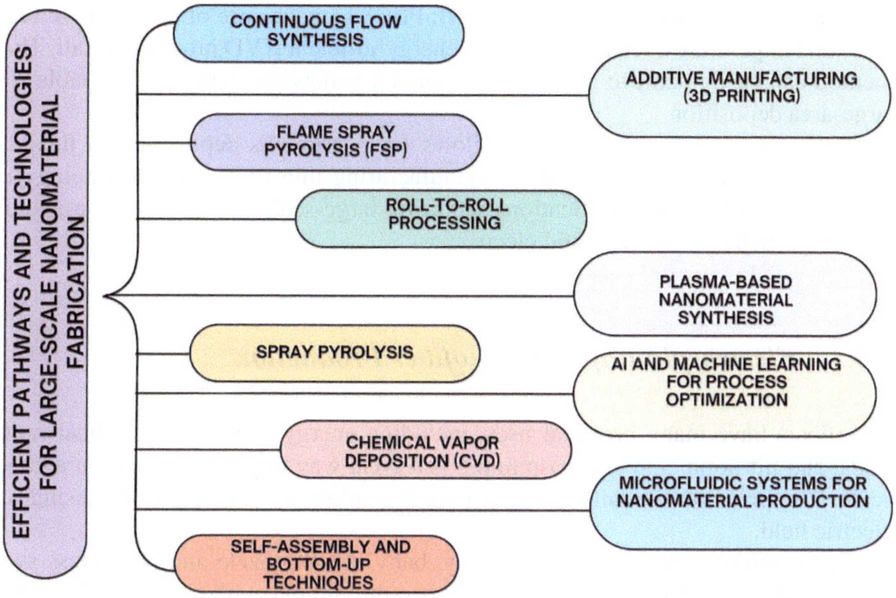

Fig. 1.5 The efficient pathways and technologies for large-scale nanomaterials fabrication

1.6.1 Continuous Flow Synthesis

The continual creation of nanomaterials is essential for large-scale manufacturing, and continuous flow synthesis provides this capability, in contrast to batch production. This approach allows for more precise regulation of reaction conditions and more reliable results each time.

Microreactors, which offer precise control over mixing, temperature, and pressure, enable the consistent synthesis of nanoparticles. Microreactors aid in the synthesis of quantum dots, metal nanoparticles, and nanostructured polymers by running several reactors in parallel, thereby increasing output volumes.

Carbon nanotubes and graphene are two examples of materials whose production requires extremely high pressure, making high-pressure continuous flow systems an absolute necessity. While decreasing reaction times, they increase uniformity of nanomaterial characteristics.

1.6.2 Chemical Vapor Deposition (CVD)

Carbon vapor deposition (CVD) is a popular method for mass-producing high-quality nanostructures and thin films, including graphene, nanowires, and carbon nanotubes (CNTs). Chemically reacting vapor-phase reactants deposit onto a substrate to form a solid layer of nanomaterials.

Enhanced CVD with Plasma (PECVD): PECVD makes use of plasma to reduce reaction temperatures, leading to a more energy-efficient CVD process overall. This method can easily achieve graphene and silicon nanowires, which are suitable for large-area deposition.

Roll-to-Roll CVD: This method allows for continuous deposition on flexible substrates, making it ideal for mass manufacturing thin films. This technology is particularly suitable for applications requiring large-scale and inexpensive manufacturing, such as solar cells and electronics.

1.6.3 Electrospinning for Nanofiber Production

Nanofibers have many practical uses, including energy storage, biomedical scaffolds, and filtration, and electrospinning is a great way to produce them on a large scale. The procedure entails drawing polymer solutions into fine fibers using an electric field.

Electrospinning has come a long way, baby! Multi-nozzle and needleless systems are now available, allowing for higher throughput without sacrificing fiber consistency. Because of this, industrial-scale nanofiber manufacturing may continue indefinitely.

Various materials employ two types of electrospinning: solution-based and melt-based. Melt electrospinning is better suited to thermoplastic polymers and requires less solvents than solution electrospinning, which is more often employed for composites and polymers.

1.6.4 Flame Spray Pyrolysis

One scalable and effective way to make catalysts, functional nanomaterials, and metal oxide nanoparticles is by flame spray pyrolysis (FSP). In this method, rapid combustion and thermal degradation cause nanoparticle formation, which entails spraying a precursor solution into a flame.

Pilot-scale FSP reactors allow for continuous nanoparticle synthesis with size and composition control. These reactors produce large quantities of materials for catalysts, ceramic nanoparticles, and energy storage components.

Industries wishing to gradually increase their nanomaterial production capacity can use modular flame spray systems, which enable them to do so by adding burners or increasing reactor size in a systematic manner.

1.6.5 Sol-Gel Processing

The sol-gel technique involves the use of a wet chemical procedure to create ceramics, hybrid nanomaterials, metal oxides, and other similar substances. We create nanomaterials by converting a liquid "sol" into a solid "gel" phase, followed by drying and calcining the mixture.

Automated Sol-Gel Reactors: By providing exact control over process parameters, these reactors allow for large-scale, continuous manufacturing of nanomaterials with consistent characteristics. Applications for this process include the production of optical materials, catalysts, and coatings.

The creation of nanomaterials with better characteristics, such as increased surface area or porosity, is crucial for applications like fuel cells and batteries. Hybrid sol-gel techniques allow for this by combining sol-gel with other scalable processes, such as spray drying or freeze drying.

1.6.6 Ball Milling and Mechanochemical Synthesis

Ball milling is one mechanical method for grinding materials to nanoscale sizes. Creating large quantities of metal nanoparticles, alloys, and nanocomposites is where it really shines.

The induction of chemical reactions during milling, a process known as mechanochemical synthesis, requires high-energy ball mills, which are required for the large-scale, consistent production of nanomaterials. Using this technology, producing massive amounts of metallic nanoparticles and nanocomposites has never been easier or more economical.

A planetary mill allows for more precise control over particle size distribution by increasing the impact force on the material through revolving containers. This method allows for the mass production of nanomaterials with specific characteristics, which have potential applications in energy storage, catalysis, and coatings.

1.6.7 Scalable Graphene Production

Several sectors highly demand graphene due to its remarkable mechanical and electrical capabilities. It is a two-dimensional substance. The complexity of graphene's fabrication processes has made it difficult to scale up manufacturing. Manufacturing on a large scale is becoming more practical due to recent innovations.

Liquid-phase exfoliation can convert bulk graphite into graphene flakes using solvents and mechanical forces. For applications such as conductive inks, batteries, and composites, liquid-phase exfoliation provides an economical and scalable method of producing graphene in large numbers.

Electrochemical exfoliation, which involves separating graphene from graphite via electrochemical reactions, is another scalable process. For electronics and sensor applications, it provides better control over graphene layer thickness and quality.

Chemically reducing graphene oxide (GO) produces reduced graphene oxide (rGO), which is a useful byproduct. This process makes it possible to make large amounts of graphene with many of its properties and uses.

1.6.8 Supercritical Fluid Technology

Supercritical fluid technology, which uses supercritical CO_2 as a solvent, facilitates both nanomaterial synthesis and processing. It provides an environmentally friendly, scalable, and energy-efficient method for making nanomaterials such as polymers, metal nanoparticles, and Nano catalysts.

To minimize the use of potentially harmful solvents, supercritical CO_2 reactors enable the synthesis of nanomaterials in a precisely regulated temperature and pressure environment. Supercritical CO_2 is an attractive approach for environmentally friendly nanomaterial synthesis because of its adjustable characteristics, which allow for optimization of particle size and surface area.

Supercritical drying not only dries aerogels and porous nanomaterials, which are particularly fragile, but also preserves their structure by preventing pore collapse. Insulation, catalysis, and medication distribution are three areas where this method shines.

1.6.9 Plasma-Based Synthesis

Plasma-based approaches use very energetic gases, or plasmas, to convert precursors into nanoparticles. Useful in semiconductors, coatings, and energy applications, these technologies may produce high-purity nanomaterials quickly and on a large scale.

Thermal plasma systems create nanoparticles by first vaporizing precursor materials in high-temperature plasmas, then cooling and condensing them. Ceramic nanoparticles, carbon nanotubes, and metal oxides are typical products of this process.

Low-Temperature Plasma Synthesis: This method provides a workable alternative to high-temperature plasma synthesis for applications that necessitate nanomaterials with low thermal degradation. It works wonders when making nanocomposites or covering surfaces with functional nanoparticles.

1.6.10 Spray Drying and Freeze Drying

Powdered nanomaterials are crucial for uses such as catalysts, food additives, and medicines; two scalable ways to create these materials are spray drying and freeze drying. Either method allows for precise control of particle size and shape.

Spray drying technologies produce nanoparticles by atomizing liquid solutions into tiny droplets. The food and pharmaceutical sectors frequently employ spray drying, as it is perfect for mass-producing nanomaterial powders with consistent particle sizes.

Freeze drying, which involves sublimating frozen ice in a solution, is one method for removing solvents from nanomaterials. Nanomaterials with high structural sensitivity, such as biological nanoparticles or aerogels, must be handled using this approach.

Innovative paths and sophisticated technologies are essential for the efficient, large-scale manufacturing of nanomaterials. These resources improve production rates, lower prices, and guarantee product uniformity. Flame spray pyrolysis, electrospinning, CVD, and continuous flow synthesis are some of the most important methods for making a lot of nanomaterials. Newer technologies, like plasma-based synthesis and supercritical fluid processing, offer more environmentally friendly and energy-efficient options. These scalable methods allow companies to move nanomaterials from the lab to the factory, where they may be mass-produced and used in a variety of commercial contexts.

References

Abd-Elsalam, K. A., Rajeshkumar, S., & Periakaruppan, R. (2022). Sustainable strategies for producing large-scale nanomaterials: A note from the editors. In *Agri-waste and microbes for production of sustainable nanomaterials* (pp. 1–13). Elsevier.

Aitken, R. J., Chaudhry, M. Q., Boxall, A. B. A., & Hull, M. (2006). Manufacture and use of nanomaterials: Current status in the UK and global trends. *Occupational Medicine, 56*(5), 300–306.

Arole, V. M., & Munde, S. V. (2014). Fabrication of nanomaterials by top-down and bottom-up approaches-an overview. *Journal of Materials Science, 1*, 89–93.

Baron, A., Szewieczek, D., & Nowosielski, R. (2007). Selected manufacturing techniques of nanomaterials. *Journal of Achievements in Materials and Manufacturing Engineering, 20*(1–2), 83–86.

Behera, A., & Chatterjee, S. (2022). Industrial scale up applications of nanomaterials recycling. In *Nanomaterials recycling* (pp. 341–361). Elsevier.

Bhanvase, B. A., Sonawane, S., Pawade, V. B., & Pandit, A. B. (Eds.). (2021). *Handbook of nanomaterials for wastewater treatment: Fundamentals and scale up issues*. Elsevier.

Chang, C.-H., Paul, B. K., Remcho, V. T., Atre, S., & Hutchison, J. E. (2008). Synthesis and post-processing of nanomaterials using microreaction technology. *Journal of Nanoparticle Research, 10*, 965–980.

Charitidis, C. A., Georgiou, P., Koklioti, M. A., Trompeta, A.-F., & Markakis, V. (2014). Manufacturing nanomaterials: From research to industry. *Manufacturing Review, 1*, 11.

Cristaldi, D. A., Yanar, F., Mosayyebi, A., García-Manrique, P., Stulz, E., Carugo, D., & Zhang, X. (2018). Easy-to-perform and cost-effective fabrication of continuous-flow reactors and their application for nanomaterials synthesis. *New Biotechnology, 47*, 1–7.

Dąbrowska, S., Chudoba, T., Wojnarowicz, J., & Łojkowski, W. (2018). Current trends in the development of microwave reactors for the synthesis of nanomaterials in laboratories and industries: A review. *Crystals, 8*(10), 379.

de Zárate, O., David, C. G.-M., Pinilla-Cienfuegos, E., Ayúcar, J. A., Griol, A., Bellières, L., Hontañón, E., Kruis, F. E., & Martí, J. (2020). Green and sustainable manufacture of ultrapure engineered nanomaterials. *Nanomaterials, 10*(3), 466.

Gacem, M. A., & Abd-Elsalam, K. A. (2022). Strategies for scaling up of green-synthesized nanomaterials: Challenges and future trends. In *Green synthesis of silver nanomaterials* (pp. 669–698). Elsevier.

GadelHak, Y., Muhammad, A., El-Azazy, M., El-Shafie, A. S., Shibl, M. F., & Mahmoud, R. (2024). Computer-aided design of large-scale nanomaterials synthesis processes: A detailed review. *ChemBioEng Reviews, 11*, e202300075.

González, C. M. O., Kharissova, O. V., González, L. T., Méndez-Rojas, M. A., Quezada, T. S., & Méndez, Y. P. (2021). Scalable synthesis of nanomaterials. In *Handbook of nanomaterials and nanocomposites for energy and environmental applications* (pp. 899–921). Springer.

Gruar, R. I., Tighe, C. J., & Darr, J. A. (2013). Scaling-up a confined jet reactor for the continuous hydrothermal manufacture of nanomaterials. *Industrial & Engineering Chemistry Research, 52*(15), 5270–5281.

Gupta, D., Boora, A., Thakur, A., & Gupta, T. K. (2023). Green and sustainable synthesis of nanomaterials: Recent advancements and limitations. *Environmental Research, 231*, 116316.

Holkar, C. R., Jain, S. S., Jadhav, A. J., & Pinjari, D. V. (2018). Scale-up technologies for advanced nanomaterials for green energy: Feasibilities and challenges. In *Nanomaterials for green energy* (pp. 433–455). Elsevier.

Jose, N. A., Kovalev, M., Bradford, E., Schweidtmann, A. M., Zeng, H. C., & Lapkin, A. A. (2021). Pushing nanomaterials up to the kilogram scale–an accelerated approach for synthesizing antimicrobial ZnO with high shear reactors, machine learning and high-throughput analysis. *Chemical Engineering Journal, 426*, 131345.

Kelesidis, G. A., & Pratsinis, S. E. (2021). A perspective on gas-phase synthesis of nanomaterials: Process design, impact and outlook. *Chemical Engineering Journal, 421*, 129884.

Khan, F. A. (2020). Synthesis of nanomaterials: Methods & technology. In *Applications of nanomaterials in human health* (pp. 15–21). Springer.

Kioumourtzoglou, S., Hof, S., Kalk, C., Toth, V., Görlin, M., Nováková, J., & Sá, J. (2024). Nanomaterials as a service (NaaS) concept: On-demand protocols for volume synthesis of nanomaterials. *Nanoscale Horizons, 9*(8), 1364–1371.

Li, C. (2010). Structure controlling and process scale-up in the fabrication of nanomaterials. *Frontiers of Chemical Engineering in China, 4*, 18–25.

Lines, M. G. (2008). Nanomaterials for practical functional uses. *Journal ofAlloys and Compounds, 449*(1-2), 242–245.

Liu, C.-J., Burghaus, U., Besenbacher, F., & Wang, Z. L. (2010). Preparation and characterization of nanomaterials for sustainable energy production. *ACS Nano, 4*, 5517–5526.

Lloyd, J. R., Byrne, J. M., & Coker, V. S. (2011). Biotechnological synthesis of functional nanomaterials. *Current Opinion in Biotechnology, 22*(4), 509–515.

Ma, C. Y., Liu, J. J., Zhang, Y., & Wang, X. Z. (2015). Simulation for scale-up of a confined jet mixer for continuous hydrothermal flow synthesis of nanomaterials. *The Journal of Supercritical Fluids, 98*, 211–221.

Mabrouk, M., Das, D. B., Salem, Z. A., & Beherei, H. H. (2021). Nanomaterials for biomedical applications: Production, characterisations, recent trends and difficulties. *Molecules, 26*(4), 1077.

Makgwane, P. R., & Ray, S. S. (2014). Synthesis of nanomaterials by continuous-flow microfluidics: A review. *Journal of Nanoscience and Nanotechnology, 14*(2), 1338–1363.

Malekzadeh, M., & Swihart, M. T. (2021). Vapor-phase production of nanomaterials. *Chemical Society Reviews, 50*(12), 7132–7249.

McCourt, K. M., Cochran, J., Abdelbasir, S. M., Carraway, E. R., Tzeng, T.-R. J., Tsyusko, O. V., & Vanegas, D. C. (2022). Potential environmental and health implications from the scaled-up production and disposal of nanomaterials used in biosensors. *Biosensors, 12*(12), 1082.

Naz, M. Y., Shukrullah, S., Ghaffar, A., Ali, K., & Sharma, S. K. (2020). Synthesis and processing of nanomaterials. In *Solar cells: From materials to device technology* (pp. 1–23). Springer.

Niu, L., Coleman, J. N., Zhang, H., Shin, H., Chhowalla, M., & Zheng, Z. (2016). Production of two-dimensional nanomaterials via liquid-based direct exfoliation. *Small, 12*(3), 272–293.

Patwardhan, S. V., Manning, J. R. H., & Chiacchia, M. (2018). Bioinspired synthesis as a potential green method for the preparation of nanomaterials: Opportunities and challenges. *Current Opinion in Green and Sustainable Chemistry, 12*, 110–116.

Pitkethly, M. J. (2004). Nanomaterials–the driving force. *Materials Today, 7*(12), 20–29.

Raliya, R., Saharan, V., Choudhary, K., Summarwar, S., Gulecha, K., Gupta, V., & Sain, P. M. (2019). Nanomaterials synthesis and characterization. In *Nanoscale engineering in agricultural management* (pp. 1–10). CRC Press.

Ramesh, K. T., & Ramesh, K. T. (2009). *Nanomaterials*. Springer.

Saldanha, P. L., Lesnyak, V., & Manna, L. (2017). Large scale syntheses of colloidal nanomaterials. *Nano Today, 12*, 46–63.

Saleh, T. A. (2022). Large-scale production of nanomaterials and adsorbents. In *Interface science and technology* (Vol. 34, pp. 167–197). Elsevier.

Sasidharan, S., Raj, S., Sonawane, S., Sonawane, S., Pinjari, D., Pandit, A. B., & Saudagar, P. (2019). Nanomaterial synthesis: Chemical and biological route and applications. In *Nanomaterials synthesis* (pp. 27–51). Elsevier.

Sebastian, V. (2022). Toward continuous production of high-quality nanomaterials using microfluidics: Nanoengineering the shape, structure and chemical composition. *Nanoscale, 14*(12), 4411–4447.

Sebastian, V., Arruebo, M., & Santamaria, J. (2014). Reaction engineering strategies for the production of inorganic nanomaterials. *Small, 10*(5), 835–853.

Souto, E. B., Silva, G. F., Dias-Ferreira, J., Zielinska, A., Ventura, F., Durazzo, A., Lucarini, M., Novellino, E., & Santini, A. (2020). Nanopharmaceutics: Part II—Production scales and clinically compliant production methods. *Nanomaterials, 10*(3), 455.

Stieberova, B., Zilka, M., Ticha, M., Frantisek Freiberg, P., Caramazana-González, J. M. K., & Lester, E. (2019). Sustainability assessment of continuous-flow hydrothermal synthesis of nanomaterials in the context of other production technologies. *Journal of Cleaner Production, 241*, 118325.

Thirugnanasambandan, T., & Gopinath, S. C. B. (2023). Laboratory to industrial scale synthesis of chitosan-based nanomaterials: A review. *Process Biochemistry, 130*, 147–155.

Tulinski, M., & Jurczyk, M. (2017). Nanomaterials synthesis methods. In *Metrology and standardization of nanotechnology: Protocols and industrial innovations* (pp. 75–98). Wiley.

Van der Heijden, A. E. D. M. (2018). Developments and challenges in the manufacturing, characterization and scale-up of energetic nanomaterials–a review. *Chemical Engineering Journal, 350*, 939–948.

von Bomhard, S., Schramm, J., Bleul, R., Thiermann, R., Höbel, P., Krtschil, U., Löb, P., & Maskos, M. (2019). Modular manufacturing platform for continuous synthesis and analysis of versatile nanomaterials. *Chemical Engineering & Technology, 42*(10), 2085–2094.

Yang, M., Yang, L., Zheng, J., Hondow, N., Bourne, R. A., Bailey, T., Irons, G., Sutherland, E., Lavric, D., & Ke-Jun, W. (2021). Mixing performance and continuous production of nanomaterials in an advanced-flow reactor. *Chemical Engineering Journal, 412*, 128565.

Zhu, J. Y., Agarwal, U. P., Ciesielski, P. N., Himmel, M. E., Gao, R., Deng, Y., Morits, M., & Österberg, M. (2021). Towards sustainable production and utilization of plant-biomass-based nanomaterials: A review and analysis of recent developments. *Biotechnology for Biofuels, 14*(1), 114.

Chapter 2
MXenes: A Promising Avenue for Sustainable Nanotechnology

Samaresh Ghosh, Subrata Kumar Maiti, Priyabrata Mandal, and Priya Goel

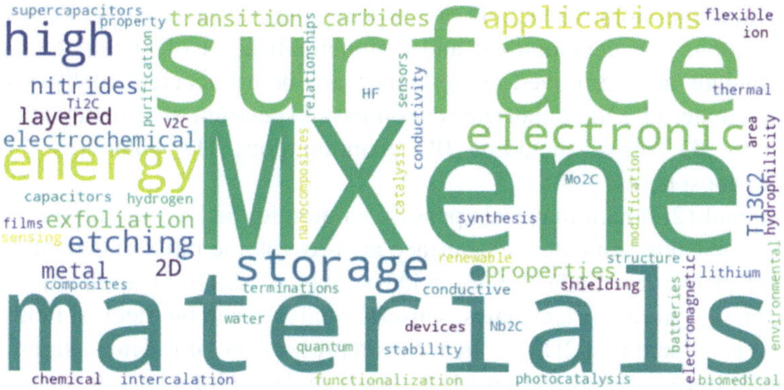

2.1 Introduction

MXene, an emerging 2D material, has attracted tremendous research attention for its fabulous properties like excellent electrical conductivity, outstanding mechanical properties, and thermal conductivity to address various challenges in material science research. MXene was discovered by the Gogotsi group at Drexel Institute in 2011. MXenes are basically transition metal nitrides or carbides or nitro carbides with $M_{n+1}X_nT_x$ formula, where **M** describes earlier transition metal, **X** denotes

S. Ghosh · S. K. Maiti
Department of Polymer and Process Engineering, IIT Roorkee, Roorkee, India

P. Mandal
Research and Technical Support Division, Permionics Membranes Pvt. Ltd, Vadodara, India

P. Goel (✉)
Department of Chemical Sciences, School of Science, GSFC University, Vadodara, India
e-mail: priya.goel@gsfcuniversity.ac.in

S. Kulkarni (ed.), *Next-Generation Nanomaterials for Sustainable Engineering*, Lecture Notes in Nanoscale Science and Technology 43,
https://doi.org/10.1007/978-3-032-05719-8_2

carbon and nitrogen, and **T** states to terminating groups, mainly hydroxyl (-OH), fluorine, and other oxygen-containing groups.

The swift advancement of high-performance technologies in the twenty-first century has been driven by the expansion of nanotechnology. Nanotechnology represents a remarkable instance of an emerging technology, providing engineered nanomaterials that possess substantial potential for the production of products with notably improved performance. These nanomaterials have surfaced as a compelling category of materials, increasingly sought after for various practical applications. Significant advancements in materials science through computing and experimentation have made it possible to modify the properties of materials at the nano-level, for example, molecular and atomic level and have opened the door to a deeper knowledge of the situation. The development and commercialization of nanomaterials with energy efficient and improved performance are being accelerated by low dimensional nanomaterials. Nanomaterials encompass a spectrum of structures that vary from 0D to 3D forms, including 0D quantum dots, 1D nanowires, 2D nanosheets, 3D nanospheres.

Two-dimensional (2D) nanomaterials possess an atomic thickness, a high density of active surface sites, an extensive surface area-to-volume ratio, and remarkable mechanical properties made them a wide variety of different applications. Particularly, in energy storage, electronics, sensors, catalysis, and biomedical fields. Consequently, it is evident that 2D nanomaterials represent one of the most significant and extensively researched classes of materials to date. Graphene, hexagonal boron nitride, phosphorene, and MXene are few examples of 2D nanomaterials.

MXene, an emerging 2D material, has attracted tremendous research attention for its fabulous properties like excellent electrical conductivity, outstanding mechanical properties, and thermal conductivity to address various challenges in material science research. MXenes are composed of multilayered sandwich structures that typically feature an odd number of layers, with carbon and/or nitrogen (X) layers positioned between metal (M) layers. They are denoted in the form $M_{n+1}X_nT_x$, which encompasses transition metals (M elements), for example, Sc, Zr, Ta, Hf, Mo, Cr, W, and Mn, along with carbon, nitrogen (X elements), and various functional surface termination groups (Tx elements). Multiple surface termination groups, for example, hydroxyl (–OH), oxygen (O), and halogens (F, Cl), give MXenes superior chemical and physical properties and make processing easier. The surface characteristics enable the combination of MXenes with suitable nanomaterials through covalent functionalization, polymerization, and self-assembly to create the MXene-based multilayer hybrid structure. Recent years have seen reviews and reports on the advancements in Mxenes-based energy storage applications, sensors, photocatalysts, water treatment and environmental remediation, photonics and optoelectronics, electromagnetic interference, and polymeric membranes (Xie, Naguib, et al., 2014).

The scientific community increasingly views MXenes as pivotal to the future of nanotechnology, driven by their expanding applications across various industries. This perspective is bolstered by recent research that demonstrates the ability to customize MXene‘s surface functional groups, thereby altering its electrical, electrochemical, and surface properties (Ko et al., 2024). Most MXenes and MXene-derived materials exhibit remarkable characteristics, with high volumetric capacitance, antimicrobial properties, electrochromic behavior, excellent optical transparency, and electronic conductivity. Consequently, MXenes not only have the potential to improve the performance of current applications but also to facilitate the development of innovative solutions.

The aim of this book chapter is to provide a comprehensive overview of current knowledge on the MXene, including synthesis, important discoveries, applications, challenges, and future directions/implications. Additionally, this chapter also incorporates the most recent research results, techniques, innovations, and applications of MXenes. It also highlights gaps and limitations in current understanding, pointing out areas that need more research. For researchers, students, and practitioners with a strong interest in MXenes, this chapter is an educational resource.

2.2 Synthesis of MXenes

2D materials are usually layered and structured, and layer-by-layer interactions are poor in nature. Removing a few layers from the precursor material is the easiest way to get a 2D structure. This removal is called etching in chemical terminology. MXene was obtained from a layered precursor named MAX phase with a strong inter-layer bond. In the MAX phase, the middle A refers to Aluminum, and it is strongly bonded with carbon and nitrogen. Hence, to remove this aluminum, there is a need for a strong etchant. The first etchant used for MXene synthesis from Ti_3AlC_2 is hydrofluoric acid HF (Anasori et al., 2015; Naguib, Mashtalir, et al., 2012; Zhou et al., 2017). As HF is very harsh in nature, an alternate way is in situ generation of HF to etch the aluminum. A mixture of HCl and LiF was applied by Ghidiu to etch the parent ternary of the MAX phase (Ghidiu et al., 2015). HF-free methods have also been developed such as (1) hydrothermal synthesis with a basic solution of NaOH (Li, Yan, et al., 2018), (2) by molten salt (Urbankowski et al., 2016), (3) electrochemical etching in diluted HCl (Sun et al., 2017). All these methods have their own merits and demerits. The hydrothermal method requires a high concentration of NaOH. The molten salt synthesis needs high temperature, and yield is poor. Electrochemical synthesis is a relatively promising one with 90% yield and better control of functionality of the final MXene. These all methods are only to synthesize carbide type MXenes. MXenes are prepared by ammonia treatment of the precursor carbide MXenes at high temperatures (Chen et al., 2025). Other fascinating synthesis methods newly developed are like thermal reduction of Ti_2SC to produce Ti_2CT_X (Guo et al., 2022).

$$2Ti_3AiC_2 + 6HF = 2Ti_3C_2 + 2AlF_3 + 3H_2 \tag{2.1}$$

$$Ti_3C_2 + 2H_2O = Ti_3C_2(OH)_2 + H_2 \tag{2.2}$$

$$2HF + Ti_3C_2 = Ti_3C_2F_2 + H_2 \tag{2.3}$$

After etching the MAX phase, multilayered MXenes resulted in various terminating functional groups on it. These functional groups are strongly bonded with H-bond and Van der Waals interactions, which resulted in several layers of MXenes being held together. Here exfoliation is performed to get a colloidal suspension with single layer MXene. Several methods applied for MXene exfoliation depending on prior etching conditions. Usually, intercalation with bulky organic moiety, cations, and shaking/sonication are the common procedures for delamination. However, the methodology for exfoliation depends on the composition of MXene. For instance, DMSO successfully works for $Ti_3C_2T_x$ and $(Mo_{0.67}Ti_{0.33})_3C_2T_z$ but does not active for other MXenes (Anasori et al., 2015; Mashtalir et al., 2013; Naguib et al., 2015). Some reports also highlighted tetrabutylammonium hydroxide as an exfoliator for different MXenes (Anasori et al., 2016; Halim et al., 2016; Urbankowski et al., 2016) Few recent studies highlighted etching-free methods like chemical vapor deposition and atomic layer assembly for Mxene synthesis. Ren et al. developed ultrathin α-Mo_2C through CVD with high stability under atmospheric conditions (Xu et al., 2015). Wang et al. also synthesized carbide and nitride by direct chemical vapor deposition technique.(Wang et al., 2023).

The theoretical study revealed there are almost 150 varieties of the MAX phase. Almost 30 different MXene samples have been reported experimentally. MXenes exist in different chemical formulas: $M_4X_3T_z$, $M_3X_2T_z$, M_2XT_z and $M_{1.33}XT_z$. MAX phase with two different transition metals named as solid solutions has expanded MXene family members to a great extent. Two different orders exist in the solid solution type MAX phase. (1) Out-of-plane order (Liu et al., 2014) where one transition metal occupies the central layer and the other element in the outer layer; (2) in-plane order (Tao et al., 2017) where two transition elements occupy the same plane.

MXene chemistry can be tuned by modifying surface functionality. Depending on the etching method (Table 2.1), terminations are =O, –OH, –F. If only HF is used for etching, the –F content is higher than when LiF/HCl is used (Hope et al., 2016).

Table 2.1 The methods, etchants at diffracted temperature of MXene

Method	Etchants	Temperature	References
Acid with fluorine	HF	Room temperature to 55 °C	Naguib et al. (2011)
	HCl + (Na, K, or NH_4F)	35–60 °C	Liu, Kan, et al. (2017)
	HCl + LiF	30–55 °C	Ghidiu et al. (2015)
	H_2O_2 + HF	40 °C	Alhabeb et al. (2018)
	NH_4HF_2	Room temperature	Karlsson et al. (2015)
Electrochemical	TMAOH/NH_4Cl	Room temperature	Yang et al. (2018)
	HCl	Room temperature	Sun et al. (2017)
Hydrothermal	NaOH	270 °C	Li, Yao, et al. (2018)
	$NaBF_4$, HCI	180 °C	Peng et al. (2018)
Lewis acid	$ZnCl_2$	550 °C	Li et al. (2019)
Molten salts	LiF + NaF + KF	550 °C	Urbankowski et al. (2016)

When metal chloride like $ZnCl_2$, used for etching surface functionality, contains –Cl (Li et al., 2019). Terminal chemistry can also be varied by differing M and X elements in $M_{n+1}X_nT_x$. Oxygen or hydroxyl functionalities can be varied by replacing C by N (Naguib et al., 2015). Post-modification to reduce –F content also investigated with TBAOH treatment, alkali treatment or by annealing.

2.3 Properties of MXenes

MXenes represent a 2D materials synthesized of transition metal nitrides, carbides, or carbonitrides, which are distinguished by their exceptional metallic and semi-metallic electronic properties. These properties arise from their layered structure, composition, and surface chemistry, positioning them as promising materials for a broad applicability in energy storage, optics, catalysis, and electronics. Below, we explore the mechanisms underlying their metallic and semi-metallic behavior, thermal and mechanical properties of different MXene based materials.

2.3.1 Electronic Properties of MXenes

2.3.1.1 Band Structure

MXenes usually display a metallic or semi-metallic band structure, attributed to the d-orbitals of transition metals present in their composition. The band structure (electronics) can be altered by varying the metal component (M) or types of surface end groups. In many of the MXenes, the conduction band originates from the d-states of the transition metal. In contrast, the valence band is typically composed of the p-states of carbon or nitrogen, and such overlap results in metallic behavior.

Numerous research studies have been undertaken to investigate the electronic band structural analysis and unique properties of this diverse family of MXenes (Kim et al., 2019). In general, functionalized MXenes are expected to exhibit metallic electronic configurations, whereas a few specific MXene are anticipated to be in the semiconducting range (Table 2.2). Band gaps of all MXenes, for example, Sc_2CT_2 (where T=F, O, OH), $Hf_2CO_2Ti_2CO_2$, and Zr_2CO_2 range from 0.20 to 1.90 eV. Among most of the MXenes, $Sc_2C(OH)_2$ is the only one with a direct band gap configuration, while the rest are reported to have indirect band gap configuration. Interestingly, biaxial strains has an significant impact on theband gap transformation. A significant amount of strain 4%, 10%, and 14% applied to indirect band gap structural MXenes, e.g., Ti_2CO_2, Zr_2CO_2, and Hf_2CO_2, respectively, can induce a transition to a direct band gap structure. Zr, Ti, and Hf are in the same group of periodic table and share the same outer s- and d-orbitals orbital electronic configuration. As a result, similar trends in the transition from metallic to semiconducting behavior have been noticed in the corresponding M_2C structures following O-group

Table 2.2 Earlier reported band gaps (in eV) of different functionalized MXenes based on some conventional and hybrid functional terminations. Type of band gaps are remarked as an antiferromagnetic and topological insulator

MXenes	Termination	PBE	HSEO6	Remarks
Sc_2C	O	1.8 (Khazaei et al., 2012), 1.86 (Lee et al., 2015), 1.84 (Kumar & Schwingenschlögl, 2016)	3.01 (Lee et al., 2015), 2.90 (Liu, Zhou, et al., 2017), 2.92 (Chandrasekaran et al., 2017)	Direct band gap
	F	1.05 (Ma et al., 2014), 1.0 (Lee et al., 2015), 1.03 (Khazaei et al., 2012)	1.88 (Lee et al., 2015), 1.64 (Ma et al., 2014), 1.84 (Liu, Zhou, et al., 2017)	
	OH	0.34 (Lee et al., 2015), 0.71 (Khazaei et al., 2014), 0.44 (Kumar & Schwingenschlögl, 2016), 0.45 (Khazaei et al., 2012)	0.71 (Lee et al., 2015),0.74 (Liu, Zhou, et al., 2017)	
	Cl	0.88 (Liu, Kan, et al., 2017)	1.64 (Liu, Kan, et al., 2017)	–
Mo_2C	O	–	–	–
	OH	0.10 (Khazaei et al., 2014)	–	–
	F	0.25 (Khazaei et al., 2014)	–	–
	Cl	0.15 (Khazaei et al., 2014)	–	–
Cr_2C	O		–	–
	F	0.22 (Khazaei et al., 2014)	3.49 (Si et al., 2015), 3.15 (You et al., 2016),	Antiferromagnetic
	OH	0.03 (Khazaei et al., 2014)	1.76 (Si et al., 2015), 1.39 (You et al., 2016)	Antiferromagnetic
	Cl	0.15 (Khazaei et al., 2014)	2.56 (Si et al., 2015)	Antiferromagnetic
Ti_2C	O	0.33 (Gandi et al., 2016), 0.17 (Khazaei et al., 2014), 0.24 (Khazaei et al., 2012)	0.88 (Xie et al., 2013), 0.92 (Zhang et al., 2016), 0.78 (Gao et al., 2017)	–
Zr_2C	O	0.95 (Gandi et al., 2016), 0.66 (Khazaei et al., 2014), 0.88 (Khazaei et al., 2012)	1.54 (Zhang et al., 2016)	–
W2C	O	0.194 (Weng et al., 2015)	0.472 (Weng et al., 2015)	Topological insulator
Hf_2C	O	0.8 (Khazaei et al., 2014), 1.00 (Khazaei et al., 2012)	1.75 (Zhang et al., 2016), 1.657 (Zha, Huang, et al., 2016)	

functionalization. Additionally, band gap of M_2CO_2 (M belongs to Zr, Ti, Hf) MXenes increases with the atomic number of the metal (M), due to the decreasing electronegativity of the metal. As an example, in $Ti_{n+1}C_nO_2$ ($n \geq 2$) MXene, the involvement of the O 2p orbital decreases as "*n*" increases, causing these MXenes to lose their semiconducting properties. Furthermore, Cr_2C and Mo_2C shows semiconducting properties when functionalized with OH, Cl, and F groups, and few double-transition metalic MXenes also exhibit a semiconducting nature with O termination.

Some MXene systems that include two distinct metal atoms can form an methodical structure, in its place of a randomly distributed one. They are classified as ordered MXenes with double-transition metal, characterized by the formulas $M'_2M''C_2$ and $M'_2M''_2C_3$, where M′ atoms constitute the outer layer and M″ atoms form the central layers. For example, in $Mo_2TiC_2T_x$, outer layers are composed of Mo, while central layers contain Ti. Mainly, this ordered structure influences the electronic characteristics of MXenes. A significant topological gap of 0.1–0.2 eV has also been described in the $Mo_2M''C_2O_2$ (M″ belongs to Ti, Hf, Zr) system, which is sufficiently large to facilitate the quantum spin Hall effect at ambient temperature (Dong et al., 2018). Actively favorable oxygen-terminated surfaces can able to protect MXenes from oxidation in ambient conditions. Additionally, M_2CO_2 (where M = W, Cr, Mo) MXenes are also classified as topological insulators. A few of these topologically non-trivial MXenes are synthesized, making them promising candidates for electronic and spintronic instruments, as well as for research in topological superconductivity. The Fermi level in MXenes is situated within the conduction band, signifying a high density of states (DOS) at this level, which characterizes the metallic behavior. This characteristic enhances their high electrical conductivity, facilitating the efficient flow of electrons.

2.3.2 Metallic Behavior

2.3.2.1 Electrical Conductivity

MXenes exhibit outstanding electrical conductivity, often more than that of graphene and certain metals too. For example, $Ti_3C_2T_x$ has reported electrical conductivities greater than 10,000 S/m, establishing an exceptional conductor. Kim et al. (Kim et al., 2017) studied molybdenum-based MXenes such as Mo_2CTx, Mo_2TiC_2Tx, and $Mo_2Ti_2C_3Tx$ to understand their thermoelectric properties. Mo_2CF_2 MXene has been computationally identified as the utmost adapted thermoelectric candidate among similar M_2CTx based MXenes. Furthermore, Mo_2TiC_2Tx, and $Mo_2Ti_2C_3Tx$ MXenes possess an ordered double-metal structure, with Mo atoms specially positioned on their surfaces, resulting in a surface nature similar to that of Mo_2CTx. Such types of MXenes were treated into free-standing, flexible papers through a straightforward vacuum filtration method. The intercalation of cationic organic molecules between the MXene nanosheets increased the interlayer spacing, which in turn reduced the electrical conductivity and Hall mobility. Annealing at 250–450 °C in an Ar/H_2 atmosphere removed the intercalated H_2O and organic species from the MXene papers, restoring their intrinsic conductive properties.

Dong et al. established the application of metallic based MXenes in flexible type triboelectric nanogenerators, which efficiently converted mechanical strain into

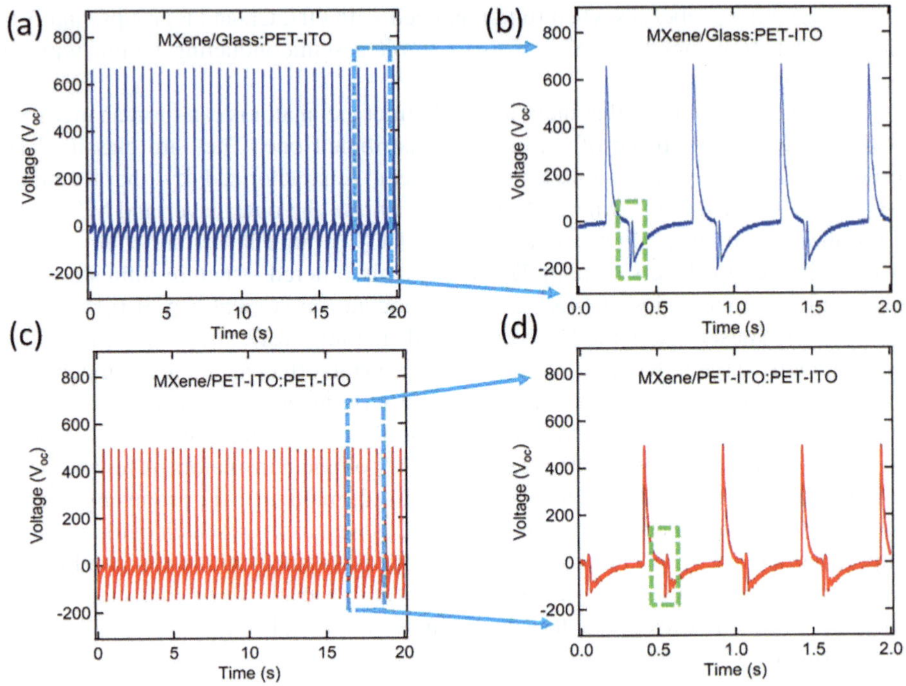

Fig. 2.1 (**a**) Application of MXene/Glass: PET-ITO TENG imperiled to a 15 N upright force at a frequency of 2 Hz, (**b**) Zoomed-in view of V_{oc}. (**c**) MXene/PET-ITO:PET-ITO as a flexible composite displayed a similar performance (**d**) Zoomed-in view of V_{oc} (Dong et al., 2018)

electrical (Fig. 2.1) (Dong et al., 2018). Anyway, Ti_3C_2Tx MXene was described to rank lower than PTFE or Teflon in the triboelectric compounds.

2.3.2.2 Temperature Dependence

The conductivity of MXenes is predominantly metallic, exhibiting minimal dependence on temperature. This stability indicates that conductivity remains consistent across a range of temperatures, making it advantageous for applications in electronic devices that function under fluctuating thermal conditions.

2.3.3 Surface Termination

Properties of MXenes. Although these groups introduce additional states into the band structure, they do not substantially hinder conductivity. Instead, they may improve the material's interaction with ions, which is advantageous for energy storage applications.

2.3.4 Semi-Metallic Behavior

Some MXenes can also display semi-metallic behavior, characterized by a small band gap typically ranging from a few hundred meV. This property can be adjusted through chemical modification, surface functionalization, or the introduction of defects into the MXene.

The semi-metallic characteristics of certain MXenes make them suitable for applications in field-effect transistors (FETs) and sensors, where even minor changes in conductivity can be detected. This sensitivity makes them ideal for applications in chemical sensing and electronic devices that demand precisely tuned electronic properties.

2.3.5 Mechanical Properties

Zha et al. showed the excellent mechanical strength of MXene (Mo_2C) based materials under biaxial strain (Zha, Yin, et al., 2016). The biaxial elastic modulus of Mo_2C MXene is 302–322 GPa, which is a little higher than MoS_2. The strong interaction between the Mo-C atoms contributes to the development of a higher modulus of elasticity. As the strain increases, the stress rises to reach the critical strain of less than 0.09, after which the MXene exhibits creep deformation. For Mo-based Mxene, the optimal strength at the critical strain is determined to be about 21 GPa. These data suggest that Mo_2C MXene is an exceptionally robust elastic 2D material, demonstrating high robustness to applied strains.

Ibrahim et al. discussed the elastic behavior and mechanical strength of different MXenes, which depend on various factors, including the number of layers and its thickness, termination functional groups, doping materials and different transition metals. The mechanical behavior of MXenes with varying compositions is highlighted in Table 2.3.

Table 2.3 Mechanical properties of different composite MXenes

Materials	Measurements	Modulus of elasticity (GPa)	References
$Ti_3C_2H_2$	VASP/PBE	419	Zha et al. (2015)
$Zr_3C_2O_2$	DFT	392.9	Fu et al. (2018)
Ti_2C	VASP	609	Yorulmaz et al. (2016)
Ti_2CO_2	Nanoindentation process	983	Zhang et al. (2017)
Ta_2C	CASTEP/Wu-Cohen	788	Si et al. (2012)
Ti_2C	MD	597	Plummer et al. (2019)
Mo_2C	VASP	312	Zhang et al. (2013)
Ti_2CO_2	VASP/PBE	745	Zha, Yin, et al. (2016)
W_2C	VASP/PBE	781.9	Peng and De (2013)

2.3.5.1 Number of Layers and Its Thickness

Lipatov et al. (2018) determined the modulus of elasticity and fracture point of mono/bi-layer of Ti_3C_2Tx flakes using atomic force microscopy (AFM) indentation. It was demonstrated that the elastic modulus for bilayer Ti_3C_2Tx flakes are twice those of monolayer MXenes. This suggests a strong interaction force among the layers, possible because of the presence of H-bonding. Single layer of $Ti_3C_2T_x$ exhibits the Young's modulus of ~333 GPa, significantly surpassing that of graphene oxide (GO) 210 GPa and several other MXenes, while, its breaking strength was measured to be 17.3 ± 1.6 GPa.

Using various numerical modeling techniques, including molecular dynamics simulations and density functional theory computations, the impact of layer thickness on the elastic and structural characteristics of various 2D MXene layers was examined. Borysiuk et al. (2015) studied the Young's modulus of $Ti_{n+1}C_n$ MXene using molecular dynamics simulation, which showed an increased value as 2-D layer thickness reduces. The simulated elastic modulus values for Ti_2C, Ti_3C_2, and Ti_4C_3 were 597, 502, and 534 GPa, respectively, for strains $\varepsilon \sim 0.01$.Furthermore, Wu et al. (2018) got similar effects on carbide ($Ti_{n+1}C_n$) and nitride-based ($Ti_{n+1}N_n$) MXenes, by density functional theory calculations, which signifies the important role of layer thickness in determining the mechanical properties of MXenes.

2.3.5.2 Functional Terminations Functional

The functional groups (–F, –O–, –OH) of MXenes significantly influence the mechanical properties, as extensively demonstrated through density functional theory calculations. Bai et al. (2016) explained that elastic stiffness is maintained only in MXenes with –O– group, whereas it deteriorates in those with F and –OH terminations. This behavior can be attributed to variations in the in-plane lattice constants of 2D MXenes caused by different termination functional group. For both Ti_2CT_2/$Ti_3C_2T_2$ MXenes, the in-plane lattice constant is shortest when the termination is

O. In contrast, F and OH terminations result in larger in-plane lattice constants, suggesting a robust interaction between the titanium and the terminating oxygen (O) atoms (Fu et al., 2016).

2.3.5.3 Transition Metal

Jin et al. (2018) utilized DFT (density functional theory) tool to study the effect of asymmetrical –F/–OH functionalization on the mechanical characterization of different monolayer of MXenes e.g., M_2X (M=Sc, V, Ti, Mn, MO, Hf, Nb; X=C, N), where X is positioned between two M layers. These results demonstrated that the surface functionalization and the mass of the transition metal both affect the mechanical characteristics. The results showed that asymmetric functionalization significantly influences the elastic properties of the materials. For pristine M_2X MXenes, the in-plane stiffness (C) of M_2C is slightly lower than that of M_2N. This difference is attributed to the extra valence electron provided by nitrogen (N) compared to carbon (C), leading to stiffer M-X bonds. However, for Mo_2X, the in-plane stiffness of Mo_2N is marginally lower than that of Mo_2C, which is due to the unique structural characteristics of Mo_2X.

2.3.5.4 Doping Effect

Doping of any molecules in the MXene intercalation has a significant role in the elastic properties determination. Chakraborty et al. (2018) compared the stress–strain diagram of pristine Ti_2C and dopped MXene layers of $Ti_2(C_{0.5}B_{0.5})$ and (Ti, V) C through the Density Functional Theory study. As the examples,V-doping led to a minor improvement, whereas B-doping notably enhanced the elastic properties by lowering the in-plane modulus of elasticity.This reduction in stiffness can be attributed to the weaker Ti-B bond relative to the Ti-C bond. Notably, the Young's modulus and in-plane stiffness of $Ti_2(C_{0.5}B_{0.5})$ decrease by about 25–27% compared to Ti_2C. Nonetheless, V-doping at the Ti sites resulted in stiffness similar to that of the undoped Ti_2C. Interestingly, the stiffness of $Ti_2(C_{0.5}B_{0.5})$ was approximately 4.2, 1.5, 1.86, and 3.1 times greater than that of various 2D materials MoS_2, graphene, h-BN, and SiC, corrospondingly, attributed to the strengthening effect of B-doping.

2.3.6 Thermal Properties

The thermal properties of MXenes, including thermal conductivity, thermal expansion, and specific heat capacity, are crucial factors that influence their performance in various applications. MXenes, especially Ti_3C_2 and Ti_2C, are recognized for their relatively high thermal conductivity compared to other 2D materials like, graphene oxide or certain polymers. This enhanced thermal conductivity is attributed to the strong metallic bonding and efficient phonon transport within the metallic layers of MXenes. However, the thermal conductivity of MXenes can be significantly affected by factors such as surface functionalization, layer thickness, and the presence of defects or point defects. Zha, Huang, et al. (2016) investigated the thermal conductivity of oxygen-functionalized Hf_2CO_2 (86.25–131.2 W m^{-1} K^{-1}) MXenes using first-principles calculations, which exhibited a higher value than MoS_2 as well as phosphorene [38 W m^{-1} K^{-1} and 12 W m^{-1} K^{-1}, respectively].

2.3.7 *Opticals Properties*

Optical response of MXenes closely depends on the electronic structural configurational behavior. Dielectric properties are one of the important characteristics to differentiate optical behavior. Dielectric constant $\omega(\varepsilon)$ of a material as a function of frequency consists of $\varepsilon(\omega)_{Re}$ (real part), and $\varepsilon(\omega)_{Im}$ (imaginary part). In general MXene structure, for example, Ti_3C_2Tx, the termination groups (Tx) can significantly influence the electronic structure and its properties, thereby playing a crucial role in determining the optical characterization. Inter- and intra-band electronic transitions play a crucial role in shaping the imaginary component of the dielectric constant, thereby impacting the material's optical absorption and its interaction with electromagnetic radiation. Thus the absorption spectrum is directly related to the total contribution of inter-band transitions within the material. In general, due to structural differences, hydroxylated and fluorinated terminations can accept only one electron, resulting in similar optical performance in the visible range. However, oxygen group requires two electrons to stabilize. Berdiyorov (2016) used computational methods to explore the functionalization-dependent optical properties of Ti_3C_2Tx. Calculations reveal that the functionalization of MXenes decreases the static dielectric function by over 50%. The oxidized Mxenes showed greater absorption in the visible light spectrum, although surface fluorination produced less absorption than virgin Ti_3C_2. In comparison to virgin Ti_3C_2, surface terminations enhance both reflectance and absorption in the ultraviolet (UV) spectrum. Oxygen termination enhances absorption and reflectivity characteristics in the visible spectrum. In contrast, Ti_3C_2 with fluorinated and hydroxylated terminations is more transparent and shows less reflection and absorption than virgin Ti_3C_2. Jhon et al. (2018) calculated the promising optical absorption and extinction coefficients of MXene Ti_3C_2 in the terahertz (THz) range. Combined with its high carrier mobility and high intrinsic charge carrier density, the results indicate that Ti_3C_2 is well-suited for terahertz (THz) detection applications.

2.4 Sustainable Applications of MXenes

2.4.1 Battery Application

MXene has a vast application area depending on chemical variety and structure. MXene has very high theoretical electrical conductivity and lots of hydrophilic groups that results in fast diffusion of charged species in between layers. MXene is used as an electrode material for batteries and supercapacitors (Perera et al., 2023). ML MXenes showed a few hundred capacity for Li and Na ions battery (Halim et al., 2016). As the M-X bond is very strong, the greater the number of M-X bonds, the harder the ion penetration in between the M-X layer, and the only possibility is penetration in between MXene layers.

Ti_2C and Ti_3C_2 have with similar surface chemistry but Ti_2C has higher gravimetric capacity than Ti_3C_2. Experimental results also established that Ti_2CT_x uptake 50% higher Li^+ than $Ti_3C_2T_X$ (Mashtalir et al., 2013; Naguib, Come, et al., 2012). The transitional element also influences capacity for example V_2CT_X has the highest Li^+ capacity among all MXenes (Naguib et al., 2013). The theoretical study revealed that low formula weight MXene has higher gravimetric capacitance (Eames & Islam, 2014). But formula weight does not merely determine specific capacity. Nb-based MXene Nb_2CT_X has a higher capacity than Ti_2CT_X, though Nb is heavier than Ti (Naguib et al., 2013). During intercalation-deintercalation of ions in side MXene layers the oxidation states of the transition element changes continuously (Xie, Naguib, et al., 2014). An added advantage of MXene as an electrode for Li-battery is its excellent electrical conductivity and formation of Li layer on MXene surface. Which further boost electrochemical performance (Xie, Naguib, et al., 2014). Subsequently, further advancement was done by hybridizing MXene with other materials or by molecular engineering of MXene electrodes in order to achieve higher capacity (Ren et al., 2016). MXene can accommodate different-sized metal ions, which enables their widespread application in the development of batteries (Xie, Dall'Agnese, et al., 2014). But the hindrance is the abundance of electrode material for other metal ions batteries. For different metal ions capacity of $M_{n+1}X_nT_x$ also varied. Figure 2.2c depicts different capacity of various metal ions in MXene based electrode. Fast diffusion of metal ions enables the high rate capability of MXenes electrodes (Wang et al., 2015). It is crucial to note that the GCD curve of MXene does not exhibit a plateau, which resembles their capacitive behavior.

2.4.2 Sodium-Ion Battery

The growing demand for wearable and portable electronics puts cost pressure on LIBs. Sodium-ion batteries have proven greater potential in the near future. It is also fourth abundant element on the Earth. Sodium is also cheap and has suitable redox potential. But Na^+ has a greater radius than Li^+ which caused sluggish diffusion of

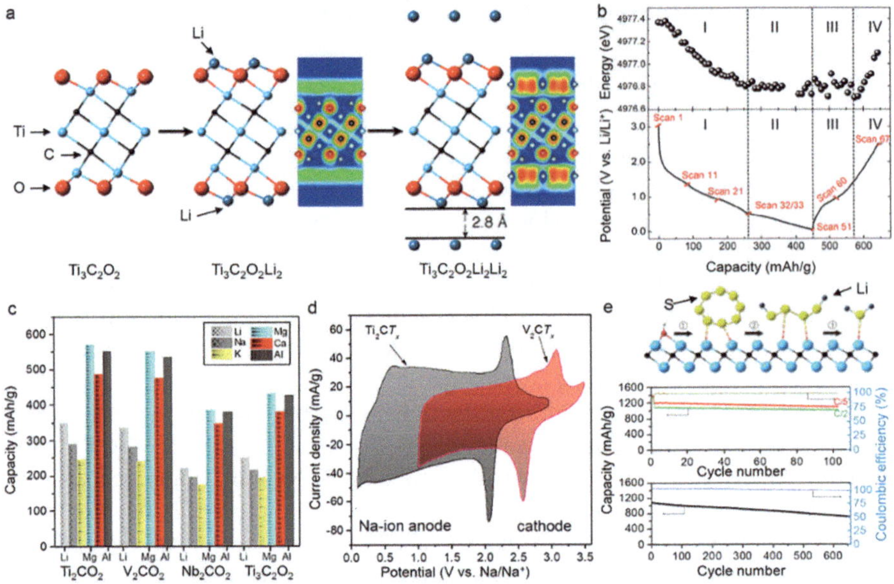

Fig. 2.2 (**a**) Insertion of Li ions in between MXene layers. (**b**) In situ XANES of Ti edge energy variation during Li^+ intercalation-deintercalation. (**c**) capacities of various ions based oxygen-containing MXene electrode. (**d**) CV of Ti_2C and V_2C for Na^+ ion electrolyte. (**e**) Performance of Ti_2C/sulfur composite in Li–S batteries

Na^+ and hence fading cycling stability. MXene with a layered structure can accommodate larger ions and open up new opportunities for sodium-ion batteries with remarkable performance. However, the compact layer inhibits the sodium-ion insertion process and limits capacity value. To overcome this challenge, mechanical grinding by assisting DMSO solvent was done by Zhang and coworkers to form a few layers of MXene (Wu et al., 2017). This modified MXene demonstrated higher capacity 267 mAh g^{-1} at 100 mA g^{-1}, and excellent stability. Another approach to enhance MXene capacity involves the creation of porous nanostructured MXene through the S loading removal technique. The porous MXene facilitates ion insertion smoothly and exposes greater active sites to improve electrochemical performance. PAN et al. reported sulfur doping of MXene improved Na ion battery performance (Li, Yan, et al., 2018). The s-doped MXene also resembled better reversibility, 121.3 mA hg^{-1} at 2 A g^{-1} and 113.9 mA hg^{-1} at 4 A g^{-1}. The other promising technology is the development of a based composite to deliver better performance. Ti_3C_2/NiCoP composite reported by Zhao et al. exhibited excellent stability, a specific capacity of 261.7 mA hg^{-1} was retained at a current density of 1 A g^{-1} for 2000 cycles (Zhao et al., 2019).

2.4.3 Potassium-Ion Battery

Potassium-ion batteries also work with similar principles to sodium and lithium-ion batteries. High abundance, low cost, and wide potential window (−2.93 V vs. SHE) make PIBs a potential storage system for next-generation electrochemical storage devices. While PIBs have numerous advantages, it also face some disadvantages. Since K^+ is larger than Na^+ and Li^+, it leads to slow diffusion and pulverized electrodes (Zhang et al., 2018). Mxene with a layered structure has been widely investigated for PIBs as an anode material. Pure Ti_3CNT_X exhibited a poor specific capacity of 32 mA hg^{-1} (Naguib et al., 2017). However, the preparation method of Mxene can enhance the specific capacity of PIBs. Dong et al. prepared nanoribbon type Mxene after treating and shaking with KOH (Dong et al., 2017). They achieved a specific capacity of 78 mA hg^{-1} at 200 mA g^{-1} current density.

2.4.4 Supercapacitor

Another promising electrochemical energy storage device is a supercapacitor. Based on the storage mechanism, there are two types of supercapacitors: electrical layer capacitors (EDLC) and pseudocapacitors. EDLC mechanism in which charge is stored by double-layer formation at the electrode–electrolyte interface, and in pseudocapacitor, the charge is stored via a surface-based redox reaction. The exceptional electrical conductivity and the variety of functional groups make MXene a potential material for supercapacitors (Awan et al., 2024). MXene layers are intercalated by different organic and metal ions. These charged species occupy electrochemically active sites on the MXene surface and participate in a charge storage mechanism. Freestanding binder free MXene electrode are much denser than carbon material or other MXene based composite electrode.

Surface chemistry of synthesized MXene plays a very crucial role in electrochemical properties. Oxygen-containing groups. It is proved that with fluorine presence of oxygen functionalities enhanced gravimetric capacitance. HF etched MXene, when treated with KOH, N_2H_4, or DMSO; the gravimetric capacitance improved two to seven times depending on electrolytes (Come et al., 2015). Other than $Ti_3C_2T_X$, other MXenes like Mo_2CT_X and $Mo_2TiC_2T_X$ also delivered high volumetric capacitance (Kim et al., 2017). Doping with a hetero atom is an current approach for enhancing the electrochemical performance of MXene. Heteroatom doping enhance electronic properties of MXene through activating charge transport property and electrochemically active site (Tian et al., 2024). Composite with carbonaceous or polymeric material with MXene resulted in higher rate capability and mechanical integrity (Dadashi Firouzjaei et al., 2025). Transparent electrodes for transparent electronics developed by combining graphene and MXene as energy storage active material. Hydrophilic polymers like PVA and PDDA (poly(diallyldimethylammonium chloride) can interact with oxygen-containing surface functionalities and impart mechanical stability to MXene-based composite electrodes (Ling et al., 2014).

2.5 Challenges and Future Perspectives

As mentioned in the above sections, MXene is a new type of 2D transition metal carbides and nitrides. It has become known for its outstanding characteristics such as high conductivity, exceptional mechanical strength, and a variety of functionalities. Yet, many obstacles prevent its broad use and market expansion, such as scalability issues, cost-efficiency concerns, and potential toxicity. Achieving the complete potential of MXene is hindered by the significant challenges in scalability and cost-effectiveness. Current synthesis techniques, which frequently require harsh chemical etching procedures, are energy-intensive, lengthy, and may result in differences in quality between batches. Improving synthesis methods to be more efficient and scalable is vital for widespread production and commercial success. Moreover, it is crucial to decrease the production cost of MXene in order to ensure its competitiveness in a wide range of applications.

Worries regarding the toxicity and environmental consequences of MXene materials have also been raised. Further investigation is required to evaluate the long-term toxicity and environmental impact of MXene despite early indications of its biocompatibility. Evaluation of both the chemicals utilized in the synthesis procedure and the possible discharge of MXene particles into the surroundings is crucial. Creating sustainable and eco-friendly synthesis techniques is crucial for reducing adverse effects. Several promising research directions are being considered to overcome these challenges and fully utilize the potential of MXene.

Investigating different methods for synthesis, like electrochemical or hydrothermal approaches, that offer greater scalability, energy efficiency, and eco-friendliness. Developing techniques to add different functional groups to MXene in order to customize its characteristics for specific purposes, such as improving its stability, compatibility with biological systems, or catalytic performance. Performing thorough studies to gain a deep understanding of the basic properties of MXene, such as its electronic structure, surface chemistry, and mechanical behavior. Creating multifunctional composites and devices involves developing techniques to seamlessly incorporate MXene with various materials like polymers, ceramics, and carbon-based materials. Creating thorough frameworks for assessing risks and regulations to promote safe and responsible usage of MXene.

By overcoming these obstacles and leveraging its distinct characteristics, MXene could transform multiple sectors and support sustainable growth.

Producing MXene typically requires aggressive chemical etching methods, which restrict its potential for widespread production and commercial success. Furthermore, MXene materials may be susceptible to oxidation and deterioration, particularly when in contact with air or moisture. Maintaining consistent quality and uniformity in the synthesis of MXene is difficult because of the inherent variability of the materials. In spite of these difficulties, the potential of MXene looks bright. Dealing with these problems will create multiple chances in different areas, like energy storage, electronics, environmental clean-up, medical uses, and composite materials. By addressing the existing difficulties and leveraging its distinct characteristics, MXene could transform different sectors and support sustainability efforts.

2.6 Conclusions

As per the above through discussion on MXene, It is perspicuous lucid that MXenes are a very versatile class of 2D materials that hold promising aproperties for a wide range of applications because of their unique properties including large surface area, high conductivity, and easy functionalization. However, in order to fully utilize their potential, some challenges need to be addressed such as stability and oxidation, particularly for applications that require long-term, sustained performance. To increase MXenes'resistance to oxidation, creative techniques must be used. Another important area that needs focus is the growth of MXene production. To enable large-scale production, cleaner and more cost-effective substitutes for the existing technologies that employ hazardous chemicals like HF must be found. Energy density and specific capacitance in energy storage can be greatly increased by increasing the surface area of MXenes.

Comparably, in order to go from small-scale laboratory research to actual use in commercial settings, membranes derived from MXene in water purification need more development and evaluation in real-world situations. To ensure MXenes'safety and biocompatibility for biomedical applications, extensive research is required. Although wearable sensors containing MXenes show potential, their hydrophilic nature and resilience in harsh conditions present challenges. Materials' mechanical properties can be enhanced and their endurance increased by using polymer reinforcement. Future research should focus on improving the synthesis processes, clarifying the relationship between the production of MXenes and their characteristics, and investigating new uses. It is essential to address these issues by carrying out targeted research and creating fresh solutions in order to fully realize the promise of MXenes and promote their broader application in many industries.

Despite the substantial advantages that MXenes offer, their successful integration into commercial and industrial applications ultimately depends on how well the current challenges are resolved. MXenes have the potential to become a key component of future technology with continued research and progress.

References

Alhabeb, M., Maleski, K., Mathis, T. S., Sarycheva, A., Hatter, C. B., Uzun, S., Levitt, A., & Gogotsi, Y. (2018). Selective etching of silicon from Ti3SiC2 (MAX) to obtain 2D titanium carbide (MXene). *Angewandte Chemie - International Edition, 57*(19), 5444–5448. https://doi.org/10.1002/anie.201802232

Anasori, B., Shi, C., Moon, E. J., Xie, Y., Voigt, C. A., Kent, P. R. C., May, S. J., Billinge, S. J. L., Barsoum, M. W., & Gogotsi, Y. (2016). Control of electronic properties of 2D carbides (MXenes) by manipulating their transition metal layers. *Nanoscale Horizons, 1*(3), 227–234. https://doi.org/10.1039/c5nh00125k

Anasori, B., Xie, Y., Beidaghi, M., Lu, J., Hosler, B. C., Hultman, L., Kent, P. R. C., Gogotsi, Y., & Barsoum, M. W. (2015). Two-dimensional, ordered, double transition metals carbides (MXenes). *ACS Nano, 9*(10), 9507–9516. https://doi.org/10.1021/acsnano.5b03591

Awan, H. T. A., Abdah, M. A. A. M., Mehar, M., Walvekar, R., Chaudhary, V., Khalid, M., & Khosla, A. (2024). MXene-polymer hybrid composites for advanced energy storage: Insights into supercapacitors and batteries. *Journal of Energy Storage, 95*, 112449. https://doi.org/10.1016/j.est.2024.112449

Bai, Y., Zhou, K., Srikanth, N., Pang, J. H. L., He, X., & Wang, R. (2016). Dependence of elastic and optical properties on surface terminated groups in two-dimensional MXene monolayers: A first-principles study. *RSC Advances, 6*(42), 35731–35739. https://doi.org/10.1039/c6ra03090d

Berdiyorov, G. R. (2016). Optical properties of functionalized Ti3C2T2 (T = F, O, OH) MXene: First-principles calculations. *AIP Advances, 6*(5), 4948799. https://doi.org/10.1063/1.4948799

Borysiuk, V. N., Mochalin, V. N., & Gogotsi, Y. (2015). Molecular dynamic study of the mechanical properties of two-dimensional titanium carbides tin+1Cn (MXenes). *Nanotechnology, 26*(26), 1–10. https://doi.org/10.1088/0957-4484/26/26/265705

Chakraborty, P., Das, T., Nafday, D., & Saha-Dasgupta, T. (2018). Manipulating the mechanical properties of Ti2C MXene: Effect of substitutional doping. *AIP Conference Proceedings, 2005*, 5050748. https://doi.org/10.1063/1.5050748

Chandrasekaran, A., Mishra, A., & Singh, A. K. (2017). Ferroelectricity, antiferroelectricity, and ultrathin 2D electron/hole gas in multifunctional monolayer MXene. *Nano Letters, 17*(5), 3290–3296. https://doi.org/10.1021/acs.nanolett.7b01035

Chen, Y., Shi, Y., Song, G., Yang, B., & Pang, H. (2025). Compositing MXene with organic coordination frameworks (MOFs, COFs, and HOFs) for enhanced electrochemical performance. *Next Materials, 6*(May), 100323. https://doi.org/10.1016/j.nxmate.2024.100323

Come, J., Black, J. M., Lukatskaya, M. R., Naguib, M., Beidaghi, M., Rondinone, A. J., Kalinin, S. V., Wesolowski, D. J., Gogotsi, Y., & Balke, N. (2015). Controlling the actuation properties of MXene paper electrodes upon cation intercalation. *Nano Energy, 17*, 27–35. https://doi.org/10.1016/j.nanoen.2015.07.028

Dadashi Firouzjaei, M., Clayton, J., Jafarian, H., Arabi Shamsabadi, A., Thakur, A., Todd, R., Nemani, S. K., Sadrzadeh, M., Elliott, M., Anasori, B., & Terry, L. (2025). A perspective on MXene-enhanced biofiltration-membrane water reuse treatment systems: A review and experimental validation. *Desalination, 593*, 118198. https://doi.org/10.1016/j.desal.2024.118198

Dong, Y., Mallineni, S. S. K., Maleski, K., Behlow, H., Mochalin, V. N., Rao, A. M., Gogotsi, Y., & Podila, R. (2018). Metallic MXenes: A new family of materials for flexible triboelectric nanogenerators. *Nano Energy, 44*, 103–110. https://doi.org/10.1016/j.nanoen.2017.11.044

Dong, Y., Wu, Z., Zheng, S., Wang, X., Qin, J., Wang, S., Shi, X., & Bao, X. (2017). Ti3 C2 MXene-derived sodium/potassium titanate nanoribbons for high-performance sodium/potassium ion batteries with enhanced capacities. *ACS Nano*. https://doi.org/10.1021/acsnano.7b01165

Eames, C., & Islam, M. S. (2014). Ion intercalation into two-dimensional transition-metal carbides: Global screening for new high-capacity battery materials. *Journal of the American Chemical Society, 136*(46), 16270–16276. https://doi.org/10.1021/ja508154e

Fu, Z., Zhang, S., Legut, D., Germann, T. C., Si, C., Du, S., Francisco, J. S., & Zhang, R. (2018). A synergetic stabilization and strengthening strategy for two-dimensional ordered hybrid transition metal carbides. *Physical Chemistry Chemical Physics, 20*(47), 29684–29692. https://doi.org/10.1039/c8cp06458j

Fu, Z. H., Zhang, Q. F., Legut, D., Si, C., Germann, T. C., Lookman, T., Du, S. Y., Francisco, J. S., & Zhang, R. F. (2016). Stabilization and strengthening effects of functional groups in two-dimensional titanium carbide. *Physical Review B, 94*(10), 1–10. https://doi.org/10.1103/PhysRevB.94.104103

Gandi, A. N., Alshareef, H. N., & Schwingenschlögl, U. (2016). Thermoelectric performance of the MXenes M2CO2 (M = Ti, Zr, or Hf). *Chemistry of Materials, 28*(6), 1647–1652. https://doi.org/10.1021/acs.chemmater.5b04257

Gao, G., O'Mullane, A. P., & Du, A. (2017). 2D MXenes: A new family of promising catalysts for the hydrogen evolution reaction. *ACS Catalysis, 7*(1), 494–500. https://doi.org/10.1021/acscatal.6b02754

Ghidiu, M., Lukatskaya, M. R., Zhao, M. Q., Gogotsi, Y., & Barsoum, M. W. (2015). Conductive two-dimensional titanium carbide "clay" with high volumetric capacitance. *Nature, 516*(7529), 78–81. https://doi.org/10.1038/nature13970

Guo, X., Gao, H., Wang, S., Yang, G., Zhang, X., Zhang, J., Liu, H., & Wang, G. (2022). MXene-based aerogel anchored with antimony single atoms and quantum dots for high-performance potassium-ion batteries. *Nano Letters, 22*(3), 1225–1232. https://doi.org/10.1021/acs.nanolett.1c04389

Halim, J., Kota, S., Lukatskaya, M. R., Naguib, M., Zhao, M. Q., Moon, E. J., Pitock, J., Nanda, J., May, S. J., Gogotsi, Y., & Barsoum, M. W. (2016). Synthesis and characterization of 2D molybdenum carbide (MXene). *Advanced Functional Materials, 26*(18), 3118–3127. https://doi.org/10.1002/adfm.201505328

Hope, M. A., Forse, A. C., Griffith, K. J., Lukatskaya, M. R., Ghidiu, M., Gogotsi, Y., & Grey, C. P. (2016). NMR reveals the surface functionalisation of Ti3C2 MXene. *Physical Chemistry Chemical Physics, 18*(7), 5099–5102. https://doi.org/10.1039/c6cp00330c

Jhon, Y. I., Seo, M., & Jhon, Y. M. (2018). First-principles study of a MXene terahertz detector. *Nanoscale, 10*(1), 69–75. https://doi.org/10.1039/c7nr05351g

Jin, W., Wu, S., & Wang, Z. (2018). Structural, electronic and mechanical properties of two-dimensional Janus transition metal carbides and nitrides. *Physica E: Low-dimensional Systems and Nanostructures, 103*(March), 307–313. https://doi.org/10.1016/j.physe.2018.06.024

Karlsson, L. H., Birch, J., Halim, J., Barsoum, M. W., & Persson, P. O. Å. (2015). Atomically resolved structural and chemical investigation of single MXene sheets. *Nano Letters, 15*(8), 4955–4960. https://doi.org/10.1021/acs.nanolett.5b00737

Khazaei, M., Arai, M., Sasaki, T., Estili, M., & Sakka, Y. (2014). Two-dimensional molybdenum carbides: Potential thermoelectric materials of the MXene family. *Physical Chemistry Chemical Physics, 16*(17), 7841–7849. https://doi.org/10.1039/c4cp00467a

Khazaei, M., Arai, M., Sasaki, T., Chung, C.-Y., Venkataramanan, N. S., Estili, M., Sakka, Y., & Kawazoe, Y. (2012). Novel electronic and magnetic properties of two-dimensional transition metal carbides and nitrides. *Advanced Functional Materials, 23*, 2185–2192.

Kim, H., Anasori, B., Gogotsi, Y., & Alshareef, H. N. (2017). Thermoelectric properties of two-dimensional molybdenum-based MXenes. *Chemistry of Materials, 29*(15), 6472–6479. https://doi.org/10.1021/acs.chemmater.7b02056

Kim, H., Wang, Z., & Alshareef, H. N. (2019). MXetronics: Electronic and photonic applications of MXenes. *Nano Energy, 60*, 179–197. https://doi.org/10.1016/j.nanoen.2019.03.020

Ko, T. Y., Ye, H., Murali, G., Lee, S. Y., Park, Y. H., Lee, J., Lee, J., Yun, D. J., Gogotsi, Y., Kim, S. J., Kim, S. H., Jeong, Y. J., Park, S. J., & In, I. (2024). Functionalized MXene ink enables environmentally stable printed electronics. *Nature Communications, 15*(1), 1–12. https://doi.org/10.1038/s41467-024-47700-y

Kumar, S., & Schwingenschlögl, U. (2016). Thermoelectric performance of functionalized Sc2 C MXenes. *Physical Review B, 94*(3), 35405. https://doi.org/10.1103/PhysRevB.94.035405

Lee, Y., Hwang, Y., & Chung, Y. C. (2015). Achieving type I, II, and III heterojunctions using functionalized mxene. *ACS Applied Materials and Interfaces, 7*(13), 7163–7169. https://doi.org/10.1021/acsami.5b00063

Li, J., Yan, D., Hou, S., Li, Y., Lu, T., Yao, Y., & Pan, L. (2018). Improved sodium-ion storage performance of Ti3C2TX MXenes by sulfur doping. *Journal of Materials Chemistry A, 6*(3), 1234–1243. https://doi.org/10.1039/c7ta08261d

Li, M., Lu, J., Luo, K., Li, Y., Chang, K., Chen, K., Zhou, J., Rosen, J., Hultman, L., Eklund, P., Persson, P. O. Å., Du, S., Chai, Z., Huang, Z., & Huang, Q. (2019). Element replacement approach by reaction with Lewis acidic molten salts to synthesize nanolaminated MAX phases and MXenes. *Journal of the American Chemical Society, 141*(11), 4730–4737. https://doi.org/10.1021/jacs.9b00574

Li, T., Yao, L., Liu, Q., Gu, J., Luo, R., Li, J., Yan, X., Wang, W., Liu, P., Chen, B., Zhang, W., Abbas, W., Naz, R., & Zhang, D. (2018). Fluorine-free synthesis of high-purity Ti3C2Tx (T=OH, O) via alkali treatment. *Angewandte Chemie - International Edition, 57*(21), 6115–6119. https://doi.org/10.1002/anie.201800887

Ling, Z., Ren, C. E., Zhao, M. Q., Yang, J., Giammarco, J. M., Qiu, J., Barsoum, M. W., & Gogotsi, Y. (2014). Flexible and conductive MXene films and nanocomposites with high capacitance. *Proceedings of the National Academy of Sciences of the United States of America, 111*(47), 16676–16681. https://doi.org/10.1073/pnas.1414215111

Lipatov, A., Lu, H., Alhabeb, M., Anasori, B., Gruverman, A., Gogotsi, Y., & Sinitskii, A. (2018). Elastic properties of 2D Ti 3 C 2 T x MXene monolayers and bilayers. *Science Advances, 4*, 491.

Liu, F., Zhou, A., Chen, J., Jia, J., Zhou, W., Wang, L., & Hu, Q. (2017). Preparation of Ti3 C2 and Ti2 C MXenes by fluoride salts etching and methane adsorptive properties. *Applied Surface Science, 416*, 781–789. https://doi.org/10.1016/j.apsusc.2017.04.239

Liu, J. H., Kan, X., Amin, B., Gan, L. Y., & Zhao, Y. (2017). Theoretical exploration of the potential applications of Sc-based MXenes. *Physical Chemistry Chemical Physics, 19*(48), 32253–32261. https://doi.org/10.1039/c7cp06224a

Liu, Z., Zheng, L., Sun, L., Qian, Y., Wang, J., & Li, M. (2014). (Cr2/3Ti1/3)3AlC2 and (Cr5/8Ti3/8)4AlC3: New MAX-phase compounds in Ti-Cr-Al-C system. *Journal of the American Ceramic Society, 97*(1), 67–69. https://doi.org/10.1111/jace.12731

Ma, Z., Hu, Z., Zhao, X., Tang, Q., Wu, D., Zhou, Z., & Zhang, L. (2014). Tunable band structures of heterostructured bilayers with transition-metal dichalcogenide and mxene monolayer. *Journal of Physical Chemistry C, 118*(10), 5593–5599. https://doi.org/10.1021/jp500861n

Mashtalir, O., Naguib, M., Mochalin, V. N., Dall'Agnese, Y., Heon, M., Barsoum, M. W., & Gogotsi, Y. (2013). Intercalation and delamination of layered carbides and carbonitrides. *Nature Communications, 4*, 1–7. https://doi.org/10.1038/ncomms2664

Naguib, M., Adams, R. A., Zhao, Y., Zemlyanov, D., Varma, A., Nanda, J., & Pol, V. G. (2017). Electrochemical performance of MXenes as K-ion battery anodes. *Chemical Communications, 53*(51), 6883–6886. https://doi.org/10.1039/c7cc02026k

Naguib, M., Come, J., Dyatkin, B., Presser, V., Taberna, P. L., Simon, P., Barsoum, M. W., & Gogotsi, Y. (2012). MXene: A promising transition metal carbide anode for lithium-ion batteries. *Electrochemistry Communications, 16*(1), 61–64. https://doi.org/10.1016/j.elecom.2012.01.002

Naguib, M., Halim, J., Lu, J., Cook, K. M., Hultman, L., Gogotsi, Y., & Barsoum, M. W. (2013). New two-dimensional niobium and vanadium carbides as promising materials for li-ion batteries. *Journal of the American Chemical Society, 135*(43), 15966–15969. https://doi.org/10.1021/ja405735d

Naguib, M., Kurtoglu, M., Presser, V., Lu, J., Niu, J., Heon, M., Hultman, L., Gogotsi, Y., & Barsoum, M. W. (2011). Two-dimensional nanocrystals produced by exfoliation of Ti 3AlC 2. *Advanced Materials, 23*(37), 4248–4253. https://doi.org/10.1002/adma.201102306

Naguib, M., Mashtalir, O., Carle, J., Presser, V., Lu, J., Hultman, L., Gogotsi, Y., & Barsoum, M. W. (2012). Two-dimensional transition metal carbides. *ACS Nano, 6*(2), 1322–1331. https://doi.org/10.1021/nn204153h

Naguib, M., Unocic, R. R., Armstrong, B. L., & Nanda, J. (2015). Large-scale delamination of multi-layers transition metal carbides and carbonitrides "mXenes". *Dalton Transactions, 44*(20), 9353–9358. https://doi.org/10.1039/c5dt01247c

Peng, C., Wei, P., Chen, X., Zhang, Y., Zhu, F., Cao, Y., Wang, H., Yu, H., & Peng, F. (2018). A hydrothermal etching route to synthesis of 2D MXene (Ti3C2, Nb2C): Enhanced exfoliation and improved adsorption performance. *Ceramics International, 44*(15), 18886–18893. https://doi.org/10.1016/j.ceramint.2018.07.124

Peng, Q., & De, S. (2013). Outstanding mechanical properties of monolayer MoS2 and its application in elastic energy storage. *Physical Chemistry Chemical Physics, 15*(44), 19427–19437. https://doi.org/10.1039/c3cp52879k

Perera, A. A. P. R., Madhushani, K. A. U., Punchihewa, B. T., Kumar, A., & Gupta, R. K. (2023). MXene-based nanomaterials for multifunctional applications. *Materials, 16*(3), 1138. https://doi.org/10.3390/ma16031138

Plummer, G., Anasori, B., Gogotsi, Y., & Tucker, G. J. (2019). Nanoindentation of monolayer tin+1CnTx MXenes via atomistic simulations: The role of composition and defects on strength. *Computational Materials Science, 157*, 168–174. https://doi.org/10.1016/j.commatsci.2018.10.033

Ren, C. E., Zhao, M. Q., Makaryan, T., Halim, J., Boota, M., Kota, S., Anasori, B., Barsoum, M. W., & Gogotsi, Y. (2016). Porous two-dimensional transition metal carbide (MXene) flakes for high-performance Li-ion storage. *ChemElectroChem, 3*(5), 689–693. https://doi.org/10.1002/celc.201600059

Si, C., Duan, W., Liu, Z., & Liu, F. (2012). Electronic strengthening of graphene by charge doping. *Physical Review Letters, 109*(22), 1–5. https://doi.org/10.1103/PhysRevLett.109.226802

Si, C., Zhou, J., & Sun, Z. (2015). Half-metallic ferromagnetism and surface functionalization-induced metal-insulator transition in graphene-like two-dimensional Cr2C crystals. *ACS Applied Materials and Interfaces, 7*(31), 17510–17515. https://doi.org/10.1021/acsami.5b05401

Sun, W., Shah, S. A., Chen, Y., Tan, Z., Gao, H., Habib, T., Radovic, M., & Green, M. J. (2017). Electrochemical etching of Ti2AlC to Ti2CT:X (MXene) in low-concentration hydrochloric acid solution. *Journal of Materials Chemistry A, 5*(41), 21663–21668. https://doi.org/10.1039/c7ta05574a

Tao, Q., Dahlqvist, M., Lu, J., Kota, S., Meshkian, R., Halim, J., Palisaitis, J., Hultman, L., Barsoum, M. W., Persson, P. O. Å., & Rosen, J. (2017). Two-dimensional Mo1.33C MXene with divacancy ordering prepared from parent 3D laminate with in-plane chemical ordering. *Nature Communications, 8*, 1–7. https://doi.org/10.1038/ncomms14949

Tian, J., Li, J., Guo, Y., Liu, Z., Liu, B., & Li, J. (2024). Oxygen vacancy mediated bismuth-based photocatalysts. *Advanced Powder Materials, 3*(4), 100201. https://doi.org/10.1016/j.apmate.2024.100201

Urbankowski, P., Anasori, B., Makaryan, T., Er, D., Kota, S., Walsh, P. L., Zhao, M., Shenoy, V. B., Barsoum, M. W., & Gogotsi, Y. (2016). Synthesis of two-dimensional titanium nitride Ti4N3 (MXene). *Nanoscale, 8*(22), 11385–11391. https://doi.org/10.1039/c6nr02253g

Wang, D., Zhou, C., Filatov, A. S., Cho, W., Lagunas, F., Wang, M., Vaikuntanathan, S., Liu, C., Klie, R. F., & Talapin, D. V. (2023). Direct synthesis and chemical vapor deposition of 2D carbide and nitride MXenes. *Science, 379*(6638), 1242–1247. https://doi.org/10.1126/science.add9204

Wang, X., Kajiyama, S., Iinuma, H., Hosono, E., Oro, S., Moriguchi, I., Okubo, M., & Yamada, A. (2015). Pseudocapacitance of MXene nanosheets for high-power sodium-ion hybrid capacitors. *Nature Communications, 6*, 1–6. https://doi.org/10.1038/ncomms7544

Weng, H., Ranjbar, A., Liang, Y., Song, Z., Khazaei, M., Yunoki, S., Arai, M., Kawazoe, Y., Fang, Z., & Dai, X. (2015). Large-gap two-dimensional topological insulator in oxygen functionalized MXene. *Physical Review B - Condensed Matter and Materials Physics, 92*(7), 1–7. https://doi.org/10.1103/PhysRevB.92.075436

Wu, D., Wang, S., Zhang, S., Yuan, J., Yang, B., & Chen, H. (2018). Highly negative Poisson's ratio in a flexible two-dimensional tungsten carbide monolayer. *Physical Chemistry Chemical Physics, 20*(28), 18924–18930. https://doi.org/10.1039/c8cp01353e

Wu, Y., Nie, P., Wang, J., Dou, H., & Zhang, X. (2017). Few-layer MXenes delaminated via high-energy mechanical milling for enhanced sodium-ion batteries performance. *ACS Applied Materials and Interfaces, 9*(45), 39610–39617. https://doi.org/10.1021/acsami.7b12155

Xie, Y., Dall'Agnese, Y., Naguib, M., Gogotsi, Y., Barsoum, M. W., Zhuang, H. L., & Kent, P. R. C. (2014). Prediction and characterization of mxene nanosheet anodes for non-lithium-ion batteries. *ACS Nano, 8*(9), 9606–9615. https://doi.org/10.1021/nn503921j

Xie, Y., Naguib, M., Mochalin, V. N., Barsoum, M. W., Gogotsi, Y., Yu, X., Nam, K. W., Yang, X. Q., Kolesnikov, A. I., & Kent, P. R. C. (2014). Role of surface structure on li-ion energy storage capacity of two-dimensional transition-metal carbides. *Journal of the American Chemical Society, 136*(17), 6385–6394. https://doi.org/10.1021/ja501520b

Xu, C., Wang, L., Liu, Z., Chen, L., Guo, J., Kang, N., Ma, X. L., Cheng, H. M., & Ren, W. (2015). Large-area high-quality 2D ultrathin Mo2C superconducting crystals. *Nature Materials, 14*(11), 1135–1141. https://doi.org/10.1038/nmat4374

Yang, S., Zhang, P., Wang, F., Ricciardulli, A. G., Lohe, M. R., Blom, P. W. M., & Feng, X. (2018). Fluoride-free synthesis of two-dimensional titanium carbide (MXene) using a binary aqueous system. *Angewandte Chemie, 130*(47), 15717–15721. https://doi.org/10.1002/ange.201809662

Yorulmaz, U., Özden, A., Perkgöz, N. K., Ay, F., & Sevik, C. (2016). Vibrational and mechanical properties of single layer MXene structures: A first-principles investigation. *Nanotechnology, 27*(33), 335702. https://doi.org/10.1088/0957-4484/27/33/335702

You, A., Be, M., & In, I. (2016). Electronic and magnetic properties of Cr2 Ge2 Se6. *Applied Physics Reviews, 546*(7657), 801106.

Zha, X. H., Huang, Q., He, J., He, H., Zhai, J., Francisco, J. S., & Du, S. (2016). The thermal and electrical properties of the promising semiconductor MXene Hf2CO2. *Scientific Reports, 6*, 1–10. https://doi.org/10.1038/srep27971

Zha, X. H., Luo, K., Li, Q., Huang, Q., He, J., Wen, X., & Du, S. (2015). Role of the surface effect on the structural, electronic and mechanical properties of the carbide MXenes. *EPL, 111*(2), 26007. https://doi.org/10.1209/0295-5075/111/26007

Zha, X. H., Yin, J., Zhou, Y., Huang, Q., Luo, K., Lang, J., Francisco, J. S., He, J., & Du, S. (2016). Intrinsic structural, electrical, thermal, and mechanical properties of the promising conductor Mo2C MXene. *Journal of Physical Chemistry C, 120*(28), 15082–15088. https://doi.org/10.1021/acs.jpcc.6b04192

Zhang, H., Fu, Z., Zhang, R., Zhang, Q., Tian, H., Legut, D., Germann, T. C., Guo, Y., Du, S., & Francisco, J. S. (2017). Designing flexible 2D transition metal carbides with strain-controllable lithium storage. *Proceedings of the National Academy of Sciences of the United States of America, 114*(52), E11082–E11091. https://doi.org/10.1073/pnas.1717219115

Zhang, H., Yang, G., Zuo, X., Tang, H., Yang, Q., & Li, G. (2016). Computational studies on the structural, electronic and optical properties of graphene-like MXenes (M2CT2, M = Ti, Zr, Hf; T = O, F, OH) and their potential applications as visible-light driven photocatalysts. *Journal of Materials Chemistry A, 4*(33), 12913–12920. https://doi.org/10.1039/c6ta04628b

Zhang, L., Yu, J., Yang, M., Xie, Q., Peng, H., & Liu, Z. (2013). Janus graphene from asymmetric two-dimensional chemistry. *Nature Communications, 4*, 1–7. https://doi.org/10.1038/ncomms2464

Zhang, Q., Mao, J., Pang, W. K., Zheng, T., Sencadas, V., Chen, Y., Liu, Y., & Guo, Z. (2018). Boosting the potassium storage performance of alloy-based anode materials via electrolyte salt chemistry. *Advanced Energy Materials, 8*(15), 1–10. https://doi.org/10.1002/aenm.201703288

Zhao, D., Zhao, R., Dong, S., Miao, X., Zhang, Z., Wang, C., & Yin, L. (2019). Alkali-induced 3D crinkled porous Ti3C2 MXene architectures coupled with NiCoP bimetallic phosphide nanoparticles as anodes for high-performance sodium-ion batteries. *Energy and Environmental Science, 12*(8), 2422–2432. https://doi.org/10.1039/c9ee00308h

Zhou, J., Zha, X., Zhou, X., Chen, F., Gao, G., Wang, S., Shen, C., Chen, T., Zhi, C., Eklund, P., Du, S., Xue, J., Shi, W., Chai, Z., & Huang, Q. (2017). Synthesis and electrochemical properties of two-dimensional hafnium carbide. *ACS Nano, 11*(4), 3841–3850. https://doi.org/10.1021/acsnano.7b00030

Chapter 3
Graphene: Properties and Applications

Sujeet Kumar Pandey, Aash Mohammad, and Rohit Kumar Singh

3.1 Introduction of Graphene

Graphene, a single layer of carbon atoms arranged in a two-dimensional lattice, has attracted various researchers' interest due to its remarkable properties and potential applications in different fields. With the unique morphology, combined with its excellent mechanical, electrical, and thermal properties, graphene as turned out to be a leading fabric in nanotechnology and materials science.

The structure and properties of graphene are composed of a single layer of carbon atoms arranged in a honeycomb lattice, characterized by sp^2 hybridization. To understand the structural integrity, coordination number and interatomic distance are very important. Graphene has a very high electrical conductivity more than that of copper. Such a higher thermal conductivity and mechanical resistance makes graphene a suitable material for electronics, energy storage and nanotechnology (Wu et al., 2023). A high surface area to volume ratio of it makes a porous structure make it essential for the advanced battery and sensor applications.

Method of synthesizing graphene has been evolved considerably, which in turn address variety of applications and scalability requirement. Synthesis of graphene involves two approaches: top-down and bottom-up approaches. Each approach possesses different advantage and disadvantages. In top-down approaches, bulk material such as graphite break down it into small layers to graphene (Mbayachi et al., 2021). In this technique, the physical separation of the graphene layers was separated from the bulk graphite, whereas in chemical exfoliation technique, electrochemical method was used to separate the graphene layer (Parvez et al., 2014). In

S. K. Pandey (✉) · A. Mohammad · R. K. Singh
Department of Chemical and Biochemical Engineering, Rajiv Gandhi Institute of Petroleum Technology, Jais, Amethi, India
e-mail: 20ce0012@rgipt.ac.in

S. Kulkarni (ed.), *Next-Generation Nanomaterials for Sustainable Engineering*, Lecture Notes in Nanoscale Science and Technology 43,
https://doi.org/10.1007/978-3-032-05719-8_3

the laser-induced graphitization laser energy transform carbon source into graphene (Movaghgharnezhad & Kang, 2024).

Furthermore, button-up technique involves building graphene from smaller precursor, at the atomic and molecular level, and this offers a better control over the quality, structure, and properties. Common bottom-up approaches involve: (1) chemical vapor deposition (CVD), at higher temperature hydrocarbon gas decomposes into metal catalyst surface (a single-layer graphene); (2) epitaxial growth, a graphene layer is grown on a silicon carbide substrate by heating it into vacuum at higher temperature, leaving behind carbon which is arranged into graphene structure. Green synthesis approach produces eco-friendly graphene using plant extract and recycled materials which reduces the production cost and also helps in addressing the environmental concern (Kausar & Ahmad, 2024; Silva et al., 2024). Although the conventional method is effective method, often faces challenges such as contamination and scalability furthermore the innovative green method revolutionizes graphene synthesis for a variety of applications.

Graphene significantly improves the water treatment procedure, which enhance the effectiveness of removing the pollutants, such as heavy metals and organic component via adsorption and photocatalysis (Rana et al., 2024). Nanopore architectures can be developed to increase the permeation performance, making them a vital resource in environmental management. In the field of corrosion, graphene as a filler material acts as a barrier against the corrosion and transport corrosion inhibitors. This new approach can solve the persistent corrosion problem of the surface (Zhang et al., 2021). Graphene-based nanomaterials are gaining popularity in the field of biomedical in the drug delivery, biosensing, and tissues engineering. The exceptional mechanical, thermal, and electrical properties of graphene show a great potential in the field of medical technologies, electronics, sensors, and energy storage applications. The challenges associated with graphene such as a toxicity and production scalability are a major obstacle in the widespread utilization of graphene in these areas.

3.2 Properties of Graphene

Graphene is a single layer's carbon atom arranged in 2-D lattice which resembles the structure of honeycomb, exhibits a range of properties which make a desirable material across various fields. Its unique structural and physical properties increased its progressive use in energy, electronics, and nanotechnology.

Graphene, a monolayer of carbon atoms organized in a two-dimensional lattice, demonstrates an exceptional range of features that render it a highly coveted material across multiple domains. The distinctive structural and physical properties that enhance its prospective uses in energy, electronics, and nanotechnology are shown in Fig. 3.1.

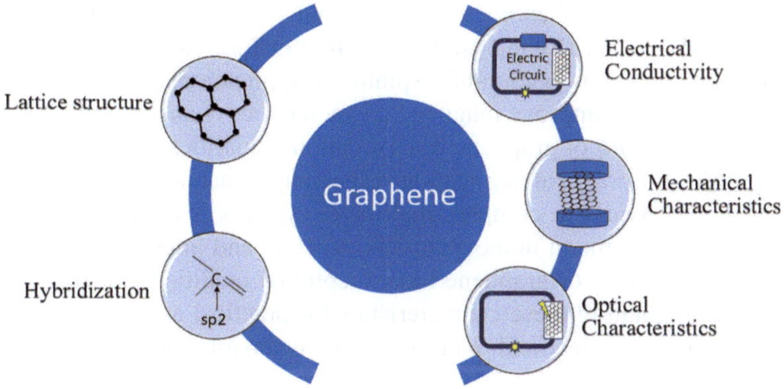

Fig. 3.1 Schematic diagram of properties of graphene

3.2.1 Structural Characteristics

As above stated, the lattice structure of graphene profoundly affects its electrical and vibrational characteristics. The numerous structural alterations such as twisting and doping increase the functionality of graphene in advance applications. The honeycomb configuration results in unique phonon frequencies that are essential for understanding its vibrational dynamics (Aziz & Khan, 2024). Graphene lattice dynamics shows an important feature, like soft mode, that is crucial for integrating graphene into electronic devices. Twisted multilayer graphene, particularly at magic angles can enhance electronic properties, such as flat bands and strong electronic correlations (Nguyen et al., 2022). Lattice relaxation in the twisted graphene configuration creates complex Moiré patterns, which affect the electron density of states and enable phenomena such as Mott insulators and superconductivity (Pantaleón et al., 2024). Doping graphene with transition metals changes its magnetic anisotropy, which is important for spintronic applications, and interactions between dopants can significantly enhance its magnetic properties (Langer et al., 2022).

The lattice structure of graphene is defined by its distinct two-dimensional honeycomb configuration, which profoundly affects its electrical and vibrational characteristics. This structure permits numerous alterations, including twisting and doping, which can augment its functionality in sophisticated applications. The lattice of graphene is a honeycomb configuration, resulting in unique phonon frequencies essential for comprehending its vibrational dynamics (Aziz & Khan, 2024). The lattice dynamics display notable characteristics, such as soft modes, which are crucial for the incorporation of graphene into electronic devices (Aziz & Khan, 2024). Twisted multilayer graphene, particularly at specific angles referred to as "magic angles," can enhance electronic properties, including flat bands and pronounced electron correlations (Nguyen et al., 2022). Lattice relaxation in twisted configurations produces intricate moiré patterns, influencing the electronic density of states and facilitating phenomena such as Mott insulators and superconductivity (Pantaleón et al., 2024). Doping graphene with transition metals modifies its magnetic anisotropy, which is essential for spintronic applications. The interplay among dopants can markedly improve magnetic characteristics (Langer et al., 2022).

The lattice structure of graphene has exceptional qualities; however, controlling these properties for practical applications especially in achieving specific electrical and magnetic behaviors in complex multilayer systems is a challenging task that requires precise adjustment in the materials structure and interaction.

The lattice structure of graphene has exceptional qualities; nonetheless, obstacles persist in regulating these characteristics for practical applications, especially in attaining specific electrical and magnetic behaviors in intricate multilayer systems.

The atomic thickness of the graphene is a crucial feature that affects the mechanical and electrical properties. The single layer of graphene has an approximate thickness of 0.34 nm, which is determined by the interlayer spacing in multilayer graphene configuration. Atomic Force Microscopy (AFM) technique measure the surface topology and thickness of graphene sheets by evaluating friction forces and employing contact mode to measure thickness perpendicular to the ripple direction (Bae et al., n.d.). X-ray diffraction is a technique that measures the diffraction patterns by passing the X-rays through the material which helps to measure the spacing between the different layers (Hess, 2020). With the recent developments graphene/boron nitride composite films with atomic thickness has a higher mechanical strength and environmental durability, while maintaining the beneficial properties of graphene (Uyor et al., 2024). The atomic thickness of graphene contributes to its exceptional permeability in membrane applications, with experiments demonstrating higher flow rates due to its ultrathin structure (Wei et al., 2017).

The atomic thickness of graphene is a crucial attribute that affects its mechanical and electrical properties. A single layer of graphene typically measures approximately 0.34 nm in thickness, a value obtained from the interlayer spacing in multilayer configurations and validated by multiple measurement techniques. Atomic Force Microscopy (AFM) is a method that enables accurate thickness assessment of graphene by evaluating friction forces and employing contact mode to measure thickness perpendicular to the ripple direction (Bae et al., n.d.). X-ray diffraction is a technique that yields precise thickness measurements for graphene and other two-dimensional materials through the analysis of interlayer spacing (Hess, 2020). Recent advancements have resulted in the creation of atomic thickness graphene/boron nitride composite films, which improve mechanical strength and environmental resilience while preserving the advantageous properties of graphene (Uyor et al., 2024). The atomic thickness of graphene enhances its remarkable permeability in membrane applications, with experimental configurations exhibiting elevated flow rates attributed to its ultrathin structure (Wei et al., 2017). The atomic thickness of graphene offers benefits for various applications, although it presents issues in handling and stability, requiring protective measures for effective use.

3.2.2 *Hybridization*

The hybridization of graphene entails the interaction between graphene and other materials or molecules, resulting in distinctive electrical and optical properties. This phenomenon is essential for the progression of applications in electronics, sensing, and materials science. Studies demonstrate that the hybridization function delineates electron interactions between impurities and the graphene lattice. This study demonstrates that varying impurity positions can result in divergent pseudo gap Kondo effects, underscoring the significance of impurity localization in the electronic characteristics of graphene (Khudhair et al., 2023). Graphene functions as a buffer layer between ferromagnetic substrates and organic molecules, exhibiting weak hybridization effects that lead to charge transfer and modified electronic states. This interaction alters the Fermi velocity and introduces localized hybrid states, demonstrating graphene's active role in hybridization (Cojocariu et al., 2024). Plasmonic hybridization, which involves the coupling of graphene plasmons with localized gold plasmons, boosts light-graphene interaction, resulting in increased efficiency in mid-infrared applications. This hybridization facilitates novel plasmonic modes, advancing sensing and nonlinear optics (Feinstein & Almeida, 2024). 3D hybrid architectures are innovative methodologies that have resulted in the development of three-dimensional graphene hybrid structures, amalgamating organic and inorganic elements. This spatially delineated design facilitates multifunctional applications, highlighting the adaptability of graphene in hybridization processes (Liu et al., 2023). Although hybridization presents considerable benefits, issues pertaining to the preservation of stability and efficiency in actual applications persist. Additional research is necessary to enhance the efficacy of these hybrid systems for practical application.

3.2.3 *Physicochemical Characteristics*

Graphene possesses exceptional physical features that render it a very desirable material across multiple domains. The distinctive structure, comprising a monolayer of carbon atoms organized in a hexagonal lattice, enhances its remarkable properties, such as superior electrical and thermal conductivity, mechanical strength, and optical clarity.

3.2.3.1 Electrical and Thermal Conductivity

Graphene demonstrates exceptional electrical and thermal conductivity, positioning it as an ideal alternative for sophisticated materials in electronics and thermal management. The study underscores multiple facets of graphene's electrical conductivity, accentuating its prospective applications. Graphene nanoplatelets (GNPs) exhibit substantial electrical conductivity, characterized by a negative derivative of electrical resistivity over various temperatures, signifying strong efficacy in electrical packaging applications (Storti et al., 2023). The distinct transport characteristics of thermal and electrical energy in graphene indicate that, although both conductivities are elevated, they react differently to external stimuli, which can be utilized in device engineering. Graphene fibers have a higher thermal conductivity, with a value of 553 W/m K in composite materials, far more than that of conventional carbon fibers (Yin et al., 2024). Bilayer graphene shows the anisotropic thermal conductivity with in-plane values around 204 W/m K, which can be further improved by structural modifications (Mohammadi et al., 2023). With the optimized gain length and stacking, the thermal conductivity of the graphene sheet has been achieved up to 3826 W/m K (Guo et al., 2023). Despite such advancement, challenges in attaining such a high conductivity in practical application is far challenging, particularly in ensuring structural integrity under varying environmental conditions.

3.2.3.2 Mechanical Characteristics

For the well-known remarkable strength, flexibility and ductility of graphene, makes it a promising material for various applications. The researcher highlights the key aspect of its mechanical behavior including its stress response, effect of particle size in the composition and the influence of particle boundaries. Graphene has a higher tensile strength (greater than 1 TPa) which makes it suitable for high-performance usage (Serna-Gutiérrez & Cordero, 2024). The mechanical properties of graphene nanoflakes depend on their aspect ratio, which is different from the traditional isotropic model (Moriche et al., 2024). The mechanical properties of graphene-reinforced composites are significantly affected by the diameter of graphene particles. Smaller the particle size generally increases the strength and ductility (Ni, 2024). Graphene improves the mechanical performance of polymer nanocomposites by increasing the Young's modulus and Poisson's ratio, indicating higher stiffness at the interface (Barakat et al., 2023). However, grain boundaries in the CVD graphite can weaken its mechanical properties, although control of these boundaries can improve the fracture strength (Heo & Han, 2023).

Multilayer graphene exhibits intricate mechanical properties under both compression and tension, uncovering mechanisms applicable for sophisticated material design (Polyakova & Baimova, 2023). Despite graphene‘s largely beneficial mechanical qualities, issues like defect control and the optimization of composite structures are essential for fully realizing its potential in practical applications.

3.2.3.3 Optical Characteristics

The optical properties of graphene are defined by distinctive characteristics arising from its electrical structure and material composition. Graphene demonstrates elevated optical conductivity, substantial light absorption, and adjustable band gaps, rendering it an attractive candidate for optoelectronic applications. The absorption coefficient of graphene is significantly elevated, especially in the parallel orientation, which facilitates light penetration (Houmad et al., 2015). The optical conductivity spectrum of twisted bilayer graphene exhibits a "redshift" as twist angles decrease, signifying a complicated interplay between its electronic structure and optical response (Liu, Liu, et al., 2024). Reduced graphene oxide (rGO) augments optical characteristics by increasing surface area and enhancing photon absorption, exhibiting a significant peak in UV-Vis absorbance at 255 nm (Ankita & Singh, 2024). The optical bandgap of graphene oxide (GO) is roughly 4 eV, decreasing to 1.9 eV upon reduction, which considerably influences its absorption properties (Tene et al., 2023). Synthesis parameters, such as drying and reduction time, affect the optical transitions and absorption coefficients of GO and rGO which improves the efficiency in energy conversion applications (Tene et al., 2023). However, despite the favorable optical properties of graphene, challenges remain in tailoring these properties for specific applications, mainly in maintaining stability and uniformity during large-scale production.

3.3 Synthesis Methods for Graphene

A variety of synthesis technique of Graphene are present which highlight its unique properties and wide range of applications. Broadly classified, the main synthesis techniques are mechanical, chemical and thermal processes are shown in Fig. 3.2.

3.3.1 Mechanical Exfoliation

Mechanical exfoliation is one of the finest methods for producing high-quality graphene sheets with minimal material damage. With the simplicity and efficiency in process, it has been frequently used technique such as electrochemical exfoliation and liquid phase exfoliation (LPE). Electrochemical exfoliation (EE) is a cost-effective and scalable approach that uses an electrolyte to mixing and separating graphite, this approach produces graphene with desired properties (Yu et al., 2022). Using eco-friendly electrolytes such as residual whey has shown promising solution in creating stable graphene dispersions without use of surfactants. Furthermore, liquid-phase exfoliation is a technique for producing graphene is large quantity (Güler et al., 2021). Scalability and environmental-friendly nature of this technique

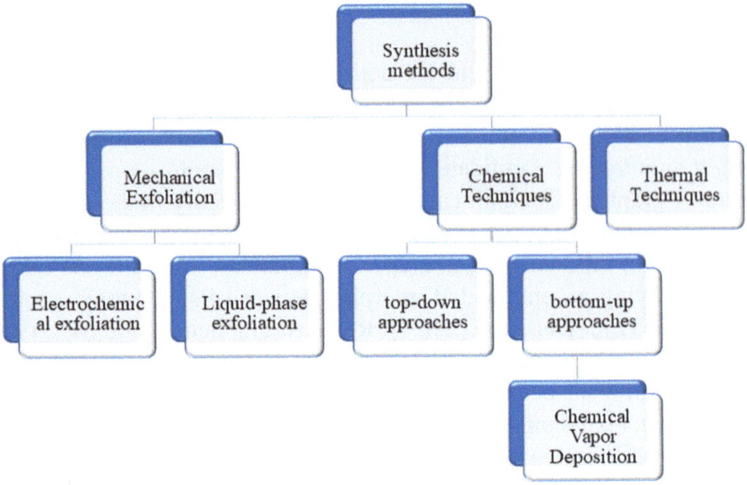

Fig. 3.2 Different methods of synthesis of graphene

make it an extremely attractive option for the growing demand of high-quality graphene (Safian et al., 2021).

3.3.2 Chemical Techniques

Chemical synthesis method include top-down and bottom-up approaches, each of which has a unique processes advantage. In a top-down technique of reducing graphene oxide considering graphite as a starting material has a large-scale graphene production, enabling efficient chemical modification and processing. In this chemical reduction method, sodium borohydride has been reduced (Khan & Shaida, 2023). In the electrocoating, mechanical splitting and electrochemical coating are used to separate graphene layers and produce the graphene sheet. The electrochemical technique employs ammonium sulphate as the electrolyte (Seenukool et al., 2023).

Bottom-up methods include the chemical vapor deposition (CVD) process, which involves the deposition of carbon atoms from gaseous precursors onto a substrate, thereby promoting the synthesis of high-quality graphene (Zhang et al., 2013). Epitaxial growth allows the synthesis of graphene on silicon carbide (SiC) substrates, where carbon atoms are thermally sublimated (Yazdi et al., 2016). Although these chemical approaches provide several avenues for graphene synthesis, problems remain in scalability and defect minimization, which are essential for practical applications.

3.3.3 Thermal Techniques

Thermal techniques for graphene synthesis have attracted considerable interest due to their efficiency and scalability. There are several methods that use thermal processes to generate high-quality graphene, each of which has its own advantages and applications. Thermal reduction is a common technique that involves heating graphene oxide (GO) to remove oxygen-containing groups, thereby improving electrical conductivity and structural integrity (Akbi et al., 2022). This process can be tailored to different temperatures and atmospheres to achieve certain properties, making it suitable for a wide range of applications. Two-step thermal transformation is a new method that uses a polymeric carbon source called polymethyl methacrylate (PMMA) to produce multilayer graphene. This method prevents total thermal dissipation and facilitates the synthesis of high-quality graphene on dielectric substrates (Huang et al., 2023). A solar-thermal chemical vapor deposition (CVD) technique using renewable energy has been established, demonstrating the ability to efficiently synthesize high-quality single-layer and bilayer graphene. This approach demonstrates the potential for sustainable manufacturing in semiconductor applications (Alghfeli & Fisher, 2024). Although thermal approaches are efficient, problems regarding scalability and homogeneity remain, requiring continued studies to enhance these techniques and investigate alternative synthesis routes.

3.4 Graphene in Environmental Applications

For environmental concern, graphene has emerged as potential material especially for the wastewater treatment applications and pollution remediations. Graphene leads with its elevated surface, mechanical toughness and superior electrical conductivity facilitate the diverse resolution for environmental concerns. The strength of graphene is substantial to enhance the water purification system. Graphene enables the extraction of heavy metals and organic molecules from the water vis photocatalytic and adsorption methods. Elevated area and distinct properties of it and its derivative facilitate the efficient extraction of heavy metals and organic molecules (Hu et al., 2018). Functionalized graphene oxide (GO) has garnered interest due to its substantial surface area and chemical reactivity, rendering it appropriate for various environmental applications, such as dye elimination and CO_2 reduction (Isah et al., 2024). Pharmaceuticals and microorganisms were found successfully removed by graphene-based materials with exceptional efficacy of wastewater treatment (Akash et al., 2022; Tiwary et al., 2024). The advancement of graphene-based nanocomposites, including Sm-doped Co-ferrite, illustrates its efficacy as an antibacterial agent and heavy metal extractor, attaining a lead removal effectiveness of up to 99.8% (Ramadan & Shafaay, 2024). Advanced functionalized graphene-cellulose composite material like aerogels improves the removal efficiency, such functionalized materials demonstrating the possibility of materials in large-scale water level of wastewater treatment (Sekwele et al., 2024). Unique characteristics of graphene enhance the performance in various treatment procedures, positioning it as an essential component in environmental research. Though the substantial benefits of graphene in water treatment, challenges remain concerning large-scale manufacturing and environmental effects, necessitating more investigation to improve its use.

3.5 Graphene in Energy Storage and Conversion

Graphene in energy storage and conversion due to its qualities, such as high surface area, electrical conductivity, and mechanical strength. Its incorporation into various energy systems improves performance and efficiency across many applications.

3.5.1 *Energy Storage Solutions*

Graphene has developed as a new material in energy storage applications, especially in batteries and supercapacitors, due to its characteristics. The high surface area, electrical conductivity, and mechanical strength improve the effectiveness of diverse energy storage systems. Graphene aerogels (GAs) enhance the energy density and stability of rechargeable batteries. Their lightweight, porous architecture promotes effective electron and ion transport, resulting in improved performance in lithium-ion, sodium-ion, and several other battery types (Sultanov et al., 2024). Graphene-based materials are essential in electrochemical energy storage devices (EESDs). Their scalability and stability make them appropriate for use in lithium-sulfur and lithium-oxygen batteries, among others (Lv et al., 2016). Recent improvements in the synthesis of disordered graphene have produced supercapacitors showing high-volumetric capacitance and energy density, hence solving issues in compact energy storage systems (Liu et al., 2024). Laser-induced graphene (LIG) provides a three-dimensional porous architecture with high electrochemical stability, making it suitable for flexible energy storage applications, such as supercapacitors and battery systems (Bai et al., 2024).

3.5.2 *Function in Energy Transformation*

Graphene in energy conversion technologies due to its qualities, such as higher electrical conductivity, mechanical strength, and thermal stability. Its applications include many energy systems, improving efficiency and performance. Graphene functions as a transparent electrode in solar cells, improving light absorption and charge transfer efficiency (Mahmoudi et al., 2018). A system that combines a graphene-based thermionic energy converter with a two-stage thermoelectric generator has shown large efficiency improvement, reaching a maximum efficiency of 54.99% and a power density of 7.93 W/cm^2 (Zhang et al., 2024). This link enables efficient waste heat recovery, indicating graphene's ability in energy applications. Graphene's features make it a special candidate for fuel cell applications since it improves catalytic performance and durability in electrodes (Madaswamy et al., 2024). Hybrid structures, including graphene, have been engineered to enhance the efficiency of fuel cell devices, hence reinforcing their significance in energy conversion. Despite the advantages of graphene, problems continue regarding scalability and cost-effectiveness for extensive commercial applications. Additional study is required to resolve these challenges and enhance graphene's abilities in energy storage and conversion technologies.

3.6 Graphene in Construction and Materials Engineering

Graphene and its derivatives are emerging as revolutionary materials in building and materials engineering, providing substantial improvements in mechanical characteristics, durability, and sustainability. Their integration into cement-based composites is notably advantageous since they tackle significant issues in contemporary building. Graphene enhances both tensile and compressive strength in concrete, with research indicating an increase of up to 12.5% in compressive strength with modest graphene incorporation (Pokorný & Ševčík, 2023). Graphene oxide (GO) improves compressive strength by 21% and abrasion resistance by 23% at optimal concentrations (Fonseka et al., 2023). The incorporation of graphene diminishes crack initiation and propagation, hence enhancing the durability of structures and increasing their resistance to environmental influences like chloride and sulfate attacks (Bheel et al., 2023). Few-layer graphene (FLG) has demonstrated a substantial reduction in electrical resistance and an enhancement in thermal conductivity, thereby fostering energy efficiency in buildings (Hamze et al., 2020). Graphene-based additives in lubricants improve mechanical efficiency, decreasing energy consumption and CO_2 emissions by up to 20% in home refrigerators and 6% in internal combustion engine automobiles (Tomanik et al., 2023). Graphene as a filler in polymer composites enhances strength, durability, and impact resistance while preserving a high strength-to-weight ratio. Its thermal conductivity facilitates efficient heat dissipation, rendering it appropriate for electronics and aerospace applications (Shahnaz et al., 2024). Notwithstanding the encouraging outcomes, additional investigation is required to comprehensively incorporate graphene into construction methodologies and material engineering, guaranteeing its practical utilization and tackling prospective scalability issues.

3.7 Graphene in Electronic Applications

Graphene has emerged as a revolutionary material in electronics, providing distinctive features that improve device performance across multiple applications. Its remarkable electrical conductivity, mechanical strength, and flexibility render it appropriate for next-generation electronic devices, including flexible electronics, sensors, and transistors. Devices function in both DC and AC modes, attaining detection limits as low as 1 attomolar for certain DNA biomarkers, thereby illustrating their efficacy in medical diagnostics (Pinzón et al., 2023). They can also identify toxins, aiding in environmental monitoring and public health. Graphene FETs (G-FETs) demonstrate exceptional sensitivity and rapid response times, rendering them suitable for the detection of biomolecules and environmental contaminants and essential for health monitoring. The integration of graphene with substances such as nanocellulose improves the electrical characteristics of flexible composites, making them appropriate for use in sensors and supercapacitors (Yang et al., 2023). Graphene polymer composites employ bioinspired techniques to replicate natural healing mechanisms, markedly enhancing autonomous repair abilities (Sabet, 2024).

The integration of stimuli-responsive molecules into the graphene matrix facilitates swift self-healing after injury, hence improving resilience (Sabet, 2024). Polydopamine/reduced graphene oxide hydrogels demonstrate elevated swelling ratios and self-healing properties, rendering them suitable for self-powered applications in wearable devices (Jiang et al., 2024). Graphene oxide-enhanced polyurethane coatings have an impressive self-healing efficiency of 97.3% post-damage, highlighting their potential for corrosion resistance (Pang et al., 2024). Self-healing polymer composites utilizing electrochemically exfoliated graphene inks exhibit enhanced healing efficiency in soft electronics, achieving a balance between self-healing properties and sensor compatibility (Şimşek et al., 2024). Layered graphene coatings exhibit self-healing and antifogging characteristics, offering an innovative solution for advanced surface coatings (Manabe & Norikane, 2023). The incorporation of graphene onto biodegradable substrates and the creation of innovative composite materials present great opportunities for enhancing flexible electronics while simultaneously addressing performance and environmental issues.

3.8 Graphene in Biomedical Applications

Graphene and its derivatives are transforming biomedical applications due to their characteristic qualities, which include enhanced electrical, mechanical, and thermal properties. These materials are being studied for various applications, ranging from biosensors to drug delivery systems, indicating their utility and ability to advance medical technologies as shown in Fig. 3.3.

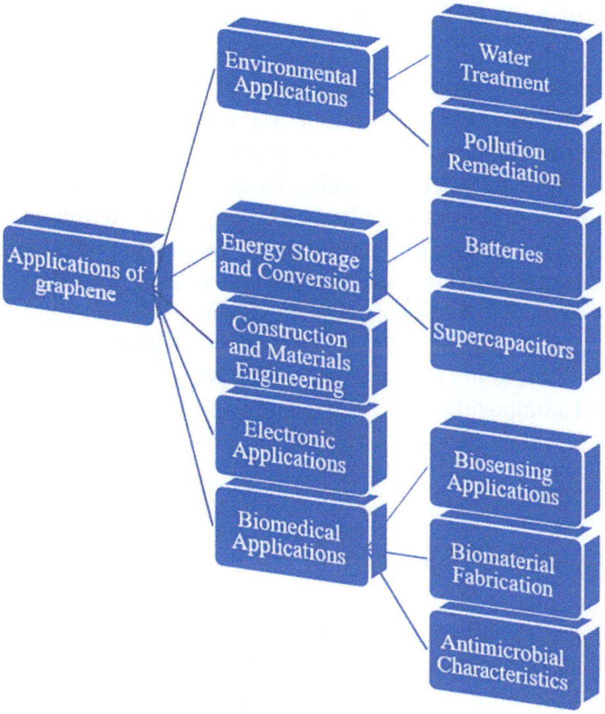

Fig. 3.3 Applications of graphene

3.8.1 Biosensing Applications

Graphene-based nanomaterials (GBNs) are very efficient in the formation of sensitive biosensors, helping illness diagnostics through the detection of pathogens and biomolecules. Their higher surface area and conductivity improve the sensitivity and precision of biosensor devices. G-FETs have been advanced via interface functionalization, improving their sensitivity for the detection of diseases such as cancer and cardiovascular illnesses (Zhao et al., 2024). Raman spectroscopy has been used to find graphene interactions with biological samples, hence increasing its utility in biosensing (Wu et al., 2024). Graphene-based biosensors help in non-invasive illness diagnosis by analyzing extracellular vesicles (Rexha et al., 2024). These sensors determine potential for environmental monitoring by detecting pollutants with excellent specificity (Rexha et al., 2024).

3.8.2 Biomaterial Fabrication

Graphene aerogels show improved drug-loading capabilities and regulated release characteristics, making them appropriate for drug delivery systems and wound dressings. GBNs are employed in tissue engineering to enhance cell production and regeneration due to their biocompatibility. Graphene oxide (GO) nanoparticles can be engineered to improve drug targeting, especially in cancer treatment, by enabling the regulated release of therapeutic chemicals (Sadeghi et al., 2023). Graphene-based nanocomposites help in gene transport, increasing the effectiveness of gene therapy (Imani et al., 2018). Graphene‘s mechanical strength and biocompatibility makes it a favorable candidate for supporting in tissue engineering, helping cell adhesion and proliferation (Kumar et al., 2023; Malisz & Świeczko-Żurek, 2023). Graphene-based composites can be designed to react to environmental stimuli, making them suitable for various tissue engineering applications (Kumar et al., 2023).

3.8.3 Antimicrobial Characteristics

Graphene and its derivatives have developed as attractive materials in opposing microbial diseases due to their typical antibacterial characteristics. Their applicability includes multiple domains, especially in facing the issues presented by multidrug-resistant organisms. Modifications to the surface of graphene enhance its antibacterial effectiveness, locating it as a feasible option for infections in medical applications. The potential of graphene in biomedicine is significant but issues related to toxicity and long-term stability must be resolved to safe and successful clinical uses. Graphene-based materials (GBMs) have antibacterial properties via multiple paths, including direct interaction with microbial membranes, resulting in structural damage and cellular mortality (Kumar et al., 2023). Graphene oxide (GO) induces oxidative stress that impairs microbial functions, while the physical characteristics of graphene can lead to mechanical wrapping and membrane stress (Bhatt et al., 2023; Kumar et al., 2023).

3.9 Challenges and Future Directions

3.9.1 Challenges

Graphene, despite its potential for diverse uses, encounters numerous substantial obstacles that impede its broad adoption. These difficulties encompass various disciplines, including material processing, synthesis, and application. Contemporary synthesis methods are expensive and dependent on non-renewable resources, resulting in elevated manufacturing costs. The shift from laboratory-scale production to industrial applications constitutes a considerable challenge. The potential hazards linked to graphene's toxicity in biomedical applications require additional investigation.

The aggregation of graphene nanoplatelets in aluminum composites results in diminished mechanical characteristics when surpassing 0.5–3 wt.% graphene (Ghaderi et al., 2024). Inadequate wetting of graphene by molten aluminum leads to uneven distribution and porosity, requiring the application of wetting agents (Ghaderi et al., 2024). The restricted catalytic efficacy of metallurgical alloys in CVD methods hinders the development of high-quality graphene films, as demonstrated by NiTi shape memory alloys (Fickl et al., 2024). High-temperature chemical vapor deposition can impair the bulk characteristics of alloys, but low-temperature techniques demonstrate potential for preserving structural integrity (Fickl et al., 2024). Graphene-based field-effect transistor biosensors encounter challenges with Debye screening, which hinders the detection of target molecules (Ono et al., 2024). Challenges encompass nonspecific adsorption and intricate detection processes, which impede practical usage (Ono et al., 2024). Biosecurity hazards and insufficient adhesion performance of graphene family nanoparticles present obstacles in medical applications (Chen et al., 2024). Electrochemical exfoliation encounters compromise between yield and quality as side reactions influence defect densities and carbon/oxygen ratios (Ummer & Kalarikkal, 2024).

3.9.2 Future Directions

Graphene research is set to experience substantial progress in multiple domains, propelled by its distinctive characteristics and prospective uses. Primary areas of emphasis are the advancement of graphene-based sensors, the establishment of a configurable bandgap, and novel fabrication methodologies for wearable electronics.

Graphene's remarkable conductivity and flexibility render it optimal for biosensors in the fields of healthcare and defense. Wearable graphene sensors are becoming essential instruments for real-time health monitoring, especially in cardiovascular diagnostics. A significant challenge persists due to the absence of a bandgap in graphene, which restricts its application in semiconductor technologies. Future research seeks to attain a tuneable bandgap via diverse methodologies, potentially establishing graphene as an alternative to silicon in electronics (Nandee et al., 2022). Progress in fabrication techniques, particularly chemical vapor deposition (CVD), is essential for the scalable production of high-quality graphene. These strategies are anticipated to improve the material's use across several sectors, including energy storage and flexible electronics (Zhang et al., 2022). Despite the great possibilities of graphene, obstacles related to scalability, affordability, and comprehension of its fundamental features must be resolved to fully exploit its promise in commercial applications.

3.10 Conclusions

Graphene has an exceptional range of features that render it a highly coveted material in numerous domains, especially in nanoelectronics. Its distinctive attributes arise from its two-dimensional configuration and atomic organization, resulting in remarkable electrical, mechanical, and thermal properties. Graphene exhibits exceptional electrical conductivity and carrier mobility, rendering it suitable for sensor and electronic device applications. The extensive specific surface area improves its efficacy in diverse sensing applications, such as magnetic and optical sensors.

Graphene shows higher mechanical strength, characterized by high tensile strength and flexibility, essential for its application in composite materials. The thickness and layering of graphene significantly affect its mechanical properties, with multilayer configurations indicating diverse behaviors in comparison to single-layer graphene. Graphene shows higher thermal conductivity which is beneficial for heat dissipation in electrical applications.

Graphene has developed as a material with a wide range of uses in electronics, biology, and environmental engineering. Its various qualities, including higher electrical conductivity, mechanical strength, and biocompatibility, make it as an important element in the development of technology and healthcare. The notable electrical properties of graphene provide its application in high-speed electronic devices, flexible screens, and solar cells. It improves energy storage systems, including lithium-ion batteries and supercapacitors, due to its higher surface area and conductivity. Graphene-based sensors are in development for artificial organs and sensory devices, such as artificial retinas and eardrums. Its antibacterial characteristics and potential for tissue support makes it promising for medical uses but additional research is required to determine safe uses of graphene. Graphene efficiently eliminates contaminants from water and enhances the sensitivity of biosensors, indicating its multifunctionality in environmental applications.

References

Akash, S., Sivaprakash, B., Rajamohan, N., Govarthanan, M., & Elakiya, B. T. (2022). Remediation of pharmaceutical pollutants using graphene-based materials - A review on operating conditions, mechanism and toxicology. *Chemosphere, 306*, 135520. https://doi.org/10.1016/j.chemosphere.2022.135520

Akbi, H., Mekki, A., Rafai, S., Touidjine, S., Boudina, N., & Sayeh, Z. B. D. (2022). Phenomenological description of the thermal reduction kinetics in graphene oxide films. *Materials Chemistry and Physics, 277*, 125477. https://doi.org/10.1016/j.matchemphys.2021.125477

Alghfeli, A., & Fisher, T. S. (2024). Sequential Bayesian-optimized graphene synthesis by direct solar-thermal chemical vapor deposition. *Scientific Reports, 14*(1), 3660. https://doi.org/10.1038/s41598-024-54005-z

Ankita, U. B., & Singh, V. (2024). *Structural and optical properties of reduced graphene oxide* (pp. 531–541). Springer. https://doi.org/10.1007/978-981-99-4878-9_74

Aziz, M. I., & Khan, Q. (2024). *Study of lattice dynamics of the graphene along highly symmetry directions* (pp. 103–111). Springer. https://doi.org/10.1007/978-981-99-4878-9_14

Bae, H. P., Jin, S. C., & Duk, H. L. (n.d.). Method for determining graphene thickness using atomic force microscope, US8753468B2.

Bai, S., et al. (2024). Laser-induced graphene: Carbon precursors, fabrication mechanisms, material characteristics, and applications in energy storage. *Chemical Engineering Journal, 493*, 152805. https://doi.org/10.1016/j.cej.2024.152805

Barakat, M., Reda, H., Chazirakis, A., & Harmandaris, V. (2023). Investigating the mechanical performance of graphene reinforced polymer nanocomposites via atomistic and continuum simulation approaches. *Polymer, 286*, 126379. https://doi.org/10.1016/j.polymer.2023.126379

Bhatt, S., Pathak, R., Punetha, V. D., & Punetha, M. (2023). Recent advances and mechanism of antimicrobial efficacy of graphene-based materials: A review. *Journal of Materials Science, 58*(19), 7839–7867. https://doi.org/10.1007/s10853-023-08534-z

Bheel, N., Mohammed, B. S., Liew, M. S., & Zawawi, N. A. W. A. (2023). Effect of graphene oxide as a nanomaterial on the durability behaviors of engineered cementitious composites by applying RSM modelling and optimization. *Buildings, 13*(8), 2026. https://doi.org/10.3390/buildings13082026

Chen, X., Zou, M., Liu, S., Cheng, W., Guo, W., & Feng, X. (2024). Applications of graphene family nanomaterials in regenerative medicine: Recent advances, challenges, and future perspectives. *International Journal of Nanomedicine, 19*, 5459–5478. https://doi.org/10.2147/IJN.S464025

Cojocariu, I., Perilli, D., Feyer, V., & Jugovac, M. (2024). Graphene-molecule hybridization at a ferromagnetic interface. *Chemistry–A European Journal*. https://doi.org/10.1002/chem.202400857

Feinstein, M. D., & Almeida, E. (2024). Hybridization of graphene-gold plasmons for active control of mid-infrared radiation. *Scientific Reports, 14*(1), 6733. https://doi.org/10.1038/s41598-024-57216-6

Fickl, B., et al. (2024). Challenges in chemical vapour deposition of graphene on metallurgical alloys exemplified for NiTi shape memory alloys. *BHM Berg-Huttenmannische Monatshefte, 169*(7), 357–365. https://doi.org/10.1007/s00501-024-01482-4

Fonseka, I., Mohotti, D., Wijesooriya, K., Lee, C.-K., & Mendis, P. (2023). Influence of graphene oxide on abrasion resistance and strength of concrete. *Construction and Building Materials, 404*, 133280. https://doi.org/10.1016/j.conbuildmat.2023.133280

Ghaderi, O., Zare, M., Niroumand, B., Church, B. C., & Rohatgi, P. K. (2024). Selected challenges in solidification processing of graphene nanoplatelets (GNPs) reinforced aluminum alloys composites. *Frontiers in Materials, 11*, 1363270. https://doi.org/10.3389/fmats.2024.1363270

Güler, Ö., Tekeli, M., Taşkın, M., Güler, S. H., & Yahia, I. S. (2021). The production of graphene by direct liquid phase exfoliation of graphite at moderate sonication power by using low boiling liquid media: The effect of liquid media on yield and optimization. *Ceramics International, 47*(1), 521–533. https://doi.org/10.1016/j.ceramint.2020.08.159

Guo, S., et al. (2023). Toward ultrahigh thermal conductivity graphene films. *2D Materials, 10*(1), 014002. https://doi.org/10.1088/2053-1583/ac9421

Hamze, S., et al. (2020). Few-layer graphene-based nanofluids with enhanced thermal conductivity. *Nanomaterials, 10*(7), 1258. https://doi.org/10.3390/nano10071258

Heo, J., & Han, J. (2023). Evaluation of mechanical properties of triple-junction-free polycrystalline graphene. *Nanotechnology, 34*(41), 415704. https://doi.org/10.1088/1361-6528/ace44e

Hess, P. (2020). Thickness of elemental and binary single atomic monolayers. *Nanoscale Horizons, 5*(3), 385–399. https://doi.org/10.1039/C9NH00658C

Houmad, M., Zaari, H., Benyoussef, A., El Kenz, A., & Ez-Zahraouy, H. (2015). Optical conductivity enhancement and band gap opening with silicon doped graphene. *Carbon, 94*, 1021–1027. https://doi.org/10.1016/j.carbon.2015.07.033

Hu, H., et al. (2018). A reduced graphene oxide quantum dot-based adsorbent for efficiently binding with organic pollutants. *ACS Applied Nano Materials, 1*(11), 6502–6513. https://doi.org/10.1021/acsanm.8b01799

Huang, Y., et al. (2023). Two-step thermal transformation of multilayer graphene using polymeric carbon source assisted by physical vapor deposited copper. *Materials, 16*(16), 5603. https://doi.org/10.3390/ma16165603

Imani, R., Mohabatpour, F., & Mostafavi, F. (2018). Graphene-based Nano-carrier modifications for gene delivery applications. *Carbon, 140*, 569–591. https://doi.org/10.1016/j.carbon.2018.09.019

Isah, M., Lawal, R., & Onaizi, S. A. (2024). CO_2 capture and conversion using graphene-based materials: A review on recent progresses and future outlooks. *Green Chemical Engineering*. https://doi.org/10.1016/j.gce.2024.09.009

Jiang, Y., Wang, D., Wang, J., Yue, L., Tang, C., & Song, J. (2024). Adhesive, self-healing polydopamine/reduced graphene oxide composite hydrogel for self-powered application. *Soft Materials, 22*(2), 119–128. https://doi.org/10.1080/1539445X.2024.2321907

Kausar, A., & Ahmad, I. (2024). Green fabrication techniques for graphene and graphene nanocomposites—State-of-the-art and modern prospects. *Materials Research Innovations, 28*(6), 478–492. https://doi.org/10.1080/14328917.2024.2326696

Khan, M. U., & Shaida, M. A. (2023). Reduction mechanism of graphene oxide including various parameters affecting the C/O ratio. *Materials Today Communications, 36*, 106577. https://doi.org/10.1016/j.mtcomm.2023.106577

Khudhair, A. M., Mohammed, M. H., Ajeel, F. N., & Mohammed, S. H. (2023). Enhancement the electronic and optical properties of the graphene nanoflakes in the present S impurities. *Chemical Physics Impact, 6*, 100154. https://doi.org/10.1016/j.chphi.2022.100154

Kumar, R., et al. (2023). Graphene-based materials for biotechnological and biomedical applications: Drug delivery, bioimaging and biosensing. *Materials Today Chemistry, 33*, 101750. https://doi.org/10.1016/j.mtchem.2023.101750

Langer, R., Mustonen, K., Markevich, A., Otyepka, M., Susi, T., & Błoński, P. (2022). Graphene lattices with embedded transition-metal atoms and tunable magnetic anisotropy energy: Implications for spintronic devices. *ACS Applied Nano Materials, 5*(1), 1562–1573. https://doi.org/10.1021/acsanm.1c04309

Liu, G., et al. (2024). Efficient fabrication of disordered graphene with improved ion accessibility, ion conductivity, and density for high-performance compact capacitive energy storage. *Advanced Science, 11*(38), 155. https://doi.org/10.1002/advs.202405155

Liu, M., Liu, Z., Cao, J. C., & Wang, C. (2024). Properties of the optical response of the twisted bilayer graphene. *Physica B: Condensed Matter, 675*, 415609. https://doi.org/10.1016/j.physb.2023.415609

Liu, X., et al. (2023). Synergistic combination of reductive covalent functionalization and atomic layer deposition—Towards spatially defined graphene-organic-inorganic heterostructures. *Angewandte Chemie International Edition, 62*(47), 14183. https://doi.org/10.1002/anie.202314183

Lv, W., Li, Z., Deng, Y., Yang, Q.-H., & Kang, F. (2016). Graphene-based materials for electrochemical energy storage devices: Opportunities and challenges. *Energy Storage Materials, 2*, 107–138. https://doi.org/10.1016/j.ensm.2015.10.002

Madaswamy, S. L., Keertheeswari, N. V., & Dhanusuraman, R. (2024). *Graphene-based fuel cells* (pp. 237–255). Springer. https://doi.org/10.1007/978-981-99-9935-4_13

Mahmoudi, T., Wang, Y., & Hahn, Y.-B. (2018). Graphene and its derivatives for solar cells application. *Nano Energy, 47*, 51–65. https://doi.org/10.1016/j.nanoen.2018.02.047

Malisz, K., & Świeczko-Żurek, B. (2023). Graphene production and biomedical applications: A review. *Crystals, 13*(10), 1413. https://doi.org/10.3390/cryst13101413

Manabe, K., & Norikane, Y. (2023). Graphene composite self-healing antifog/frost-resist transparent coatings with zwitter-wettability. *Surfaces and Interfaces, 42*, 103363. https://doi.org/10.1016/j.surfin.2023.103363

Mbayachi, V. B., Ndayiragije, E., Sammani, T., Taj, S., Mbuta, E. R., & Ullah Khan, A. (2021). Graphene synthesis, characterization and its applications: A review. *Results in Chemistry, 3*, 100163. https://doi.org/10.1016/j.rechem.2021.100163

Mohammadi, R., Ghaderi, M. R., & Hajian, E. (2023). A molecular dynamics simulation study of in- and cross-plane thermal conductivity of bilayer graphene. *Materials, 16*(20), 6714. https://doi.org/10.3390/ma16206714

Moriche, R., et al. (2024). Influence of graphene-based nanostructures processing routes and aspect ratio in flexural strength and fracture mechanisms of 3Y-TZP-matrix composites. *Ceramics International, 50*(11), 19217–19227. https://doi.org/10.1016/j.ceramint.2024.03.021

Movaghgharnezhad, S., & Kang, P. (2024). Laser-induced graphene: Synthesis advances, structural tailoring, enhanced properties, and sensing applications. *Journal of Materials Chemistry C, 12*(19), 6718–6742. https://doi.org/10.1039/D3TC04677J

Nandee, R., Chowdhury, M. A., Shahid, A., Hossain, N., & Rana, M. (2022). Band gap formation of 2D materialin graphene: Future prospect and challenges. *Results in Engineering, 15*, 100474. https://doi.org/10.1016/j.rineng.2022.100474

Nguyen, V. H., Hoang, T. X., & Charlier, J.-C. (2022). Electronic properties of twisted multilayer graphene. *Journal of Physics: Materials, 5*(3), 034003. https://doi.org/10.1088/2515-7639/ac6c4a

Ni, G. (2024). Effect of graphene particle diameter on mechanical properties of graphene composite material. *Journal of Physics Conference Series, 2706*(1), 012014. https://doi.org/10.1088/1742-6596/2706/1/012014

Ono, T., Okuda, S., Ushiba, S., Kanai, Y., & Matsumoto, K. (2024). Challenges for field-effect-transistor-based graphene biosensors. *Materials, 17*(2), 333. https://doi.org/10.3390/ma17020333

Pang, W., et al. (2024). Graphene oxides enhanced polyurethane based composite coating with long term corrosion resistance and self-healing property. *European Polymer Journal, 207*, 112825. https://doi.org/10.1016/j.eurpolymj.2024.112825

Pantaleón, P. A., Sainz-Cruz, H., & Guinea, F. (2024). Designing moiré patterns by shearing. *ACS Nano, 18*(42), 28575–28584. https://doi.org/10.1021/acsnano.4c08302

Parvez, K., et al. (2014). Exfoliation of graphite into graphene in aqueous solutions of inorganic salts. *Journal of the American Chemical Society, 136*(16), 6083–6091. https://doi.org/10.1021/ja5017156

Pinzón, E., et al. (2023). Graphene biosensors operated in DC transistor and AC electrochemical modes for DNA sensing. In *2023 IEEE sensors* (pp. 1–4). IEEE. https://doi.org/10.1109/SENSORS56945.2023.10324951

Pokorný, J., & Ševčík, R. (2023). The influence of graphene admixture on strength properties of cement-based composites. *Journal of Physics Conference Series, 2628*(1), 012025. https://doi.org/10.1088/1742-6596/2628/1/012025

Polyakova, P. V., & Baimova, J. A. (2023). Mechanical properties of graphene networks under compression: A molecular dynamics simulation. *International Journal of Molecular Sciences, 24*(7), 6691. https://doi.org/10.3390/ijms24076691

Ramadan, R., & Shafaay, A. S. (2024). Graphene-based Sm-doped co-ferrite for environmental applications. *Journal of Materials Science: Materials in Electronics, 35*(19), 1331. https://doi.org/10.1007/s10854-024-12980-z

Rana, K., Kaur, H., Singh, N., Sithole, T., & Siwal, S. S. (2024). Graphene-based materials: Unravelling its impact in wastewater treatment for sustainable environments. *Next Materials, 3*, 100107. https://doi.org/10.1016/j.nxmate.2024.100107

Rexha, J., Perta, N., Romagnoli, A., Cappetta, E., Guescini, M., & Di Marino, D. (2024). Exploring the potential of graphene field-effect transistors in biosensing for health and environment. *Biophysical Journal, 123*(3), 162. https://doi.org/10.1016/j.bpj.2023.11.1089

Sabet, M. (2024). Unveiling advanced self-healing mechanisms in graphene polymer composites for next-generation applications in aerospace, automotive, and electronics. *Polymer-*

Plastics Technology and Materials, 63(15), 2032–2059. https://doi.org/10.1080/25740881.2024.2365282
Sadeghi, M. S., et al. (2023). Graphene oxide nanoarchitectures in cancer therapy: Drug and gene delivery, phototherapy, immunotherapy, and vaccine development. *Environmental Research, 237*, 117027. https://doi.org/10.1016/j.envres.2023.117027
Safian, M. T., Umar, K., & Mohamad Ibrahim, M. N. (2021). Synthesis and scalability of graphene and its derivatives: A journey towards sustainable and commercial material. *Journal of Cleaner Production, 318*, 128603. https://doi.org/10.1016/j.jclepro.2021.128603
Seenukool, N., et al. (2023). Graphene oxide synthesis methods based on electrochemical exfoliation pencil core. *Materials Today Proceedings*. https://doi.org/10.1016/j.matpr.2023.04.665
Sekwele, K. G., Tichapondwa, S. M., & Mhike, W. (2024). Cellulose, graphene and graphene-cellulose composite aerogels and their application in water treatment: A review. *Discover Materials, 4*(1), 23. https://doi.org/10.1007/s43939-024-00097-3
Serna-Gutiérrez, A., & Cordero, N. A. (2024). Mechanical properties of small quasi-square graphene nanoflakes. *Crystals, 14*(4), 314. https://doi.org/10.3390/cryst14040314
Shahnaz, T., et al. (2024). Graphene-based nanoarchitecture as a potent cushioning/filler in polymer composites and their applications. *Journal of Materials Research and Technology, 28*, 2671–2698. https://doi.org/10.1016/j.jmrt.2023.12.108
Silva, W. C. H., Zafar, M. A., Allende, S., Jacob, M. V., & Tuladhar, R. (2024). Sustainable synthesis of graphene oxide from waste sources: A comprehensive review of methods and applications. *Materials Circular Economy, 6*(1), 23. https://doi.org/10.1007/s42824-024-00117-w
Şimşek, B., Ruhkopf, J., Plachetka, U., Rademacher, N., Belete, M., & Lemme, M. C. (2024). Silver nanoparticle-assisted electrochemically exfoliated graphene inks coated on PVA-based self-healing polymer composites for soft electronics. *ACS Applied Materials & Interfaces, 16*(6), 7838–7849. https://doi.org/10.1021/acsami.3c17851
Storti, E., Lojka, M., Lencová, S., Hubálková, J., Jankovský, O., & Aneziris, C. G. (2023). Synthesis and characterization of graphene nanoplatelets-containing fibers by electrospinning. *Open Ceramics, 15*, 100395. https://doi.org/10.1016/j.oceram.2023.100395
Sultanov, F., Tatykayev, B., Bakenov, Z., & Mentbayeva, A. (2024). The role of graphene aerogels in rechargeable batteries. *Advances in Colloid and Interface Science, 331*, 103249. https://doi.org/10.1016/j.cis.2024.103249
Tene, T., et al. (2023). Temperature-dependent optical properties of oxidized graphenes. *Nanomaterials, 13*(15), 2263. https://doi.org/10.3390/nano13152263
Tiwary, S. K., Singh, M., Chavan, S. V., & Karim, A. (2024). Graphene oxide-based membranes for water desalination and purification. *npj 2D Materials and Applications, 8*(1), 27. https://doi.org/10.1038/s41699-024-00462-z
Tomanik, E., Christinelli, W., Souza, R. M., Oliveira, V. L., Ferreira, F., & Zhmud, B. (2023). Review of graphene-based materials for tribological engineering applications. *Eng, 4*(4), 2764–2811. https://doi.org/10.3390/eng4040157
Ummer, F. C., & Kalarikkal, N. (2024). *Electrochemical exfoliation techniques for graphene synthesis: Challenges and future opportunities* (pp. 321–338). Springer. https://doi.org/10.1007/978-981-97-2128-3_12
Uyor, U. O., Popoola, A. P. I., & Popoola, O. M. (2024). Chemically functionalized graphene–boron nitride/polypropylene nanocomposites with enhanced nanomechanical and anti-wear properties. *Chemistry Africa, 7*(3), 1595–1604. https://doi.org/10.1007/s42250-023-00824-y
Wei, G., Quan, X., Chen, S., & Yu, H. (2017). Superpermeable atomic-thin graphene membranes with high selectivity. *ACS Nano, 11*(2), 1920–1926. https://doi.org/10.1021/acsnano.6b08000
Wu, Q., Jiang, S., Zhang, Y., He, H., & Wang, X. (2023). The application of graphene in the field of electronics and electricity. In *2023 international seminar on computer science and engineering technology (SCSET)* (pp. 752–756). IEEE. https://doi.org/10.1109/SCSET58950.2023.00169
Wu, W., Ranasinghe, J. C., Chatterjee, A., & Huang, S. (2024). Recent advances on Raman spectroscopy of graphene: Towards biosensing applications. *Materials Chemistry and Physics, 318*, 129281. https://doi.org/10.1016/j.matchemphys.2024.129281

Yang, H., et al. (2023). Nanocellulose-graphene composites: Preparation and applications in flexible electronics. *International Journal of Biological Macromolecules, 253*, 126903. https://doi.org/10.1016/j.ijbiomac.2023.126903

Yazdi, G., Iakimov, T., & Yakimova, R. (2016). Epitaxial graphene on SiC: A review of growth and characterization. *Crystals, 6*(5), 53. https://doi.org/10.3390/cryst6050053

Yin, Q., et al. (2024). Highly thermally conductive composites of graphene fibers. *Composites. Part A, Applied Science and Manufacturing, 185*, 108290. https://doi.org/10.1016/j.compositesa.2024.108290

Yu, Q., et al. (2022). Electrochemical synthesis of graphene oxide from graphite flakes exfoliated at room temperature. *Applied Surface Science, 598*, 153788. https://doi.org/10.1016/j.apsusc.2022.153788

Zhang, J., Peng, H., Zhang, H., Maruyama, S., & Lin, L. (2022). The roadmap of graphene: From fundamental research to broad applications. *Advanced Functional Materials, 32*(42), 8378. https://doi.org/10.1002/adfm.202208378

Zhang, R., Yu, X., Yang, Q., Cui, G., & Li, Z. (2021). The role of graphene in anti-corrosion coatings: A review. *Construction and Building Materials, 294*, 123613. https://doi.org/10.1016/j.conbuildmat.2021.123613

Zhang, Y., Zhang, L., & Zhou, C. (2013). Review of chemical vapor deposition of graphene and related applications. *Accounts of Chemical Research, 46*(10), 2329–2339. https://doi.org/10.1021/ar300203n

Zhang, Z., Huang, Y., & Lu, S. (2024). Graphene thermionic energy converter integrated with two-stage thermoelectric generator for energy cascade utilization. *Applied Thermal Engineering, 254*, 123867. https://doi.org/10.1016/j.applthermaleng.2024.123867

Zhao, W., et al. (2024). Sensitivity-enhancing strategies of graphene field-effect transistor biosensors for biomarker detection. *ACS Sensors, 9*(6), 2705–2727. https://doi.org/10.1021/acssensors.4c00322

Chapter 4
Nanocellulose, Chitosan-Based Nanomaterials: A Renewable Resource

Pradnya Dilip Rao and Ameya Bipin Karapurkar

4.1 Introduction

4.1.1 Renewable Nanomaterials: Overview

Naturally renewing sources such as plant extracts, agricultural residues, and bio-based precursors provide considerable advantages in the synthesis of renewable nanomaterials. These resources are available abundantly and hence are inexpensive. Therefore, renewable nanomaterials are valuable for scaled applications including catalysts, pharmaceuticals, solar cells, electronics, and many others (Mazari et al., 2021). Renewable materials obtained from natural sources contain functional groups that serve as reducing agents, stabilizers, or base templates that allow for precise control over nanoparticle synthesis and characteristics. Biomass-derived lignin is used in stabilizing nanoparticles which increase catalytic activity and dispersion (Morsali et al., 2022). Renewable nanomaterials are important for achieving sustainable development by minimizing dependency on conventional finite resources and promoting environmental-friendly technology. Natural fibers (e.g., nanocellulose) and biopolymers (e.g., polyhydroxyalkanoate, Chitosan) are employed in various nanomaterial synthesis because they are ecological and reduce carbon emissions (Asunis et al., 2020). Nanomaterials developed using renewable materials enable applications in areas such as medicine delivery, water purification, and biosensors (Zhang et al., 2023).

P. D. Rao (✉) · A. B. Karapurkar
Department of Chemistry, The University of Maine, Orono, ME, USA

S. Kulkarni (ed.), *Next-Generation Nanomaterials for Sustainable Engineering*, Lecture Notes in Nanoscale Science and Technology 43,
https://doi.org/10.1007/978-3-032-05719-8_4

4.1.2 *Sustainability Importance in Nanomaterials*

The synthesis of nanomaterials using a sustainable approach is paramount for tackling environmental problems. To achieve this, it is advised that renewable resources be used more frequently in production because they ensure a replenishable supply of raw materials. In recent years, various initiatives have been helping to reduce greenhouse gas emissions, energy consumption, and toxic byproducts via eco-friendly synthesis methods that use natural precursors and agricultural waste. Industrial demand for nanoparticles is growing in all sectors, from the consumer market to the business-to-business model, and it is more necessary than ever to implement sustainable methods to limit environmental impact. It is crucial to increase recycling and material reintegration in the manufacturing process to improve resource efficiency. Circular economy approaches, such as recycling and biodegradable design, address environmental concerns about nanowaste. Balancing sustainability and technological advancement support both environmental resilience and economic growth. Nanotechnology, by utilizing sustainable processes and renewable resources, can address global issues while also promoting environmental harmony.

4.2 Types and Sources of Nanocellulose and Chitosan-Based Nanomaterials

4.2.1 *Types and Sources of Nanocellulose*

Nanocellulose is derived from plants, agricultural byproducts, and various organic animal waste materials. The term "nanocellulose"refers to various cellulose nanoparticles, including cellulose nanocrystals (CNCs), cellulose nanofiber (CNFs), and bacterial, microbial, and cellulose nanowhiskers (CNW). The cellulose polymer backbone is composed of repeated cellobiose units connected by β-1,4-glycosidic linkages. Its structure has crystalline and amorphous regions; the crystalline regions are very organized, while the amorphous areas are more amenable to chemical or biological alteration (Rojas, 2016). CNCs and CNFs are the two main forms of nanocellulose. CNFs are created by mechanical disintegration. CNFs have spaghetti-like entangled polymeric networks, while CNCs are extracted through chemical treatments like acid hydrolysis. CNCs are stiff rod-like structures (George & Sabapathi, 2015). The surfaces of CNFs and CNCs are rich in hydroxyl groups that can be functionalized for various applications. These hydroxyl groups can be esterified or crosslinked to convert the hydrophilic nature of CNFs and CNCs to hydrophobic materials, which makes them suitable for the polymer and plastic industries. A third kind of nanocellulose, known as bacterial cellulose, is made of bacteria, including Gluconacetobacter xylinus. Bacterial cellulose (BC) is valued for both its potential for use in biomedical applications and its exceptional mechanical properties. However, BC production is less scalable due to certain limitations (Reshmy et al., 2021). Today, nanocellulose is employed in various applications, such as coatings, medical scaffolds, rheology modifiers, and nanocomposites.

4.2.2 Sources of Chitosan

Chitosan is a biopolymer consisting of Chitin monomer. Chitin is a byproduct of the seafood industry. These crustacean wastes of shrimp, crabs, and lobsters which contain 20–30% chitin and 30–50% calcium carbonate by weight provide a cost-effective and environmental-friendly raw material for the large-scale synthesis of chitosan (Chakravarty & Edwards, 2022). Some species, such as Homarus and Nephrops lobsters have a high chitin content of up to 75%, which makes them perfect for Chitosan production (Muthu et al., 2021). In some cases, byproducts of Procambarus clarkii crayfish, which have a chitin content of roughly 20–23%, are desirable because of their accessibility and low cost (Chen et al., 2024). Fungi can also be used to manufacture chitosan, albeit the yields are less because of fermentation efficiency and challenges. Additionally, scientists are developing new methods for producing chitosan from various sources like insects and mollusks like squid and clams (Terkula Iber et al., 2022). However, due to issues with availability and acceptance, these materials are employed less frequently.

Even if it appears that crustacean waste is the most common source of raw materials for the manufacturing of chitosan, it is paramount to investigate underutilized sources such as marine sponges, krill, and algae in order to diversify supply chains and provide environmental advantages as sustainability becomes more important.

4.3 Production of Nanocellulose and Chitosan-Based Nanomaterials

4.3.1 Cellulose Nanocrystal Production

The production of cellulose nanocrystals (CNCs) involves selective hydrolysis of the amorphous regions of cellulose fibers while retaining the crystalline domains (see Fig. 4.1). The standard approach uses acid hydrolysis, most commonly with concentrated sulfuric acid (approximately 64–65% by weight), to break glycosidic bonds in the amorphous sections and isolate the rigid crystalline nanostructures. Although sulfuric acid is the most commonly used acid, other acids like hydrochloric acids and phosphoric acids are also used to synthesize CNC. With the selection of acids other key factors such as temperature (typically 30–70 °C), reaction duration, and acid concentration are carefully controlled to optimize yield and material properties (George & Sabapathi, 2015). Once the hydrolysis step is completed

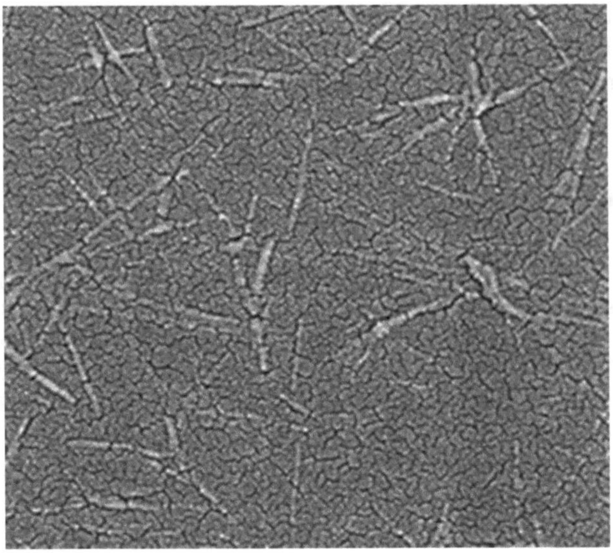

Fig. 4.1 Cellulose nanocrystals

residual acids and other byproducts are removed through purification steps like washing, centrifugation, and dialysis. Further sonication and homogenization are generally performed to achieve a uniform CNCs dispersion.

Though this technique yields CNCs characterized by a high aspect ratio and negatively charged sulfate ester groups that enhance stable aqueous suspensions, it also raises substantial environmental issues attributed to excessive water consumption and the creation of acidic waste. To address these environmental concerns, scientists are exploring different techniques for synthesizing CNCs, such as employing oxidizing agents instead of acids, microwave or ultrasound-assisted hydrolysis, high-pressure hydrolysis, and enzymatic methods (Mohd Jamil et al., 2022; Rovera et al., 2018). Renewable resources such as agricultural residues or industrial byproducts frequently produce a greater yield of CNCs, but the characteristics of the CNCs depend on the source and the hydrolysis conditions (Jonoobi et al., 2015). CNCs are also called cellulose nanowhiskers.

4.3.2 Cellulose Nanofiber Production

The production of cellulose nanofibers (CNFs) requires meticulously planned pretreatment and mechanical methods to harvest nanoscale fibers from plant-derived cellulose sources while preserving their structural integrity and nanoscale characteristics. Generally, the procedure starts by eliminating non-cellulosic elements like lignin and hemicellulose through chemical methods such as alkali or acid hydrolysis. After hydrolysis, mechanical techniques like high-pressure homogenization,

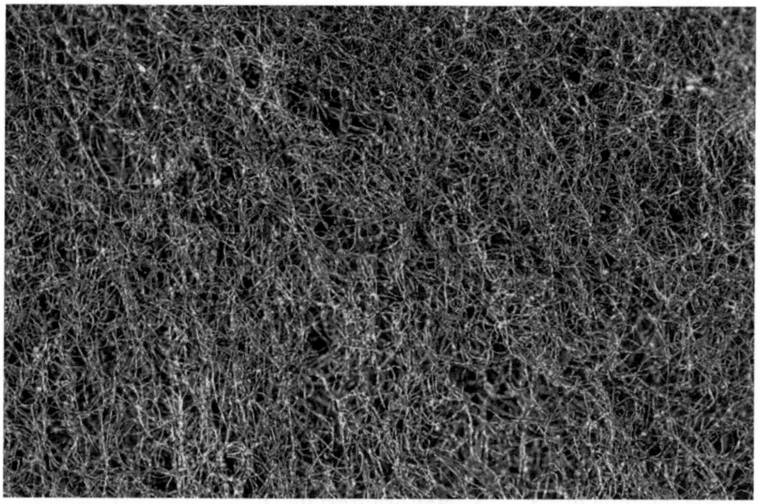

Fig. 4.2 Cellulose nanofibers

microfluidization, grinding, or ultrasonication disintegrate the cellulose into nanoscale fibers using strong shear forces (Siddiqui et al., 2011). Nonetheless, these methods can require a lot of energy and frequently produce microfibril clusters instead of completely separated fibers (See Fig. 4.2). Alternative techniques have been created to reduce energy usage, including cryo-crushing or chemical pretreatments (Chakraborty et al., 2005). For instance, 2,2,6,6-Tetramethylpiperidin-1-oxyl (TEMPO)-driven oxidation has been developed to alter the cellulose surface chemically, adding carboxyl groups that enhance fibrillation efficiency, and lower energy use. This chemical pretreatment also improves the characteristics of the resulting fibers (Isogai, 2013).

Similarly, enzymatic treatments using cellulase enzymes may also be integrated to selectively degrade amorphous regions of cellulose. The degradation of amorphous regions makes the fibers easier to separate. Unlike cellulose nanocrystals (CNCs), which are sensitive to cellulose sources, the properties of CNFs are more influenced by the chosen pretreatment and processing techniques (Zheng, 2014).

The production processes for CNF and CNC can be seamlessly incorporated into current production lines. For example, the biorefinery and winemaking industries generate large amounts of lignocellulosic waste, making it advantageous for them to combine CNC and CNF production (García et al., 2016).

4.3.3 Production of Chitosan and Its Nanomaterials

The process of creating chitosan nanomaterials consists of several steps aimed at controlling their size, shape, and surface properties. Figure 4.3 illustrates the general process of making the chitosan biopolymer and nanomaterials. Chitosan

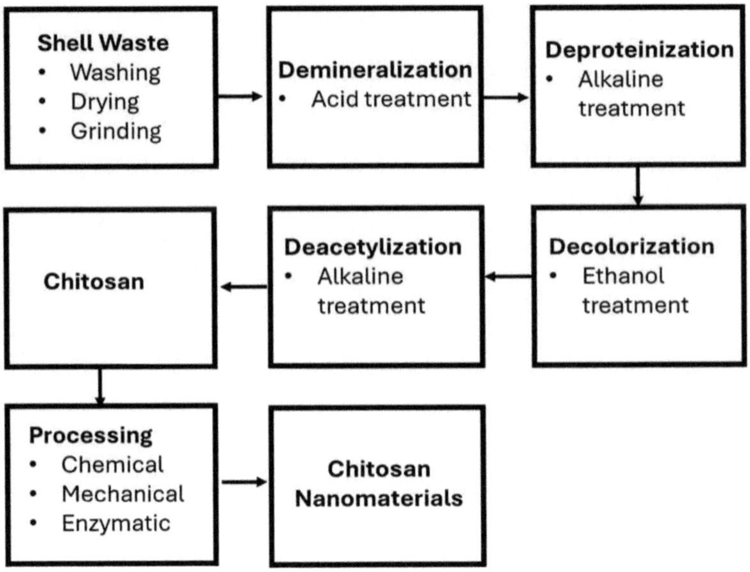

Fig. 4.3 General production process of Chitosan—Adapted (Pratiwi et al., 2023)

(Fig. 4.4) is obtained from byproducts of crustaceans waste. The extraction method can be categorized as either chemical or biological. The chemical extraction of chitosan consists of three main steps: demineralization, deproteinization, and deacetylation. During the first demineralization step, acid such as hydrochloric acid (HCl) is utilized to eliminate calcium carbonate. Although HCl is effective, in certain situations, a more environmental-friendly option like acetic acid may be preferred, as it yields soluble byproducts such as calcium acetate, which is less toxic and more compatible with microbial digestion. The efficiency of calcium carbonate removal is influenced by the reaction time and the concentration of the acid. The second stage of deproteinization uses bases such as sodium hydroxide (NaOH) or other environmental-friendly options like potassium hydroxide (KOH), which produces agriculturally useful byproducts. The final stage requires the application of strong bases like NaOH to eliminate acetyl groups from chitin, a process referred to as deacetylation. Elements such as the concentration of the base, duration of the reaction, and temperature influence the extent of deacetylation. The resulting chitosan copolymer is comprised of N-acetylglucosamine and glucosamine units in its structure (El Knidri et al., 2018).

Alongside chemical extraction, there is an eco-friendly approach to extracting Chitosan from crustacean byproducts using biological methods, such as enzymatic and fermentation processes. Enzymatic approaches use Proteinases enzymes derived from microbial or fish resources to perform deproteinization and deacetylase enzymes obtained from bacteria or genetically modified cultures to perform deacetylation at lower temperatures (25–59 °C), avoiding the harsh chemicals and high temperatures utilized in chemical processes (Mohan et al., 2022).

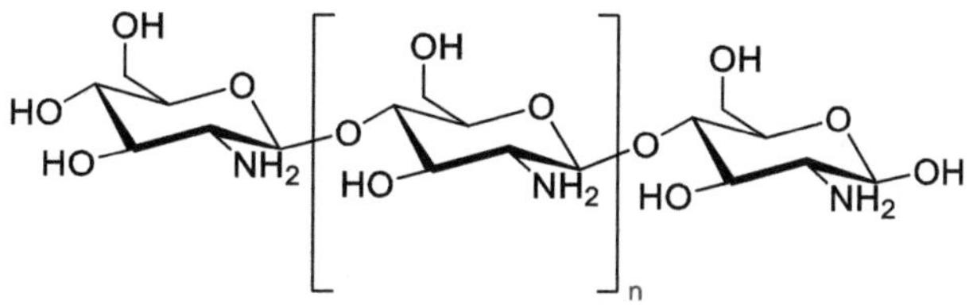

Fig. 4.4 Chitosan biopolymer

Although enzymatic approaches have a lesser environmental effect, they possess obstacles such as high enzyme costs, and poor reaction efficiency necessitating additional chemical treatments for refinement. Additional research is needed to overcome efficiency limitations and optimize these methods for industrial-scale applications (Kou et al., 2021).

Once Chitosan is extracted by either chemical or biological method, it is converted into nanomaterials through either chemical or mechanical methods. Various chemical processes exist for the synthesis of chitosan nanomaterials. Methods such as ionic gelation using polyanionic crosslinkers such as sodium tripolyphosphate (Desai, 2016) and emulsion-droplet coalescence with organic solvents and surfactants (Riegger, Bäurer, et al., 2018) are the most prevalent. Mechanical methods involve sonication (Tang et al., 2003), spray drying (Grenha et al., 2005), and high-pressure homogenization (Riegger, Kowalski, et al., 2018). There is still research going on developing scalable, environmental-friendly processes for the synthesis of chitosan nanomaterials.

4.4 Nanocellulose and Chitosan-Based Nanomaterial Properties

4.4.1 Properties of Cellulose Nanocrystals

The properties of cellulose nanocrystals (CNCs) are influenced by the extraction processes and the source of the raw material. For example, CNCs made from corncobs or pineapple leaves usually have better aspect ratios than those made from potato peels or rice husks (Owoyokun et al., 2021). To obtain the necessary CNC properties, such as aspect ratio, surface charge, thermal stability, and size, it is paramount to use the right raw material and hydrolysis/extraction techniques.

CNC stability in the suspension is also influenced by the type of acids used in the extraction procedure. CNCs produced by sulfuric acid hydrolysis are generally more stable in suspension than those extracted with phosphoric or hydrochloric acid (Pawcenis et al., 2022). CNCs extracted using sulfuric acid have negatively charged sulfate groups thus they show more stability in the suspension. Higher temperatures and longer hydrolysis durations can be used to further enhance suspension stability and surface charge in the sulfuric acid hydrolysis method (Kusmono et al., 2020).

CNC typically has a rod-like shape, with diameters between 5 and 20 nm and lengths between 100 and 500 nm. Depending on the extraction process, different techniques can produce spherical or semi-spherical CNCs. For instance, as the aspect ratio rises, CNC films exhibit increased tensile strength. CNCs with greater aspect ratios are typically made from materials with better crystallinity and using sulfuric acid hydrolysis. Nonetheless, increased sulfate levels from sulfuric acid hydrolysis may decrease the thermal stability of nanocrystals (Bangar et al., 2022). Therefore, it is crucial to specify desired product attributes before synthesizing CNCs.

4.4.2 Properties of Cellulose Nanofiber

CNFs production heavily depends on the chemical and mechanical processes used than on the source of raw materials. Nonetheless, some raw materials, such as industrial crops, seem to be easier to defibrillate and use less energy during processing. Other than energy consumption raw material sources do not significantly affect the CNFs properties. Differences in the amount of lignin, hemicelluloses, or other components like pectin may be the cause of variations in the energy required to convert cellulose into cellulose nanofibers (Yu et al., 2021).CNF is being used in many applications, and one of them is a reinforcement material in composites; hence, its mechanical properties are crucial. The production method significantly affects the mechanical behavior of nanofiber or their composites. Microfluidization, disk refining, and ultra-refining each produce cellulose nanofibrils (CNFs) with distinct tensile strength and modulus characteristics (Banvillet et al., 2023). Enzymatic pretreatment also increases the Young's modulus (Nie et al., 2018). Furthermore, TEMPO-mediated nanofibers tend to have improved mechanical properties, this is attributed to the fact that TEMPO nanofibrils, which are uniform and have a high aspect ratio, result in stronger films due to better inter-fibril hydrogen bonding and network structure.

The characteristics of cellulose nanofibers (CNFs) are strongly influenced by the amount of hemicellulose present in the raw materials. When hemicelluloses are removed by enzymatic hydrolysis, the elastic modulus of CNFs is increased, but the adhesion forces between fibers may be weakened (Dias et al., 2022).

When considering the thermal properties of CNF, materials with a higher crystalline index tend to be more thermally stable. For instance, sources with high crystallinity, such as kenaf (81.5%, 361 °C) and jute (88.6%, 325 °C), showed high degradation temperatures (Karimi et al., 2014). In contrast, less crystalline materials like soy hulls (69.6%, 290 °C) and rice straw (62%, 310 °C) started degrading at low temperatures (Alemdar & Sain, 2008).

See Table 4.1 for a side-by-side comparison of the properties of cellulose nanocrystals (CNCs) and cellulose nanofibrils (CNFs).

Table 4.1 Comparison of properties of cellulose nanocrystal and cellulose nanofibers

	Cellulose nanofibers (CNFs)	Cellulose nanocrystals (CNCs)
Source	Produced from natural cellulose fibers (e.g., wood pulp)	Produced from cellulose fibers using chemical processes
Production method	Mechanical and/or enzymatic treatment	Chemical hydrolysis
Morphology	Flexible, spaghetti-like fibers	Rigid, rod-like nanoparticles
Purity	May contain residual lignin, hemicellulose, and impurities	Higher purity, mostly crystalline
Dimensions	Width: 5–20 nm, length: several micrometers	Length: 100–1000 nm
Surface charge	Neutral or variable	Typically negatively charged
Regions in structure	Crystalline, para crystalline, amorphous	Crystalline, para crystalline
Aspect ratio	>10	>5, <50
Other abbreviations	Nano-fibrillated cellulose (NFC), cellulose nanofibers (CNFs), micro fibrillated cellulose (MFC)	Nanocrystalline cellulose (NCC) Cellulose nanowhiskers (CNW)

4.4.3 Chitosan Nanomaterials Properties

Chitosan is a multifunctional polysaccharide recognized for its solubility, ability to adhere to mucosal surfaces, capacity to form films, and behavior as a polyelectrolyte. The presence of surface amino groups allows it to effectively chelate metals, effectively binding heavy metals under both neutral and acidic environments (Varma et al., 2004). Due to its polycationic characteristics, chitosan demonstrates various chemical and biological functions, such as promoting regenerative effects on gum tissue (Arora et al., 2023), aiding osteoblast formation for bone growth (Lee et al., 2002), and exhibiting antifungal and anti-cholesterolic properties. Consequently, it is considered highly valuable for tissue engineering and related fields. Chitosan is a copolymer made up of units of glucosamine and N-acetyl-glucosamine. Its behavior in acid-base conditions is determined by its degree of acetylation which also determines the solubility of chitosan.

4.5 Nanocellulose and Chitosan-Based Nanomaterials Applications

4.5.1 Cellulose Nanocrystal Applications

The exceptional qualities like strength, biodegradability, and availability of surface functional groups have made cellulose nanocrystals (CNCs), a promising material for various applications. Even at low concentrations, CNCs are frequently added to polymer matrices for polymer reinforcement to enhance their mechanical and thermal characteristics. CNCs can effectively absorb stress due to their nanoscale size, which makes them perfect for improving building and automobile materials (Grishkewich et al., 2017). CNCs are biocompatible and biodegradable which makes them promising candidates in biomedical applications such as tissue engineering and bioimaging.

CNC-based materials offer an advantage in localized and targeted drug delivery with reduced side effects (Akhlaghi et al., 2013). CNC composites are under investigation for energy device applications such as recyclable solar cells and energy storage systems due to CNC's mechanical robustness and ease of surface functionalization. This makes it adaptable to blend with both hydrophilic and hydrophobic materials during the formulation and manufacturing stage (Zhou et al., 2013). CNC also creates new opportunities for adaptive materials, including haptic technology, electronics, and drug delivery systems, combining thermoresponsive and photoresponsive polymers.

4.5.2 Cellulose Nanofiber Applications

Cellulose nanofibers (CNFs) are useful for various applications due to their exceptional mechanical, thermal, and functional properties. CNFs are produced from renewable biomass which provides a sustainable substitute for traditional polymers. When bioactive materials like essential oils are added, CNFs offer strong, environmental-friendly food packaging choices that prolong shelf life and maintain food quality (Wagh et al., 2023). CNFs have a large surface area which makes them suitable for applications such as water treatment which need effective filtering and adsorption capacity. They effectively remove organic contaminants, heavy metals, and oils (Ma et al., 2011). In comparison to conventional techniques, these CNF-based membranes are more eco-friendly and show promise in reverse osmosis, nanofiltration, and ultrafiltration systems. The exceptional strength-to-weight ratio and biocompatibility of CNFs make them useful in environmental cleaning, biomedical, automotive, and aerospace (Sharma et al., 2023). For instance, they reinforce composites and can be used in medical environments. CNFs are also ideal for paper and textiles, and they are gaining traction in newer industries like flexible electronics and energy storage (Zhang et al., 2022). Due to their adaptability and eco-friendliness, CNFs are positioned to play a significant role in the transition to green technologies and sustainable materials. The rising number of research papers and patents demonstrates the growing contribution of CNFs to bio-based and sustainable economies as well as their capacity to drive innovation in materials science. By aligning with circular economy principles, CNFs have the potential to replace non-renewable materials across various industries and spur sustainable development. Refer to Table 4.2 for a list of nanocellulose applications across various industries.

Table 4.2 Applications of CNF and CNCs

Application	Description
Reinforcement in composites	Improves mechanical strength and stiffness of composite materials
Barrier films	Enhances barrier properties for packaging to protect against gases and moisture
Biomedical devices	Used in medical implants and devices due to biocompatibility and strength
Drug delivery systems	Facilitates controlled and targeted release of pharmaceuticals
Stabilizers in emulsions	Acts as stabilizers in emulsions for food and cosmetic industries
Electronic displays	Used in flexible and transparent electronic displays
Sensors	Improves sensitivity and performance in various types of sensors
Paper and textiles	Enhances durability and functionality of paper products and textiles
Wastewater treatment	Used in removal of contaminants and pollutants from water
3D printing	Utilized as a printable material for advanced manufacturing techniques

4.5.3 Applications of Chitosan-Based Nanomaterials

Chitosan nanomaterials have numerous uses in various sectors. This is primarily attributed to their chemical properties and adjustable physicochemical characteristics. Their potential applications include drug delivery systems, adhesives, antioxidants, biosensors, food packaging solutions, membranes, and medical uses. The structure of chitosan allows for significant chemical alterations, leading to new features. In healthcare and drug delivery, chitosan nanomaterials are crucial due to their capacity to encapsulate active ingredients and manage their release. Their mucoadhesive properties make them excellent candidates for delivering medications through nasal, oral, and ocular routes (Kumar et al., 2023). Chitosan nanomaterials are vital in tissue engineering, functioning as scaffolds that enhance cell adhesion, proliferation, and differentiation and support the regeneration of tissues and organs (Croisier & Jérôme, 2013). The cationic nature of chitosan, stemming from its positive charge, enables it to interact electrostatically with the negatively charged bacterial cell wall, compromising its integrity, leading to cellular leakage, and ultimately resulting in cell death. This mechanism is responsible for the antibacterial properties of chitosan nanomaterials, which are beneficial in wound healing and infection control (Li & Zhuang, 2020). In agriculture, chitosan nanoparticles facilitate the targeted delivery of fertilizers and pesticides, aiding in plant growth and serving as biocontrol agents (Kashyap et al., 2015). Additionally, these nanoparticles are utilized in the food industry to enhance food packaging by inhibiting microbial growth and reducing oxygen and moisture permeability, thereby prolonging shelf life. Their film-forming abilities also make them suitable for edible coatings on fresh produce (Wang et al., 2018). Chitosan nanoparticles exhibit versatility and possess the potential to drive advancements in biotechnology, cosmetics,

Table 4.3 Applications of Chitosan

Application	Description
Drug delivery	Used to formulate controlled and targeted drug delivery systems
Wound healing	Promotes faster healing and has antimicrobial properties
Water treatment	Acts as a flocculant to remove contaminants from water
Food preservation	Used as an antimicrobial and biodegradable food packaging material
Agriculture	Enhances plant growth and protects against pathogens
Cosmetics	Utilized in skincare and haircare products for its film-forming and moisturizing properties
Tissue engineering	Serves as a scaffold material for tissue regeneration
Bioplastics	Used to create biodegradable plastic alternatives
Medical devices	Used in the development of biocompatible medical implants and devices
Biomedical applications	Involved in various applications like gene therapy, antibacterial coatings, and more

environmental studies, biosensors, water purification, and nanomedicine. A list of applications for chitosan nanomaterials is presented in Table 4.3.

4.6 Applications of Nanocellulose and Chitosan Nanomaterial Composites

It is challenging to make composites of nanocellulose and chitosan as nanocellulose is hydrophilic and chitosan is an amphiphilic polymer due to the presence of both hydrophilic (amine and hydroxyl groups) and hydrophobic (acetyl groups) sites. However, surface-modified nanocellulose and chitosan composites are important in various sectors. Surface modification can create tunable surfaces and provide benefits in creating nanocellulose and chitosan nanomaterials synergy. This synergy helps in creating stable composites. Both biopolymers are biodegradable and play an important role in packaging industries (Costa et al., 2021). These composites provide films with good oxygen and oil and grease barriers with great mechanical properties that are suited for applications like edible coatings and bioplastics. In healthcare, these composites enhance strength and antibacterial activity, making them helpful for biosensors and skin regeneration (Zamel et al., 2023). In the textile industry, these hybrid composites offer strength benefits and moisture management properties, making this composite ideal for high-performance, ecological choice in textile materials' development (El-Shafei et al., 2019). They are also used to make 3D-printed structures for buildings, drug delivery, flexible and wearable technology, and robots, all of which encourage sustainable innovation.

4.7 Environmental Benefits of Nanocellulose and Chitosan-Nanomaterials

4.7.1 *Biodegradability and Biocompatibility*

Nanocellulose and chitosan are capable of biodegradability and biocompatibility, which makes them ideal for use in sustainable and biomedical applications. Nanocellulose can be fully degraded through enzymatic and microbial processes in nature into environmentally harmless byproducts like glucose and water. This significantly reduces the environmental impact when replacing non-degradable materials in packaging, textiles, or composites. Enzymes like lysozyme break down chitosan from chitin into non-toxic residues like glucosamine. This avoids buildup or toxicity in biological systems, which is especially advantageous in biomedical applications.

Nanocellulose and chitosan are naturally biocompatible, they can interact with live tissues without triggering adverse immunological reactions. Nanocellulose shows excellent mechanical strength and has the ability to mimic the extracellular matrix in tissue engineering. This provides a scaffold for cell adhesion and growth. While, chitosan can offer antimicrobial and hemostatic properties, which make it suitable for wound dressings and drug delivery systems. The cationic characteristics of chitosan increase cellular adhesion and absorption by improving its interaction with negatively charged cell membranes and macromolecules. Together both biopolymers hold significant potential to expand their uses in drug delivery and regenerative medicine. To promote environmentally sustainable solutions, both materials must be able to break down and integrate into natural cycles without leaving any hazardous leftovers. These characteristics put them at the forefront of creating novel, environmental-friendly biopolymers.

4.7.2 *Waste Valorization and Circular Economy*

Chitosan and nanocellulose are important for waste valorization and they strengthen the circular economy. These biopolymers provide sustainable alternatives for conventional petroleum-based materials as they are made from renewable and industrial waste materials. Nanocellulose is extracted from lignocellulosic biomass, agricultural waste, and byproducts like sawdust and paper waste. Further extracted nanocellulose is used to create high-value products for applications like packaging, construction, and polymer composites. Conversion of byproducts from the seafood processing industry and fungal cell walls make it possible to produce chitosan. This chitosan is a valuable biopolymer for bioplastics, water treatment, biomedical, and bio-sensor applications.

Utilizing these waste materials to synthesize biopolymers such as nanocellulose and chitosan reduces landfill impact and generates high-end functional products. In a circular economy, nanocellulose and chitosan encourage the development of biodegradable products that can be reintegrated into natural ecosystems. For example, nanocellulose films show the potential to replace single-use plastics (Shanmugam et al., 2019), whereas Chitosan serves as biodegradable carriers for fertilizers and pesticides in agriculture (Riseh et al., 2023). Nanocellulose and chitosan are great examples of how bio-based materials may propel waste management and innovation. By turning waste streams into high-performance materials, it is possible to achieve a sustainable circular economy.

4.7.3 Reducing Reliance on Petroleum-Based Products

Chitosan and nanocellulose are playing a key role in the reduction of reliance on petroleum-based products. They provide bio-based and sustainable substitutes for a range of businesses. These nanostructures help improve the barrier qualities which makes them perfect for making films and coatings. They have shown high levels of oxygen, moisture, and grease resistance, which are essential for replacing petroleum-based polymers in industrial and food packaging applications.

Thus, with the combined use of chitosan and nanocellulose to create bio-based composite materials, it is possible to replace the synthetic alternatives. These hybrid materials have shown improved performance in various applications while supporting a lower carbon footprint. These advancements support the transition to circular economy models that are emphasizing biodegradability and recyclability. These biopolymer composites have played an important role in enhancing the developments in the manufacturing of biomaterials.

Advancements in nanotechnology have further honed their characteristics to meet specific industry requirements for applications such as controlled drug release in medicines or high tensile strength for automotive parts. Industries are progressively substituting petroleum-based products with sustainable materials such as nanocellulose and chitosan to solve global concerns like resource depletion, plastic pollution, and greenhouse gas emissions.

4.8 Conclusion

4.8.1 Future Outlook on Renewable Nanomaterials

Renewable nanomaterials have an exciting and promising future ahead of them since they are leading the development in sustainable innovation in several industries. Developments in nanotechnology have made it possible to precisely manipulate these materials' characteristics, leading to the creation of multipurpose solutions. Renewable nanomaterials, such as nanocellulose and chitosan nanomaterials, are being investigated in the medical profession for antimicrobial coatings, wound healing, and drug delivery systems. In the electronics industry, they have the potential to be used in flexible substrates, conductive inks, and biodegradable sensors. Renewable nanomaterials are necessary for creating a circular economy since they reduce carbon emissions, resource utilization, and waste generation. These materials support global sustainability goals by promoting the development of renewable energy technologies that improve sustainability and performance, in applications such as enhanced supercapacitors, batteries, and solar cells. Collaborative research efforts involving academia, industry, and government accelerate innovations by addressing scalability, cost-effectiveness, and material performance challenges across diverse conditions. The industry will undergo a revolution when modern production techniques, machine learning, and artificial intelligence are combined to enable more efficient processes and faster design cycles for renewable nanomaterials. Legislators across the globe have shown support for such environmental-friendly products and technologies. By striking a balance between technological innovation, environmental preservation, and economic competitiveness, these materials can contribute to sustainable development. However, issues including biodiversity preservation, ethical sourcing, and industrial homogeneity need to be addressed. Life cycle assessments and standardized green synthesis procedures are essential to achieve moral and environmental criteria.

4.8.2 Role in Reducing Environmental Impact

The development of eco-friendly products has increased the use of renewable resources and is supporting closed-loop systems which can reduce one's ecological footprint. Bionanomaterials can help reduce the environmental footprints of many types of businesses and applications. This can help pave the way for a sustainable future in which ecological preservation and technological advancement coexist.

References

Akhlaghi, S. P., Berry, R. C., & Tam, K. C. (2013). Surface modification of cellulose nanocrystal with chitosan oligosaccharide for drug delivery applications. *Cellulose, 20*, 1747–1764.

Alemdar, A., & Sain, M. (2008). Isolation and characterization of nanofibers from agricultural residues—Wheat straw and soy hulls. *Bioresource Technology, 99*, 1664–1671.

Arora, S., et al. (2023). Role of chitosan hydrogels in clinical dentistry. *Gels, 9*, 698.

Asunis, F., et al. (2020). The dairy biorefinery: Integrating treatment processes for cheese whey valorisation. *Journal of Environmental Management, 276*, 111240.

Bangar, S. P., et al. (2022). Surface modifications of cellulose nanocrystals: Processes, properties, and applications. *Food Hydrocolloids, 130*, 107689.

Banvillet, G., et al. (2023). Cellulose nanofibril production by the combined use of four mechanical fibrillation processes with different destructuration effects. *Cellulose, 30*, 2123–2146.

Chakraborty, A., Sain, M., & Kortschot, M. (2005). Cellulose microfibrils: A novel method of preparation using high shear refining and cryocrushing. *Holzforschung, 59*, 102–107.

Chakravarty, J., & Edwards, T. A. (2022). Innovation from waste with biomass-derived chitin and chitosan as green and sustainable polymer: A review. *Energy Nexus, 8*, 100149.

Chen, B., et al. (2024). Red swamp crayfish (Procambarus clarkii) as a growing food source: Opportunities and challenges in comprehensive research and utilization. *Food, 13*, 3780.

Costa, S. M., et al. (2021). Active natural-based films for food packaging applications: The combined effect of chitosan and nanocellulose. *International Journal of Biological Macromolecules, 177*, 241–251.

Croisier, F., & Jérôme, C. (2013). Chitosan-based biomaterials for tissue engineering. *European Polymer Journal, 49*, 780–792.

Desai, K. G. (2016). Chitosan: Nanoparticles prepared by ionotropic gelation: An overview of recent advances. *Critical Reviews in Therapeutic Drug Carrier Systems, 33*, 107–158.

Dias, M. C., et al. (2022). Influence of hemicellulose content and cellulose crystal change on cellulose nanofibers properties. *International Journal of Biological Macromolecules, 213*, 780–790.

El Knidri, H., Belaabed, R., Addaou, A., Laajeb, A., & Lahsini, A. (2018). Extraction, chemical modification and characterization of chitin and chitosan. *International Journal of Biological Macromolecules, 120*, 1181–1189.

El-Shafei, A. M., Adel, A. M., Ibrahim, A. A., & Al-Shemy, M. T. (2019). Dual functional jute fabric biocomposite with chitosan and phosphorylated nano-cellulose (antimicrobial and thermal stability). *International Journal of Biological Macromolecules, 124*, 733–741.

García, A., Gandini, A., Labidi, J., Belgacem, N., & Bras, J. (2016). Industrial and crop wastes: A new source for nanocellulose biorefinery. *Industrial Crops and Products, 93*, 26–38.

George, J., & Sabapathi, S. N. (2015). Cellulose nanocrystals: Synthesis, functional properties, and applications. *Nanotechnology, Science and Applications, 8*, 45–54.

Grenha, A., Seijo, B., & Remuñán-López, C. (2005). Microencapsulated chitosan nanoparticles for lung protein delivery. *European Journal of Pharmaceutical Sciences, 25*, 427–437.

Grishkewich, N., Mohammed, N., Tang, J., & Tam, K. C. (2017). Recent advances in the application of cellulose nanocrystals. *Current Opinion in Colloid & Interface Science, 29*, 32–45.

Isogai, A. (2013). Wood nanocelluloses: Fundamentals and applications as new bio-based nanomaterials. *Journal of Wood Science, 59*, 449–459.

Jonoobi, M., et al. (2015). Different preparation methods and properties of nanostructured cellulose from various natural resources and residues: A review. *Cellulose, 22*, 935–969.

Karimi, S., Tahir, P. M., Karimi, A., Dufresne, A., & Abdulkhani, A. (2014). Kenaf bast cellulosic fibers hierarchy: A comprehensive approach from micro to nano. *Carbohydrate Polymers, 101*, 878–885.

Kashyap, P. L., Xiang, X., & Heiden, P. (2015). Chitosan nanoparticle based delivery systems for sustainable agriculture. *International Journal of Biological Macromolecules, 77*, 36–51.

Kou, S. G., Peters, L. M., & Mucalo, M. R. (2021). Chitosan: A review of sources and preparation methods. *International Journal of Biological Macromolecules, 169*, 85–94. https://doi.org/10.1016/j.ijbiomac.2020.12.005

Kumar, S., et al. (2023). Chitosan: Applications in drug delivery system. *Mini-Reviews in Medicinal Chemistry, 23*, 187–191.

Kusmono, Listyanda, R. F., Wildan, M. W., & Ilman, M. N. (2020). Preparation and characterization of cellulose nanocrystal extracted from ramie fibers by sulfuric acid hydrolysis. *Heliyon, 6*, e05486.

Lee, J.-Y., et al. (2002). Enhanced bone formation by controlled growth factor delivery from chitosan-based biomaterials. *Journal of Controlled Release, 78*, 187–197.

Li, J., & Zhuang, S. (2020). Antibacterial activity of chitosan and its derivatives and their interaction mechanism with bacteria: Current state and perspectives. *European Polymer Journal, 138*, 109984.

Ma, H., Burger, C., Hsiao, B. S., & Chu, B. (2011). Ultra-fine cellulose nanofibers: New nanoscale materials for water purification. *Journal of Materials Chemistry, 21*, 7507.

Mazari, S. A., et al. (2021). Nanomaterials: Applications, waste-handling, environmental toxicities, and future challenges—A review. *Journal of Environmental Chemical Engineering, 9*, 105028.

Mohan, K., et al. (2022). Green and eco-friendly approaches for the extraction of chitin and chitosan: A review. *Carbohydrate Polymers, 287*, 119349.

Mohd Jamil, N. A., et al. (2022). Isolation of cellulose nanocrystals from banana peel using one-pot microwave and mild oxidative hydrolysis system. *Nanomaterials, 12*, 3537.

Morsali, M., Moreno, A., Loukovitou, A., Pylypchuk, I., & Sipponen, M. H. (2022). Stabilized lignin nanoparticles for versatile hybrid and functional nanomaterials. *Biomacromolecules, 23*, 4597–4606.

Muthu, M., et al. (2021). Crustacean waste-derived chitosan: Antioxidant properties and future perspective. *Antioxidants, 10*, 228.

Nie, S., et al. (2018). Enzymatic pretreatment for the improvement of dispersion and film properties of cellulose nanofibrils. *Carbohydrate Polymers, 181*, 1136–1142.

Owoyokun, T., Pérez Berumen, C. M., Luévanos, A. M., Cantú, L., & Lara Ceniceros, A. C. (2021). Cellulose nanocrystals: Obtaining and sources of a promising bionanomaterial for advanced applications. *Biointerface Research in Applied Chemistry, 11*, 11797–11816. https://doi.org/10.33263/BRIAC114.1179711816

Pawcenis, D., Leśniak, M., Szumera, M., Sitarz, M., & Profic-Paczkowska, J. (2022). Effect of hydrolysis time, pH and surfactant type on stability of hydrochloric acid hydrolyzed nanocellulose. *International Journal of Biological Macromolecules, 222*, 1996–2005.

Pratiwi, R. D., et al. (2023). Eco-friendly synthesis of chitosan and its medical application: From chitin extraction to nanoparticle preparation. *ADMET DMPK*. https://doi.org/10.5599/admet.1999

Reshmy, R., et al. (2021). Bacterial nanocellulose: Engineering, production, and applications. *Bioengineered, 12*, 11463–11483.

Riegger, B. R., Bäurer, B., Mirzayeva, A., Tovar, G. E. M., & Bach, M. (2018). A systematic approach of chitosan nanoparticle preparation via emulsion crosslinking as potential adsorbent in wastewater treatment. *Carbohydrate Polymers, 180*, 46–54.

Riegger, B. R., Kowalski, R., Hilfert, L., Tovar, G. E. M., & Bach, M. (2018). Chitosan nanoparticles via high-pressure homogenization-assisted miniemulsion crosslinking for mixed-matrix membrane adsorbers. *Carbohydrate Polymers, 201*, 172–181.

Riseh, R. S., Vazvani, M. G., & Kennedy, J. F. (2023). The application of chitosan as a carrier for fertilizer: A review. *International Journal of Biological Macromolecules, 252*, 126483.

Rojas, O. J. (2016). *Cellulose chemistry and properties: Fibers, nanocelluloses and advanced materials* (Vol. 271). Springer International Publishing.

Rovera, C., et al. (2018). Enzymatic hydrolysis in the green production of bacterial cellulose nanocrystals. *ACS Sustainable Chemistry & Engineering, 6*, 7725–7734.

Shanmugam, K., Doosthosseini, H., Varanasi, S., Garnier, G., & Batchelor, W. (2019). Nanocellulose films as air and water vapour barriers: A recyclable and biodegradable alternative to polyolefin packaging. *Sustainable Materials and Technologies, 22*, e00115.

Sharma, S., Kumar, K., & Thakur, N. (2023). A comprehensive review on analysis of functionalization techniques and techno-innovative uses of nano-cellulosic materials. *Industrial Crops and Products, 206*, 117632.

Siddiqui, N., Mills, R. H., Gardner, D. J., & Bousfield, D. (2011). Production and characterization of cellulose nanofibers from wood pulp. *Journal of Adhesion Science and Technology, 25*, 709–721.

Tang, E. S. K., Huang, M., & Lim, L. Y. (2003). Ultrasonication of chitosan and chitosan nanoparticles. *International Journal of Pharmaceutics, 265*, 103–114.

Terkula Iber, B., Azman Kasan, N., Torsabo, D., & Wese Omuwa, J. (2022). A review of various sources of chitin and chitosan in nature. *Journal of Renewable Materials, 10*, 1097–1123.

Varma, A. J., Deshpande, S. V., & Kennedy, J. F. (2004). Metal complexation by chitosan and its derivatives: A review. *Carbohydrate Polymers, 55*, 77–93.

Wagh, R. V., Khan, A., Priyadarshi, R., Ezati, P., & Rhim, J.-W. (2023). Cellulose nanofiber-based multifunctional films integrated with carbon dots and anthocyanins from Brassica oleracea for active and intelligent food packaging applications. *International Journal of Biological Macromolecules, 233*, 123567.

Wang, H., Qian, J., & Ding, F. (2018). Emerging chitosan-based films for food packaging applications. *Journal of Agricultural and Food Chemistry, 66*, 395–413.

Yu, S., et al. (2021). Nanocellulose from various biomass wastes: Its preparation and potential usages towards the high value-added products. *Environmental Science and Ecotechnology, 5*, 100077. https://doi.org/10.1016/j.ese.2020.100077

Zamel, D., et al. (2023). Regenerated cellulose and composites for biomedical applications. In *Regenerated cellulose and composites* (pp. 265–311). Springer. https://doi.org/10.1007/978-981-99-1655-9_10

Zhang, B., Wang, L., Zhang, C., & Wu, S. (2022). High-performance cellulose nanofiber-derived composite films for efficient thermal management of flexible electronic devices. *Chemical Engineering Journal, 439*, 135675.

Zhang, Y., Poon, K., Masonsong, G. S. P., Ramaswamy, Y., & Singh, G. (2023). Sustainable nanomaterials for biomedical applications. *Pharmaceutics, 15*, 922.

Zheng, H. (2014). Production of fibrillated cellulose materials-effects of pretreatments and refining strategy on pulp properties department of forest products technology professorship fiber products technology. www.aalto.fi

Zhou, Y., et al. (2013). Recyclable organic solar cells on cellulose nanocrystal substrates. *Scientific Reports, 3*, 1536.

Part II
Next Generation Nanomaterials for Specific Sustainable Engineering Applications

Chapter 5
Exploring Quantum Dots for Photonics Devices

Rajat Kumar Goyal, Nidhi Bharadwaj, and M. Chandrasekhar

5.1 Introduction

Alexei Ekimov, Moungi Bawendi, and Louis Brus were recognized with the Nobel Prize in 2023 for their discovery and development of quantum dots (QDs) in the chemistry domain. This field of research is rich with promising discoveries in terms of both fundamental science and applications that are currently being developed and those that will be developed in the future. Over the past four decades, QDs have captivated and captivated the attention of numerous generations of scientists. With the growth of semiconductor technology, numerous new sensors and electrical devices are being fabricated and utilized in various applications. Optoelectronic and photonic systems have gained significant attention in image sensors, displays, neuromorphic devices, and healthcare tools technologies.

Silicon-based CMOS technology is always the first choice due to the development of compact, rapid, and cost-effective photonic devices. Conventional silicon-based devices have certain constraints that limit their performance in advanced electronic and photonic systems. Si-based technology has constraints like poor

---

R. K. Goyal
Interdisciplinary Research Division, Indian Institute of Technology, Jodhpur, Rajasthan, India

Department of Physics and Astronomy, National Institute of Technology,
Rourkela, Odisha, India

N. Bharadwaj
Interdisciplinary Research Division, Indian Institute of Technology, Jodhpur, Rajasthan, India

M. Chandrasekhar (✉)
Department of Physics and Astronomy, National Institute of Technology,
Rourkela, Odisha, India

Department of Physics, CVR College of Engineering, Hyderabad, Telangana, India

S. Kulkarni (ed.), *Next-Generation Nanomaterials for Sustainable Engineering*,
Lecture Notes in Nanoscale Science and Technology 43,
https://doi.org/10.1007/978-3-032-05719-8_5

electron mobility, thermal performance issues, inefficiency in light transmission, insulation challenges, crystal lattice mismatch, limitation in scaling, etc. Consequently, researchers seek novel, adaptable, and responsive materials to address these difficulties. Nanoscale materials exhibit significant potential for developing next-generation photonic devices, including high sensitivity and multi-functionality. In recent years, significant progress has been made in developing new photonic devices using materials like amorphous oxide semiconductors, organic semiconductors, and low-dimensional nanomaterials. Amorphous oxide semiconductors offer high carrier mobility, good reliability, and low off-current. However, their wide optical bandgap limits visible light absorption, which poses challenges for their application in conventional photonic devices. Organic semiconductors would be a promising candidate for photonic applications because of relatively simple processing procedures, low-temperature requirements, and the wide range of sensing possibilities; however, practical application in commercial sensors was restricted due to lower carrier mobility and poor stability in ambient conditions. QDs represent zero-dimensional nanostructures and have attracted researchers because of sensing features and multifunctionality. Quantum confinement occurs within these materials when the particle size is less than the de Broglie wavelength, resulting in a higher density of states. This enables highly photosensitive behavior with smaller feature sizes, potentially reducing pixel size and enhancing sensor resolution for more compact systems. By combining QDs with other materials, it is possible to overcome the limitations of single-material devices, enhancing sensitivity and optoelectronic functionality. Hybrid devices combining QDs with high-performance semiconductors can extend their applications and improve overall performance.

QDs are vivid practical examples of the "particle in a box" concept of quantum mechanics. QD's journey began in the late 1970s with Alexei Ekimov. While working in the Soviet Union, he found size-dependent optical properties in copper chloride nanocrystals (Ekimov & Onushchenko, 2023). Alexander Efros and Louis Brus later developed theories to explain the effects of size quantization. Alexander Efros described a general theory for optical absorption of a broad class of semiconductors (Efros, n.d.). Louis Brus noticed the quantization effect in cadmium sulfide nanocrystals (Rossetti et al., 1983) and used theoretical models to describe them (Brus, 1983, 1984). From here, developments in QDs started by perspective of theoretical and experimental point of view. An important milestone came in 1993 by Christopher Murray, David Norris, and Moungi Bawendi. They demonstrated developed highly controlled colloidal nanocrystals of cadmium chalcogenides by a hot injection method (Murray et al., 1993) This technique involves a rapid mixing of chemical precursors at high temperatures which allows precise control over nanocrystal nucleation and growth. The method produced highly crystalline QDs with superior optical properties and low defect densities compared to previous approaches. Later advancements like core/shell structures and shape control (e.g., rods, platelets) further enhanced the functionality of nanocrystals for applications such as polarized displays (Ithurria & Dubertret, 2008; Peng et al., 2000).

In parallel with synthesis, colloidal QDs have shown significant potential in bio-imaging (Bruchez et al., 1998), sensing (Medintz et al., 2005), light-emitting diodes (LEDs) (Colvin et al., 1994), solar cells (Nozik, 2002), lasers (Guan et al., 2020), detectors (Konstantatos et al., 2006), and more (Meinardi et al., 2014). Bawendi's group grow nanocrystals of other materials such as group 4, 4–6, and 3–5. InP-based QDs offer a fully compatibility with regulations of hazardous materials (ROHS) which is crucial for large-scale production of display technology. After continuous developments and progress, metal halide perovskites has emerged (still a very hot topic now a days) as an exceptional light emitters and less affected by defects compared to traditional QDs. $CsPbBr_3$ QDs (since 2001) offer an effective strategy (adjusting the ionic compositions especially for halide ions) for tuning optical properties via ionic composition which leads to significant breakthroughs in optoelectronics (Akkerman et al., 2015; Nedelcu et al., 2015).

Colloidal QDs are now used in quantum light-emitting diodes (QLEDs) and biological labeling, in lasers, detectors, and solar concentrators. Biomedical applications could benefit from near infrared (NIR) QDs for imaging and cancer treatment. QDs are capable of emitting single photons which shows an advancement for quantum technologies. Bawendi's group recently observed indistinguishable single photons from $CsPbBr_3$ at low temperatures which has a strong implication for quantum computing, quantum teleportation, and quantum key distributions. (Kaplan et al., 2023) Another study on InP QDs shows a long coherence times that is a basic need for quantum technologies (Proppe et al., 2023). QDs are also called "artificial atoms" due to their atomic-like electronic states (Banin et al., 1999) opening new frontiers in quantum technologies.

In this chapter, we will briefly summarize fundamentals properties of QDs Quantum Confinement Effects, size-dependent optical properties, band gap tenability, and absorption and emission characteristics in Sect. 5.2. In Sect. 5.3, we explored a wide range of photonic device applications (photodiodes, light-emitting devices, phototransistors, photoconductors, solar cells, and memory devices) including their performance characteristics. In Sect. 5.4, we briefly mentioned challenges and future expects of QDs. In the end, we conclude the whole book chapter.

5.2 Fundamentals of QDs

QDs are nanometers-sized semiconductor particles, typically ranging from 2 to 10 nm in size. They are made from various semiconductor materials such as cadmium selenide, cadmium sulphide, silicon, indium arsenide, lead sulphide, and lead selenide. In this we will discuss about quantum confinement effect in details and further size-dependent optical properties, band gap tunability, absorption, and emission characteristics.

Quantum confinement is a fundamental phenomenon in nanoscale materials and confined systems. Quantum mechanics is foundational framework in physics that describe how matter and energy behave on atomic and subatomic level. Quantum mechanics is based on wave–particle duality nature, according to this particle can exhibit both particle-like and wave-like properties. Particle's de Broglie wavelength (λ) is inversely related to their momentum (p). When the size of the confinement approaches the de Broglie wavelength, the wave-like nature of particles becomes dominant. This occurs when the spatial dimensions of the material are reduced to sizes comparable to or smaller than the wavelength of the particles, such as electrons. In bulk materials, electrons can occupy a continuous range of energy levels due to the large number of atoms. However, in nanoscale materials, the limited size imposes boundary conditions that restrict the possible standing wave patterns, leading to quantized energy levels. This results in discrete energy levels, like steps on a ladder, rather than a smooth gradient. Energy levels are shown in Fig. 5.1.

The bandgap of a material is the energy gap between valence band and conduction band. It behaves differently in nanoscale structures. While the bandgap in bulk materials is generally constant, it increases as the size of semiconductor nanoparticles or QDs decreases due to quantum confinement effects (Ramalingam et al., n.d.; Ramalingam et al., 2020). Consequently, electronic transitions that influence optical properties occur at higher energies (shorter wavelengths) compared to bulk materials.

To understand quantum confinement, it is important to know about QDs because these effects can be observed in QDs. QDs are made up of electrons or electron–hole pairs, also known as excitons. When an electron is excited (by any methods like absorption, electroluminescence, etc.), this leaves behind a positively charged "hole" at the lower-energy level, creating an electron–hole pair. Sometimes, the electron and hole form a bound state due to their attraction, creating a quasi-particle called an exciton. The strength of the bond between them is determined by the exciton binding energy (Mueller & Malic, 2018).

Excitons are made up of an electron and a hole, and the distance between them is called the exciton Bohr radius. Each semiconductor material has its own exciton Bohr radius, which helps us categorize excitons into two types: Frenkel and Wannier (Shiau & Combescot, 2023). Frenkel excitons have a small radius like the size of a unit cell, while Wannier excitons have a larger radius that is bigger than the unit cell. When the particle's size is much larger than the de Broglie wavelength, it behaves as free particle, and its energy states remain unchanged. The bandgap also stays in its original position.

However, when the size of the particle shrinks to the nanoscale, the energy spectrum becomes discrete, leading to a shift in the bandgap, which depends on particle size as shown in Fig. 5.1. This results in a blue shift in the emitted light as the particle size decreases. This happens when the electron and exciton are confined in a space near the exciton Bohr radius (Ramalingam et al., 2020; Zhang et al., 2022). There are three confinement regimes based on the exciton Bohr radius:

(a) Weak confinement: The crystallite radius is larger than the exciton Bohr radius.
(b) Moderate confinement: The crystallite radius is similar to the exciton Bohr radius.
(c) Strong confinement: The crystallite radius is smaller than the exciton Bohr radius.

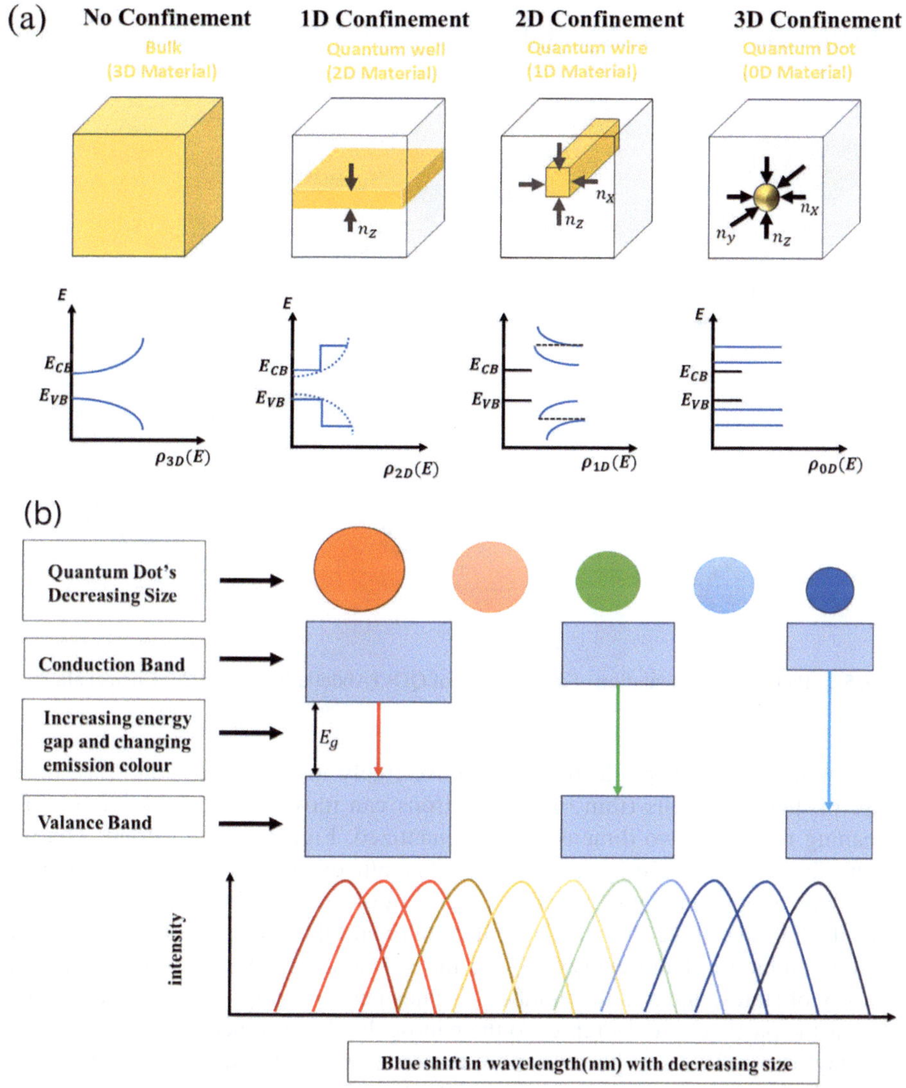

Fig. 5.1 (**a**) Representation of quantum confinement with corresponding energy levels and density of states of the QMs in different dimensions, (**b**) Variation in bandgap energy with size of the QMs (Goyal, 2024)

As the semiconductor particle gets smaller and approaches or becomes smaller than the exciton radius, the bandgap (Eg) increases, causing a blue shift in the light emitted. This property has various applications in fields like quantum wells, superlattice devices, nonlinear optical materials, photocatalysis, and imaging systems (Brus, 1984).

Based on the dimension, confinement can be classified in three categories as shown in Fig. 5.2. For instance, when electrons are confined in one direction

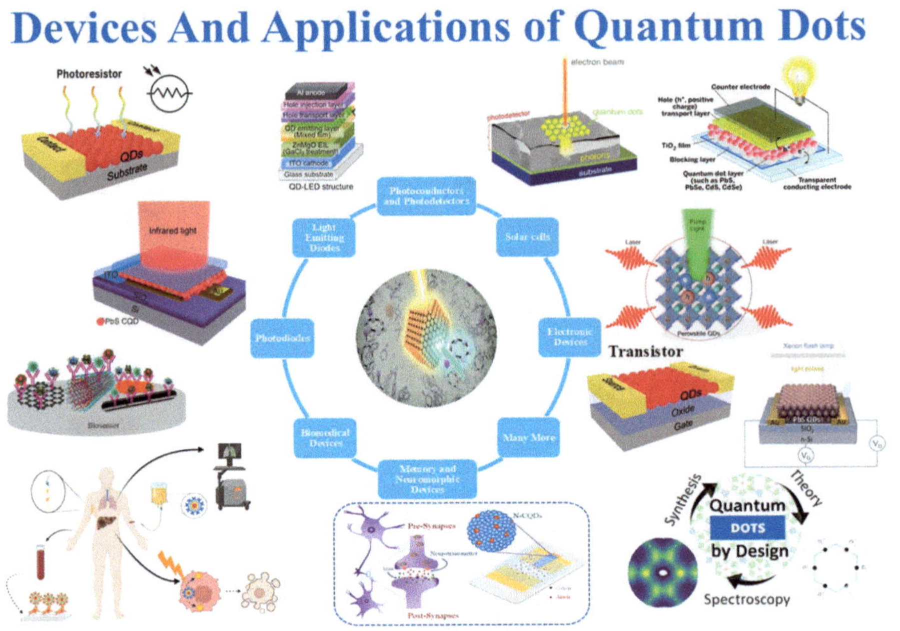

Fig. 5.2 Pictorial representation of applications of QDs (Abdellatif et al., 2022; Chen et al., 2020; Kagan et al., 2016)

(quantum wires), where electrons can move freely in one dimension (1D). In the case of quantum wells (thin films), electrons can move in two dimensions (2D), meaning the other two dimensions are quantized. Figure 5.1a illustrates confinement, showing their energy levels and a schematic example of broken symmetry. It also depicts how the density of states varies in one-dimensional (1D), two-dimensional (2D), and three-dimensional (3D) confined QDs materials. Here, E_{CB} refers to the conduction band energy, E_{VB} to the valence band energy, ρ represents the density of states, and n is the number of states per unit space. Figure 5.1b explains that as the particle size decreases to the nanoscale, the confined dimensions shrink, causing energy levels to become more discrete, which increases the band gap and raises the band gap energy (Edvinsson, 2018).

For nanoparticles, quantum confinement has a big impact on how they absorb and emit light. The energy levels become fixed, making the gap between them larger, which causes the light absorbed or emitted to have higher energy. This leads to stronger fluorescence, luminescence, and the ability to change colors, making these materials useful in things like displays, lasers, and biological imaging.

In bulk materials, conductivity is mostly calculated by the number of free-charge carriers. But in a confined system, the energy levels become discrete, which means a gap between the highest occupied and lowest unoccupied states. So, the charge carriers become slow and conductivity decreases. Quantum tunneling is a phenomenon that occurs when particles pass through an energy barrier that is higher than their total energy. This phenomenon is explained by Quantum mechanics (QM). According to QM, a particle has wave-like properties, and its wave function may penetrate the barrier height. The probability of tunneling depends on the width, height of the barrier and the energy of the particle. This phenomenon does not follow classical mechanics rules. This phenomenon enables carriers to pass through energy barriers and results in unique conductivity characteristics. Quantum tunneling is crucial in various semiconductor devices e.g. tunnel diodes and scanning tunneling microscopes (Anirban, 2022).

5.3 QDs-Based Photonic Devices

To enhance light absorption and charge transfer efficiency using highly photosensitive materials is essential for improving the performance of photonic device. A key strategy is the hybridization of various semiconducting materials, which has been employed to create multifunctional optoelectronic devices and optimize interfacial engineering for more efficient charge transfer in innovative device structures. QD-based photonic systems and devices have significantly impacted the advancements of light-emitting devices, photodiodes, phototransistors, solar cells, memory technologies, etc., as shown in Fig. 5.2. In this section, we will discuss the detailed application of QDs.

5.3.1 QDs for Light-Emitting Diodes (LEDs)

Due to quantum confinement phenomena, QDs have developed high-quality QD light-emitting applications. By precisely controlling the chemical composition of QD elements, it is possible to fine-tune the spectral properties for efficient light emission, which is a distinguished key feature of QD light-emitting devices (QLEDs). Due to their inorganic crystalline structure, QDs hold great promise for advanced light-emitting devices. The inorganic crystalline structure of QDs supports enhanced efficiency, stability, operational lifetime, and current density by overcoming limitations seen in small-molecule and polymer LEDs. Specifically, CdSe QDs are commonly used in visible light QLEDs, while lead chalcogenide QDs are preferred for infrared emission. Research on CdSe-based QLEDs has shown significant progress through optimization of core/shell configurations with advancements in ligand engineering, and the incorporation of effective charge transport and blocking layers.

In 2014, a study by Dai et al. explored CdSe/CdS core–shell QD-based red LEDs with an insulating layer between QDs. They achieved up to 20.5% EQE, which decreased up to 15.1% over 100,000 h of operational lifetime at 100 cd/m^2 (Dai et al., 2014). Song et al. later improved this with $Zn_{1-x}Cd_xSe$ QDs and observed

30.9% EQE and 1,800,000 h of lifetime at 100 cd/m^2. Pradhan et al. used PbS QD-based LEDs with advanced passivation, reducing trap density. They observed 7.9% EQE over 60% photoluminescence efficiency and improvement in performance. Gao et al. incorporated PbS QDs into a perovskite matrix and achieved 8.1% EQE with extended stability. Won et al. developed InP-based QLEDs and reached 21.4% EQE with 100,000 cd/m^2 brightness over 1000,000 h of lifetime, a comparable study to Cd-based QLEDs. Ligand exchange and shell engineering further improved efficiency by reducing defects. Wijaya et al. used near-infrared-emitting QLEDs based on heavy metal-free In(Zn)As–In(Zn)P–GaP–ZnS QDs and reached 4.6% EQE with 75% photoluminescence efficiency. Xu et al. studied perovskite QLEDs and reported 18.7% EQE with 75 cd/A current efficiency for $CsPbBr_3$ QDs. Dong et al. introduced a bipolar-shell approach and achieved 22% EQE for green devices with over 90% photoluminescence quantum yield and enhanced stability. Yang et al. studied QD-based patterns with 1400 pixels per inch for full-color large-area displays and achieved 14.6% EQE using photo-patterned QD films. These advancements promise improvements in efficient, stable, and environmentally friendly QLED technology.

Recent advancements in QLED technology have been achieved by utilizing well-designed QDs and optimizing synthesis techniques. By incorporating innovative structures with balance carrier transport layers, QLEDs have reached new levels of photonic performance and achieved peak external quantum efficiencies (EQEs) up to 21% in the visible range. Song et al. observed alloyed $Zn_{1-x}Cd_xSe$ QD-based red LED and achieved a 30.9% EQE peak with decreasing efficiency of up to 25% over an operational lifetime exceeding 1,800,000 h at a luminance of 100 cd m^{-2} (Song et al., 2019). This was achieved through precise shell growth with employing a ZnSe intermediate layer and an outer ZnS layer, along with surface ligand modification using 2-ethylhexane-1-thiol to enhance hole transport which leads to more balanced charge injection. Additionally, Pradhan et al. demonstrated that infrared LEDs based on PbS QDs could reach EQEs of 7.9% and achieve photoluminescence quantum efficiencies above 60%, which showcases their potential for efficient infrared emission (Pradhan et al., 2018).

A 2023 study by Chen Ping reported that the performance of blue QLED based on the toxic Cd, Pb can be enhanced by using orthorhombic boron nitride (BN). They achieved 31.3% photoluminescence quantum yield (PLQY) and electroluminescence at 437 nm with turn on voltage, maximum luminance, and maximum current density as 9 V, 6.55 Cd m^{-2}, and 34.42 mA cm^{-2}, respectively (Chen, Ko, et al., 2023). Dezhang Chen has demonstrated the advancements in $CsPbI_3$-based QD LEDs by reducing long chain ligands (Chen, Yang, et al., 2023) The best performance was shown by cysteine-passivated quantum dots with a dual passivation effect. Shiho Terada developed a sustainable method to develop SiQD LED using a colloidal solution of decyl-passivated SiQD derived from rice husks containing 20%wt SiO_2 to address the growing environmental concerns. It showed orange-red photoluminescence at 680 nm wavelength with 21% quantum yield (Terada et al., 2022).

5.3.2 *Photodiodes*

Photodetectors can be constructed either using diode or transistor configurations. Phototransistors generally have slower photo-response times yet exhibit high photoconductive gain, whereas photodiodes have faster response times with relatively lower photoconductive gain. Extensive research has been carried out on several low-dimensional materials as substitutes for conventional amorphous and polycrystalline silicon in photodiodes, to improve sensitivity and multifunctionality in optoelectronic devices. QDs are distinguished among many material possibilities due to their solution processability, enhanced absorption coefficients, and spectrum tunability, making them especially beneficial for advanced applications.

Tang et al. used HgTe QDs to create a flexible infrared photodiode with 7.5×10^{10} Jones specific detectivity and 260-ns reaction time (Tang et al., 2019). The device's photo-responsivity and specific detectivity remained steady even after 1000 cycles of concave bending and varying bending radii, allowing for infrared imaging applications in various bending situations. Zheng et al. developed a flexible photodetector with $CsPbI_3$ perovskite QDs and single-walled carbon nanotubes for large-area wearable optoelectronic devices (Zheng et al., 2019). The device maintained constant photocurrent despite bending radii as tiny as 2 mm and 1000 bending cycles. Shen et al. demonstrated a flexible photodetector using $CsPbI_3$ perovskite QDs and a CsBr/KBr treatment (Shen et al., 2020) After 1600 bending cycles and bending up to 60°, the photo-induced current retained more than 93% of its initial value. These flexible photodetectors have exceptional tensile strength and stability, ensuring good photonic response and resilience even after repeated bending and stretching.

A recent study by Di Yunxiang enhanced the performance of PbS colloidal QD photodiodes by introducing two-dimensional titanium carbide ($Ti_3C_2T_x$) MXenes film as a transparent conducting electrode and using interface engineering between MXenes and PbS layer. It showed a high specific detectivity of 5.51×10^{12} cmW^{-1} $Hz^{1/2}$ with a 140 dB dynamic response range and 0.76 MHz bandwidth at 940 nm in the self-powered state (Di et al., 2024).

In 2023, a study by Wei Gong revealed that multi-interface engineering could enhance the performance of PbS Carbon Quantum Dot (CQD) near-infrared (NIR) photodiodes. The defect density and dark current ($\sim$70 nA cm^{-2}) can be reduced by using high-quality n-type CQD inks and gentle ligands cross-linked p-type CQD solid films, forming the well-passivated interface of the photoactive layer. The photogenerated charge carrier extraction efficiency can be increased by constructing a carrier-selective layer of LiF and MoO_x in the electrode. This resulted in ultrahigh detectivity up to 1.42×10^{12} (Gong et al., 2023).

5.3.3 Photovoltaics Devices (Solar Cell)

Solution-processed solar cells have sparked widespread interest due to their lower production costs when compared to regular crystalline silicon solar cells. QDs stand out from other materials proposed for third-generation solar cells due to their size-dependent tunable bandgap, which enables broad-spectrum light absorption. McDonald et al. introduced solution-processed PbS QD solar cells in 2005 that used MEH-PPV to detect infrared beyond 800 nm. Their device achieved a peak open-circuit voltage of 0.36 V, a short-circuit current of 350 nA, and 0.006% internal quantum efficiency (McDonald et al., 2005).

Since then, solar cell performance has improved a lot by modifying device architectures and materials. For example, in 2011, Semonin et al. created PbSe QD solar cells with a high exterior quantum efficiency of 114% and internal quantum efficiency of 130% (Semonin et al., 2011) The conductive QDs layer was formed by a stepwise ligand exchange method involving 1,2-ethanedithiol and hydrazine. This technique enabled the efficient collection of charge carriers and increases both photocurrent and power conversion efficiency (PCE) by up to 30%. Crisp et al. utilized a similar method to develop PbS and PbSe QDs solar cells with metal halide salts. They increased conductivity by reducing carbon content and changing band locations and observed a PCE of more than 7% (Crisp et al., 2015). Furthermore, employing donor–acceptor phase separation via p-n junctions promoted efficient exciton dissociation and reduced recombination which improves overall device performance. Yang et al. created a unique approach for bulk heterojunctions by mixing QDs and creating well-aligned donor and acceptor phases and result in a high PCE of 10.4% (Yang et al., 2017). Kim et al. have demonstrated advancements in QD-based infrared solar cells by utilizing sodium acetate for facet-specific passivation by improved colloidal stability and photophysical characteristics with an external quantum efficiency of over 70% at the excitonic peak (Kim et al., 2019).

Over the last decade, major efforts have been made to improve PbS QD solar cells via material optimization, surface ligand engineering, and structural improvements, resulting in a PCE of more than 11%. Zhang et al. demonstrated PbS QD solar cells using a bulk-heterojunction interlayer of PTB7-Th: $PC_{71}BM$, reaching a PCE of 12.02% (Zhang et al., 2020). Yuan et al. also investigated tandem solar cells, which combined dye sensitized and PbS QD solar cells and resulted in a 12.37% increase in PV efficiency (Yuan et al., 2020). Perovskite QD-based solar cells have received a lot of attention recently. In 2011, Im et al. created highly efficient perovskite $(CH_3NH_3)PbI_3$ QD-sensitized solar cells with a PCE of 6.54% (Im et al., 2011). Although the redox electrolyte limits these cells' stability, techniques to improve their long-term stability are being studied. For example, surface passivation with cesium salts has been demonstrated to enhance electron coupling, short-circuit current density, and overall performance by 14.1%. In 2020, a study by Hao et al. showed perovskite QD-based solar cells with a 16.6% PCE and reduced fault density (Hao et al., 2020). Despite advancements, there are still obstacles in constructing extremely stable and efficient solar cells with optimized electron and hole transport layers. To achieve high-efficiency QD-based solar cells, more advances in surface treatments, doping, and defect state passivation are required to lengthen carrier lifetimes.

In 2022, Abinaya synthesized cadmium sulfide (CdS) QDs in three solvents (distilled water, ethanol, and isopropyl alcohol) and found QDs in range from 7.5 to 9 nm. Solvents influence the optical properties of the solar cell due to the quantum confinement. Among all the solvents, CdS QDs prepared in ethanol exhibited the highest power conversion efficiency of 0.68% (Abinaya et al., 2022). A study by Meng Xing in 2022 proposed the use of perovskite $CsPbI_3$ QDs as an interfacial layer between PbS QD active later and organic polymer HTL can improve charge extraction and reduce charge recombination thereby enhancing the photovoltaic performance from 10.5% to 12.32% (Meng et al., 2022).

In 2023, a study by Jun Ma revealed that in situ growth MoS_2 QDs can be used as carrier-selective interface material in silicon heterojunction solar cell leading to improving photovoltaic conversion efficiency up to 23% (Ma et al., 2023). Furthermore, a study by Ziqiang Sun focused on colloidal QDs that are free from toxic heavy metals, positioning them as promising candidates for next-generation solar cells due to their solution processability, tunable bandgaps, and wide absorption range. However, inherent defects in these QDs capture photo-generated carriers hinder device performance. This study presents silver-gold-selenide (AgAuSe) QDs as eco-friendly photo-absorbers with defect states treated and further enhanced by an added layer of gold nanoparticles. This interfacial modification led to a power conversion efficiency of 4.14% (Sun et al., 2023).

5.3.4 Photoconductors and Photodetectors

Photoconductors typically rely on a single type of charge carrier (unipolar), while photodiodes facilitate the movement of electrons and holes. Studies indicate that ligand exchange and chemical modifications can enhance charge carrier mobility and photoconductivity, thus improving photoconductor responsivity and detectivity. For example, in 2006, Konstantatos et al. demonstrated solution-processed lead sulfide (PbS) QDs photoconductors achieving responsivities exceeding 10^3 A/W and specific detectivity surpassing 1.8×10^{13} Jones at 1300 nm in the infrared spectrum (Konstantatos et al., 2006). Ligand engineering, such as the use of butylamine to crosslink semiconductor QDs, improves film conductivity and reduces noise from trap sites at QD interfaces, resulting in fast carrier transit times of 500 ns at 100 V bias. Keuleyan et al. developed mercury telluride (HgTe) QD photoconductors achieving detectivities greater than 10^{10} Jones in the mid-wavelength infrared region (3-5 μm) (Keuleyan et al., 2011). Dodecanethiol-quenched HgTe QDs have improved optoelectronic characteristics by reducing ligand molecule vibrations over 3 μm, which previously hindered charge separation. New ligands, such as $In_2Se_4^{2-}$, enhance photoconductors' optical performance. For example, CdSe

QD-based photoconductors capped with these complexes exhibited specific detectivities above 10^{13} Jones and responsivities of 260 A/W at 450 nm.

Phototransistors, like photoconductors, can operate in n-type (electron-conducting) and p-type (hole-conducting) circuits. QDs function as a semiconducting bridge between the source and drain electrodes in phototransistors. These QDs' conductivity is influenced by gate voltage and light exposure. However, while QD-based phototransistors offer certain advantages, their application has been restricted. This is due to exciton quenching, which lowers free carrier generation by promoting exciton recombination. Several novel solutions, including hybrid phototransistors, have been offered to overcome this problem. These devices use QDs as the light-absorbing and carrier-generating layer, together with different semiconductors that serve as carrier transport layers, increasing photosensitivity and broadening the spectrum of applications.

Hwang and colleagues designed a hybrid phototransistor by integrating lead sulfide (PbS) QDs with amorphous indium gallium zinc oxide (a-IGZO). This setup achieved an impressive photo-responsivity exceeding 10^6 A/W and a specific detectivity of 10^{13} Jones for infrared light at 1000 nm (Hwang et al., 2016). Traditionally, a-IGZO thin-film transistors (TFTs) have been prominent in high-resolution active-matrix displays and sensors due to their high electron mobility, uniformity, and stability. Furthermore, their adaptability to solution processing makes a-IGZO TFTs well-suited for cost-effective, large-area electronic and optoelectronic applications. In Hwang's study, PbS QDs were layered atop a-IGZO TFTs; photo-excited electrons from the QDs were transferred to the a-IGZO layer's conduction band, which produce a notable shift in the TFTs' threshold voltage.

Cho et al. took a similar technique, designing hybrid phototransistors with QDs implanted between amorphous oxide semiconductors (AOSs). The QDs and AOSs' enhanced interface area allowed for efficient charge transfer, leading to strong photo-responsivity (~10^4 A/W), specific detectivity (8.1×10^{13} Jones), and a photoconductive gain of 10^7. This design, which included a monolayer of QDs, permitted efficient charge transfer by decreasing trap sites inside the QD layer. In addition, an electro-optical NOT gate was demonstrated for multispectral sensing at 500 and 570 nm wavelengths.

Materials such as graphene and Transition Metal Dichalcogenides (TMDCs) have emerged as interesting possibilities for QD hybridization. Molybdenum disulphide (MoS_2), a TMDC, has gained popularity for optoelectronic applications due to its high carrier mobility, adjustable bandgap based on layer thickness, and mechanical stability. Konstantatos et al. demonstrated hybrid photodetectors (monolayer/bilayer of graphene covered with a thin film of colloidal QDs) with responsivities above 10^7 A/W and specific detectivity over 7×10^{13} Jones from short-wavelength infrared to visible (Konstantatos et al., 2012). Kufer et al. developed MoS_2/PbS QD hybrid phototransistors with a high photo-responsivity of 10^6 A/W and specific detectivity of 7×10^{14} Jones (Kufer et al., 2015). The strong electrical interaction between MoS_2 and QDs improved optoelectronic capabilities that outperformed those of individual materials.

5.3.5 Memory and Neuromorphic Devices

QDs have intriguing properties, making them ideal for developing high-performance memory and neuromorphic computer devices. Solution-processable QD semiconductors offer options for scalable neuromorphic devices due to their excellent optoelectronic performance, stability, customizable size attributes, low cost, and large-area manufacturing potential. Several studies have looked at QD-based neuromorphic devices that can be included in flash memory, resistive random-access memory (RRAM), and photonic synaptic devices.

Memory platforms offer an essential feature to erase, which is needed for high-performance and low-power devices quickly. For example, a study by Jeong et al. showed rapid photo-induced recovery in CdSe QDs floating gates using low-intensity light (Jeong et al., 2018). They modified the surface using ligands like octadecyl phosphonic acid and perfluorinated thiol by improving charge transfer between QDs, reducing erasing times and resulting in dynamic switching behaviors. Another study by Wang et al. used all-inorganic $CsPbBr_3$ perovskite QDs in photonic flash memory and important property of flash memory, i.e., optical charge trapping using exciton separation at heterostructure interfaces (Wang et al., 2018). These optical qualities also apply to synaptic properties such as short-term plasticity (STP) and long-term plasticity (LTP) in artificial neuromorphic systems. Various QD memristors have tremendous potential in scientific computing and efficient neuromorphic applications. In a study by Hu et al. phosphorene/ZnO nanoparticle hybrid heterojunction was used to create stable broadband photonic RRAM with fast charge transfer (Hu et al., 2018). Moreover, biocompatible materials-based bio-inspired neuromorphic devices have gained attention due to their potential in smart, environmentally friendly optoelectronics. Carbon-based devices that offer flexibility and optoelectronic capabilities have been explored for bioinspired memory electronics. For example, a study by Lv et al. photoactive bio-memories with carbon dots and silk showed increased photonic resistive switching by ultraviolet irradiation (Lv et al., 2018) Li et al. investigated ZnO and PbS QD-based heterostructures in neuromorphic devices mimicking bio-synaptic activity by spectral-dependent carrier excitation (Li et al., 2019). The combined features of QDs and metal-oxide materials allowed the simulation of synaptic behaviors, enabling the construction of artificial neural networks (ANNs) with efficient photonic neuromorphic structures. Due to easy colloidal production and strong photonic switching properties of QD, they have become low-cost fundamental materials for neuromorphic devices and high-performance optoelectronic memory. Shuming Ye explored Mn_xGe_{1-x} QDs for spintronic devices. In this study, $Mn_{0.05}Ge_{0.95}$ QDs/graphene heterostructures were prepared by ion beam sputtering, which revealed high hole transport properties and a Curie temperature of 342 K. This work provides a cost-effective approach to integrate Mn_xGe_{1-x} QDs with 2D materials, advancing spintronic device research (Ye et al., 2024).

The integration of flexible photonic devices into various applications has governed significant attention due to their potential for small, portable, and functional information technologies. Flexible devices due to soft and thin platforms, they can attach to and interact with diverse surfaces such as skin and textiles which makes them ideal for wearable optoelectronics with high mechanical stability. Flexible QD-based photonic devices have attracted interest for their low cost, flexibility, durability, and exceptional optical properties. Integration of Complementary Metal-Oxide-Semiconductor (CMOS) technology with QD photodetectors has emerged as a key advancement which combines the well-established CMOS platform with the unique optical properties of QDs and enables the creation of compact, cost-effective, and high-performance optoelectronic systems for flexible imaging and sensing. Up-conversion photodetectors (which transform lower-energy photons into higher-energy visible photons) have also become important for flexible applications particularly for detecting infrared wavelengths. QD-based designs showing excellent efficiency and mechanical durability. The development of QD-based flexible lasers and amplifiers has seen similar interest due to their optical output and amplification capabilities. QD-based flexible photonic devices have shown promise due to their tenable wavelength capabilities, mechanical flexibility, and resilience in demanding environments, offering opportunities for advanced, efficient, and portable communication systems. Furthermore, QLED technology has revolutionized display and imaging systems, significantly improving color accuracy and performance. QDs play a critical role in advancing the capabilities of displays like QD-OLEDs and QD-μLEDs. Photocatalysts (use light to accelerate chemical reactions/organic transformations) generate electron–hole pairs when light is absorbed (energy should be equal to or greater than their bandgap), which helps in redox reactions at the surface and enables various applications. Plasmonic photocatalysis (Plasmonic metal nanoparticles with semiconductor) enhances light absorption and photocatalytic efficiency under visible light (Raja-Mogan et al., 2020) The localized surface plasmon resonance (LSPR) of metallic nanoparticles can be adjusted by controlling their dimensions and morphology, which allows enhanced photocatalytic performance. Photocatalysis has wide applications like hydrogen production, degradation of organic pollutants, self-cleaning surfaces, and antimicrobial coating, and many more. Table 5.1 details the different types of QDs and information about their size range, bandgap, emission wavelength, surface ligands, optical properties, and potential applications in displays, bio-imaging, LEDs, photocatalysis, and solar cells.

5.4 Challenges and Future Directions

Nowadays, QDs emerge as a valuable material for various up-to-date applications. It has several challenges and limitations. This section briefly mentions some common challenges and future directions for QDs.

One of the most significant challenges for QDs is the difficulty synthesizing high-quality, uniform QDs at a large scale. There are several issues in the synthesis of uniform shape and size of QDs, which shows batch-to-batch variability. This affects the efficiency and consistency of QDs. The optical and electronic properties

Table 5.1 Properties and applications of various quantum dots (QDs)

S. No.	Type of QD	Size (nm)	Bandgap (eV)	Emission wavelength (nm)	Surface ligands	Optical properties	Applications
1.	CdSe (cadmium selenide)	2–6	1.7–2.4	450–640	Oleic acid, TOPO	High quantum yield, tunable emission	Displays, bio-imaging, LEDs
2.	CdS (cadmium sulfide)	2–10	2.4–3.8	400–600	Thiols, amines	High stability, tunable emission	Photocatalysis, LEDs, solar cells
3.	PbS (lead sulfide)	3–10	0.3–1.5	800–2000	Oleylamine, thiols	Infrared emission, narrow bandgap	Infrared detectors, photovoltaics
4.	PbSe (lead selenide)	2–10	0.4–1.6	1000–2000	Fatty acids, amines	Strong NIR emission	NIR imaging, sensors, thermoelectrics
5.	InP (indium phosphide)	2–8	1.4–2.0	450–700	Fatty acids, amines	Low toxicity, broad emission tuning	Bio-imaging, solid-state lighting, displaysl
6.	InAs (indium arsenide)	2-8	0.3–1.2	800–1600	Thiols, alkyls	NIR emission, narrow bandgap	Infrared imaging, sensors
7.	CdTe (cadmium telluride)	2–8	1.5–3.0	500–800	Thiols, carboxylates	Tunable emission, wide bandgap	Biosensors, imaging, lasers
8.	ZnSe (zinc selenide)	1–6	2.6–3.7	380–500	Carboxylates, phosphines	High quantum yield, UV emission	UV lasers, LEDs, photodetectors
9.	ZnS (zinc sulfide)	2–10	3.7–4.2	300–400	Thiols, amines	UV emission, high stability	UV LEDs, sensors, bio-labeling
10.	Ag_2S (silver sulfide)	2–10	0.9–1.1	900–1300	Thiols, fatty acids	Near-IR emission, biocompatible	Bio-imaging, NIR imaging
11.	Ag_2Se (silver selenide)	3–10	0.1–1.2	1000–2000	Fatty acids, phosphines	NIR absorption	Infrared detectors, bio-imaging
12.	$CuInS_2$ (copper indium sulfide)	3-12	1.5–2.5	600–1000	Thiols, fatty acids	Broad absorption, high stability	Photovoltaics, bio-imaging

(continued)

Table 5.1 (continued)

S. No.	Type of QD	Size (nm)	Bandgap (eV)	Emission wavelength (nm)	Surface ligands	Optical properties	Applications
13.	$CuInSe_2$ (copper indium selenide)	3–12	1.0–1.5	700–1100	Thiols, fatty acids	NIR emission, broad tuning	Solar cells, NIR sensors
14.	ZnO (zinc oxide)	2–10	3.3–4.0	380–450	Oleic acid, phosphates	UV emission, high exciton binding energy	UV detectors, biosensors, solar cells
15.	Graphene QDs (GQDs)	3–10	2.1–5.5	400–700	Hydroxyl, carboxyl	High photostability, low toxicity	Sensors, bio-imaging, energy storage
16.	Si (silicon) QDs	2–5	1.5–2.1	600–900	Si-OH, alkyls	Biocompatible, tunable emission	Bio-imaging, LEDs, sensors
17.	SnO_2 (tin oxide)	3–7	3.6–4.0	300–400	Phosphonic acids	UV emission, wide bandgap	UV sensors, photocatalysis
18.	TiO_2 (titanium dioxide)	5–20	3.0–3.2	320–400	Fatty acids, phosphates	UV absorption, high stability	Photocatalysis, solar cells
19.	MoS_2 (molybdenum disulfide)	2–10	1.2–1.9	600–1000	Thiols, alkyl chains	Strong absorption, high carrier mobility	Transistors, optoelectronics
20.	WS_2 (tungsten disulfide)	2–10	1.2–2.0	600–1000	Thiols, alkyl chains	High photoluminescence	Bio-imaging, photovoltaics
21.	NiO (nickel oxide)	5–15	3.6–4.0	300–400	Oleylamine, thiols	UV emission, wide bandgap	UV detectors, solar cells
22.	Mn-doped ZnS (manganese)	2–7	2.8–4.1	580–620	Thiols, fatty acids	Orange emission, high stability	Bio-labeling, imaging, optoelectronics
23.	FeS_2 (iron pyrite)	5–20	0.9–1.2	800–1000	Oleylamine	Narrow bandgap, NIR absorption	Photovoltaics, sensors
24.	Co_3O_4 (cobalt oxide)	3–12	2.0–2.6	450–500	Fatty acids, phosphates	Broad absorption, catalytic properties	Sensors, energy storage

25.	YVO_4: Eu (yttrium vanadate)	3–10	3.0–3.6	610–630	Oleic acid	Strong red emission, phosphorescent	Displays, lighting, bio-imaging
26.	$BaTiO_3$ (barium Titanate)	10–30	3.0–3.2	400–500	Fatty acids	Ferroelectric, high stability	Capacitors, sensors, photovoltaics
27.	$CsPbBr_3$ (perovskite)	5–15	2.0–2.3	500–550	Alkyl chains, amines	High quantum yield, tunable emission	LEDs, displays, solar cells
28.	$CsPbI_3$ (perovskite)	5–15	1.5–1.8	600–700	Alkyl chains, amines	High quantum efficiency, red emission	Displays, optoelectronics
29.	$CsPbCl_3$ (perovskite)	5–15	3.0–3.5	400–450	Alkyl chains, amines	High quantum yield, blue emission	Displays, optoelectronics
30.	BaS (barium sulfide)	2–10	3.0–3.5	300–400	Phosphonic acids	UV emission, high thermal stability	UV lasers, photocatalysis *BaS (barium sulfide
31.	SrS (strontium sulfide)	3–12	3.5–4.2	300–450	Fatty acids	UV emission, wide bandgap	UV sensors, lighting
32.	$NaYF_4$:Yb,Er (sodium yttrium fluoride)	10–20	2.4–2.6	550–700	Oleic acid, thiols	Strong up-conversion luminescence	Bio-imaging, sensing, lighting
33.	KCl (potassium chloride)	5–15	5.0–5.5	200–300	Phosphonic acids	UV transparency, high stability	UV optics, scintillators
34.	Cu_2O (copper oxide)	5–15	5.0–5.5	200–300	Phosphonic acids	UV transparency, high stability	UV optics, scintillators
35.	$ZnWO_4$ (zinc tungstate)	5–15	4.2–4.6	250–350	Oleic acid, phosphates	UV emission, high thermal stability	Scintillators, UV detectors
36.	LiF (lithium fluoride)	3-10	12.0–14.0	200	Phosphonic acids	Strong UV transparency, high bandgap	UV optics, scintillators

(continued)

Table 5.1 (continued)

S. No.	Type of QD	Size (nm)	Bandgap (eV)	Emission wavelength (nm)	Surface ligands	Optical properties	Applications
37.	EuS (europium sulfide)	2-8	1.6–1.9	600–700	Fatty acids, thiols	Strong red emission, magnetic properties	Magnetic imaging, displays
38.	EuTe (europium telluride)	2-10	1.8–2.2	500–600	Oleic acid, phosphonic acids	Red emission, magneto-optical properties	Displays, bio-imaging, sensors
39.	$NaGdF_4$ (sodium gadolinium fluoride)	5-15	2.6–2.8	500–700	Thiols, fatty acids	Up conversion emission, low toxicity	Bio-imaging, sensing, lighting
40	Y_2O_3:Eu (yttrium oxide:Eu)	3-10	3.0–3.5	600–700	Oleylamine	Strong red emission, high stability	Displays, lighting
41.	$TbPO_4$ (terbium phosphate)	5-12	5–12	400–500	Oleylamine	Strong green emission, high stability	Displays, lighting, bio-imaging l
42.	ZrS_2 (zirconium disulfide)	2-10	1.3–1.6	600–900	Oleic acid, thiols l	Strong NIR emission, high stability	NIR imaging, sensors
43.	NiS (nickel sulfide)	3-10	1.1–1.3	800–1000	Oleylamine	Strong NIR absorption, narrow bandgap	Infrared sensors, photovoltaics
44.	V_2O_5 (vanadium oxide)	5-15	2.2–2.5	500–600	Thiols, phosphonic acids	Broad absorption, high stability	Photovoltaics, photocatalysis
45.	Al_2O_3 (aluminum oxide)	3-10	6.0–6.5	200–300	Phosphonic acids	UV transparency, high stability	UV optics, high-power electronics
46.	InGaN (indium gallium nitride)	3-12	2.0–3.4	400–600	Oleylamine, thiols	Tunable emission, high stability	LEDs, displays, lasers
47.	InSb (indium Antimonide)	3-12	0.2–0.3	1000–5000	Oleic acid, thiols	Strong NIR emission, low bandgap	Infrared detectors, sensors
48.	Cr_2O_3 (chromium oxide)	5-10	3.2–3.8	400–500	Fatty acids	Strong green emission, high stability	Displays, sensors, optoelectronics
49.	$BiVO_4$ (bismuth vanadate)	5-15	2.4–2.6	500–700	Oleylamine	High photocatalytic activity	Photocatalysis, solar cells

50.	SiC (silicon carbide)	3-8	2.2–2.7	500–700	Alkyl chains, phosphonic acids	High thermal stability, tunable emission	Sensors, power devices, LEDs
51.	AlN (aluminum nitride)	5-20	6.0–6.3	200–300	Oleic acid, thiols	Strong UV absorption, high stability	UV detectors, high-power electronics
52.	Ge (germanium)	2-10	0.7–1.6	1200–1600	Oleylamine, thiols	Near-IR emission, high mobility	Photodetectors, sensors, transistors
53.	MnO (manganese oxide)	3-10	3.5–4.2	300–400	Thiols, fatty acids	High stability, UV emission	UV sensors, photocatalysis
54.	SnS (tin sulfide)	5-20	1.3–1.5	600–800	Oleylamine, amines	NIR absorption, narrow bandgap	Photovoltaics, optoelectronics
55.	HgTe (mercury telluride)	5-20	0.1–0.4	1000–3000	Thiols, fatty acids	Strong NIR absorption	NIR imaging, thermoelectrics
56.	NiSe (nickel selenide)	3-10	1.1–1.3	800–1000	Oleylamine, phosphates	NIR emission	Infrared imaging, sensors
57.	ZnTe (zinc telluride)	3-10	2.2–2.4	500–700	Phosphines, thiols	High quantum yield, tunable emission	Photodetectors, lasers
58.	HfO_2 (hafnium oxide)	5-10	5.2–6.0	200–400	Oleic acid, phosphonic acids	UV transparency, high thermal stability	UV detectors, gate dielectrics
59.	ZrO_2 (zirconium oxide)	5-10	5.0–6.1	200–400	Oleylamine, phosphates	UV absorption, high stability	UV optics, photocatalysis
60.	Cu_2S (copper sulfide)	3-10	1.0–1.3	600–900	Oleylamine, thiols	NIR absorption, high conductivity	Solar cells, thermoelectrics
61.	Fe_3O_4 (magnetite)	5-15	2.1–2.5	500–600	Thiols, fatty acids	Magnetic properties, tunable emission	Magnetic imaging, biomedicine

(continued)

Table 5.1 (continued)

S. No.	Type of QD	Size (nm)	Bandgap (eV)	Emission wavelength (nm)	Surface ligands	Optical properties	Applications
62.	TiN (titanium nitride)	5-20	3.2–3.4	300–400	Phosphonic acids	High stability, UV emission	Photocatalysis, UV detectors
63.	GaN (gallium nitride)	3-10	3.4–3.6	350–450	Oleylamine, thiols	UV and blue emission, high thermal stability	LEDs, lasers, high-power electronics
64.	Bi_2S_3 (bismuth sulfide)	5-15	1.3–1.5	600–1000	Oleylamine, thiols	NIR absorption, high stability	Photovoltaics, sensors Sb2S3 (antimony sulfide)
65.	Sb_2S_3 (antimony sulfide)	5-15	1.5–1.7	600–800	Oleylamine, thiols	Narrow bandgap, strong absorption	Photovoltaics, sensors
66.	In_2O_3 (indium oxide)	5-10	3.0–3.5	350–450	Oleic acid, thiols	High conductivity, UV absorption	Transparent electronics, sensors
67.	CeO_2 (cerium oxide)	3-8	3.0–3.4	300–400	Oleylamine, phosphonic acids	UV absorption, high stability	UV detectors, photocatalysis
68.	$LaPO_4$:Eu (lanthanum phosphate)	3-10	2.7–3.5	600–700	Oleic acid, phosphates	Red emission, high stability	Displays, bio-imaging
69.	YAG:Ce (yttrium aluminum garnet)	5-15	2.4–2.6	500–600	Oleylamine	Strong yellow emission, phosphorescent	Lighting, displays

of QDs highly depend on their size and shape. With variations in size and shape, we get inconsistent performance from devices. The synthesis process of QDs involves complex chemical reactions and expensive precursors. This raises the cost of synthesis and makes it expensive on a large scale. Another challenge in synthesis arises from the necessity of maintaining accurate temperature control. Batch-to-batch variability affects the commercial ability of QDs, especially in display and imaging technologies. A major challenge is creating artificial solids without unique properties in regular crystals. (Rainò et al., 2018) While chemists have made significant progress in QDs synthesis, but challenges remain in understanding the nucleation, surface chemistry, and defects. Particularly, colloidal synthesis of NIR materials like InAs or InSb lags due to difficult chemistry and expensive optical tools. Advances are needed in synthesis methods, such as molten salt environments (Srivastava et al., 2018), and in improving the stability of metal halide QDs, a rapidly growing research area.

Another challenge is the toxic nature of most used QDs [cadmium selenide (CdSe) and lead sulphide (PbS)]. This issue is especially important for biomedical applications where biocompatibility and safety are essential. If QDs (especially Cd-based) are released into the environment, they can persist and accumulate in ecosystems. Long-term environmental problems include heavy metal poisoning of water and soil, which can contaminate the food chain. QDs are generally not biodegradable. Nowadays, synthesizing biodegradable and environmental-friendly QDs is a significant research challenge.

Unstability under certain environmental conditions is another issue in QDs. When light is exposed to QDs, they degrade over time and reduce their luminescence efficiency. This phenomenon is known as photobleaching. This limits their usage in long-term display or bio-imaging applications. QDs (chalcogenides such as CdSe) can be oxidized in air or water. Oxidation affects their surface chemistry, which affects optical characteristics and optoelectronic device performance. Some QDs have poor thermal stability. QDs can be degraded and lose functionality in electrical devices with fluctuating operational temperatures.

The quantum yield is a measure of the efficiency of a photophysical process. It depends on the number of photons emitted and absorbed. Quantum yield significantly depends on the material and synthesis process. High quantum yield is essential for displays, solar cells, and LEDs. Some QDs have low quantum yields, which limits their applications. Photovoltaic and optoelectronic applications require efficient charge separation and transmission. Some QDs have low recombination time which can limit their efficiency in applications by preventing efficient charge carrier extraction. QD size can influence the bandgap but fine-tuning it for optimal performance across devices often involves a compromise in terms of efficiency, stability, and toxicity.

To prevent the degradation of QDs, people do surface passivation, but this process has its own challenges, like incomplete/uniform passivation, defect introduction, environmental instability, and many more. QDs have a high surface area-to-volume ratio, which leads to a significant number of surface defects. Surface defects can act as trap sites for charge carriers, reducing the efficiency of QD-based devices. For stability and solvent solubility, QDs are frequently coated with organic ligands. These ligands can reduce the conductivity of QDs when used in electronic device applications. Ligand exchange mechanisms are another approach in which insulating ligands are replaced with conductive ones. This process is difficult and influences QDs optical characteristics.

Nowadays, QDs are the frontier materials in multi-domain technologies, which show great potential in next-generation displays, lighting technology, energy, medicine, and environmental monitoring. QLEDs are common nowadays. Future advancements are expected to improve energy efficiency, flexibility, and durability for enabling applications such as foldable screens, wearable displays, and transparent panels. Compared to LEDs, QD-based lighting systems provide more accurate color and energy efficiency. QD solar cells (QDSCs) can improve solar power efficiency through larger light spectrum absorption and improved quantum effects. Hybrid systems (perovskite-QD solar cells) combine the best properties of both materials, which promise significant improvements in solar performance. Perovskite and graphene QDs increase their potential also. Functionalized QDs are being developed for targeted drug delivery and advanced diagnostics. QD-based biosensors offer possibilities in early disease detection and personalized medicine. Future quantum computers may use QDs as qubits due to their scalability and compatibility with semiconductor technology. QDs are being developed for pollution sensors and hydrogen generation photocatalysis.

5.5 Conclusion

In conclusion, due to the quantum confinement effect, QDs represent a transformative element in the field of photonic devices. Their size-dependent behavior allows for tuneable band gaps and precise light absorption and emission control. This unique capability has made QDs essential for various applications, from electronics and photonics to advanced sensors and biomedical devices. This book chapter highlights the significant breakthrough in QD research and the promising future of their integration into emerging technologies. As QDs evolve, they will undoubtedly play a pivotal role in advancing next-generation energy systems, display technologies, and beyond, offering exciting possibilities for innovation across various industries. Nonetheless, several challenges must be resolved to advance the science of QDs further. Despite the limitations, the potential of QDs is promising, and they are anticipated to contribute significantly to the progress of multiple domains. QDs represent a promising research domain with the potential to yield substantial progress across multiple disciplines. This comprehensive resource provides key insights into the fundamentals and future directions of QDs for researchers and professionals alike.

References

Abdellatif, A. A. H., Tawfeek, H. M., Younis, M. A., Alsharidah, M., & Al Rugaie, O. (2022). Biomedical applications of quantum dots: Overview, challenges, and clinical potential. *International Journal of Nanomedicine, 17*, 1951–1970. https://doi.org/10.2147/IJN.S357980

Abinaya, S., Vinoth, G., & Kadiresan, M. R. (2022). Stimulus effect of solvents on cadmium sulfide quantum dots prepared for solar cell application. *Journal of Inorganic and Organometallic Polymers and Materials, 32*(12), 4540–4548. https://doi.org/10.1007/S10904-022-02438-2/TABLES/2

Akkerman, Q. A., D'Innocenzo, V., Accornero, S., Scarpellini, A., Petrozza, A., Prato, M., & Manna, L. (2015). Tuning the optical properties of cesium lead halide perovskite nanocrystals by anion exchange reactions. *Journal of the American Chemical Society, 137*(32), 10276–10281. https://doi.org/10.1021/JACS.5B05602

Anirban, A. (2022). 40 years of scanning tunnelling microscopy. *Nature Reviews Physics, 4*(5), 291–291. https://doi.org/10.1038/s42254-022-00462-2

Banin, U., Cao, Y. W., Katz, D., & Millo, O. (1999). Identification of atomic-like electronic states in indium arsenide nanocrystal quantum dots. *Nature, 400*(6744), 542–544. https://doi.org/10.1038/22979

Bruchez, M., Moronne, M., Gin, P., Weiss, S., & Alivisatos, A. P. (1998). Semiconductor nanocrystals as fluorescent biological labels. *Science, 281*(5385), 2013–2016. https://doi.org/10.1126/SCIENCE.281.5385.2013

Brus, L. E. (1983). A simple model for the ionization potential, electron affinity, and aqueous redox potentials of small semiconductor crystallites. *The Journal of Chemical Physics, 79*(11), 5566–5571. https://doi.org/10.1063/1.445676

Brus, L. E. (1984). Electron–electron and electron-hole interactions in small semiconductor crystallites: The size dependence of the lowest excited electronic state. *The Journal of Chemical Physics, 80*(9), 4403–4409. https://doi.org/10.1063/1.447218

Chen, D., Ko, P. K., Li, C. H. A., Zou, B., Geng, P., Guo, L., & Halpert, J. E. (2023). Amino acid-passivated pure red CsPbI3 quantum dot LEDs. *ACS Energy Letters, 8*(1), 410–416. https://doi.org/10.1021/ACSENERGYLETT.2C02243

Chen, J., Du, W., Shi, J., Li, M., Wang, Y., Zhang, Q., & Liu, X. (2020). Perovskite quantum dot lasers. *InfoMat, 2*(1), 170–183. https://doi.org/10.1002/INF2.12051

Chen, P., Yang, S., Liu, F., Jiang, Y., Wang, Y., Huang, Y., Hu, J., & Chen, L. (2023). Blue-emitting orthorhombic boron nitride quantum dots and quantum-dot light-emitting diodes. *Advanced Photonics Research, 4*(3), 2200344. https://doi.org/10.1002/ADPR.202200344

Colvin, V. L., Schlamp, M. C., & Alivisatos, A. P. (1994). Light-emitting diodes made from cadmium selenide nanocrystals and a semiconducting polymer. *Nature, 370*(6488), 354–357. https://doi.org/10.1038/370354a0

Crisp, R. W., Kroupa, D. M., Marshall, A. R., Miller, E. M., Zhang, J., Beard, M. C., & Luther, J. M. (2015). Metal halide solid-state surface treatment for high efficiency PbS and PbSe QD solar cells. *Scientific Reports, 5*, 9945. https://doi.org/10.1038/SREP09945

Dai, X., Zhang, Z., Jin, Y., Niu, Y., Cao, H., Liang, X., Chen, L., Wang, J., & Peng, X. (2014). Solution-processed, high-performance light-emitting diodes based on quantum dots. *Nature, 515*(7525), 96–99. https://doi.org/10.1038/NATURE13829

Di, Y., Ba, K., Chen, Y., Wang, X., Zhang, M., Huang, X., Long, Y., Liu, M., Zhang, S., Tang, W., Huang, Z., Lin, T., Shen, H., Meng, X., Han, M., Liu, Q., & Wang, J. (2024). Interface engineering to drive high-performance MXene/PbS quantum dot NIR photodiode. *Advanced Science, 11*(6), 2307169. https://doi.org/10.1002/ADVS.202307169

Edvinsson, T. (2018). Optical quantum confinement and photocatalytic properties in two-, one- and zerodimensional nanostructures. *Royal Society Open Science, 5*(9), 180387. https://doi.org/10.1098/RSOS.180387

Efros, A. (n.d.). *Interband light absorption in semiconductor spheres*. Retrieved October 1, 2024, from https://www.researchgate.net/publication/279890805

Ekimov, A. I., & Onushchenko, A. A. (2023). Quantum size effect in three-dimensional microscopic semiconductor crystals. *JETP Letters, 118*(1), S15–S17. https://doi.org/10.1134/S0021364023130040/FIGURES/3

Gong, W., Wang, P., Deng, W., Li, J., Li, W., Li, J., Chen, Z., Li, J., & Zhang, Y. (2023). Ultrahigh detectivity from multi-interfaces engineered near-infrared colloidal quantum dot photodetectors. *IEEE Transactions on Electron Devices, 70*(7), 3668–3674. https://doi.org/10.1109/TED.2023.3276730

Goyal, R. K. (2024). *Exploring quantum materials & applications: A review*. Retrieved from https://arxiv.org/abs/2404.17594v1

Guan, J., Sagar, L. K., Li, R., Wang, D., Bappi, G., Wang, W., Watkins, N., Bourgeois, M. R., Levina, L., Fan, F., Hoogland, S., Voznyy, O., De Pina, J. M., Schaller, R. D., Schatz, G. C., Sargent, E. H., & Odom, T. W. (2020). Quantum dot-plasmon lasing with controlled polarization patterns. *ACS Nano, 14*(3), 3426–3433. https://doi.org/10.1021/ACSNANO.9B09466

Hao, M., Bai, Y., Zeiske, S., Ren, L., Liu, J., Yuan, Y., Zarrabi, N., Cheng, N., Ghasemi, M., Chen, P., Lyu, M., He, D., Yun, J. H., Du, Y., Wang, Y., Ding, S., Armin, A., Meredith, P., Liu, G., & Wang, L. (2020). Ligand-assisted cation-exchange engineering for high-efficiency colloidal Cs1xFAxPbI3 quantum dot solar cells with reduced phase segregation. *Nature Energy, 5*(1), 79–88. https://doi.org/10.1038/S41560-019-0535-7

Hu, L., Yuan, J., Ren, Y., Wang, Y., Yang, J. Q., Zhou, Y., Zeng, Y. J., Han, S. T., & Ruan, S. (2018). Phosphorene/ZnO Nano-heterojunctions for broadband photonic nonvolatile memory applications. *Advanced Materials, 30*(30), 1801232. https://doi.org/10.1002/ADMA.201801232

Hwang, D. K., Lee, Y. T., Lee, H. S., Lee, Y. J., Shokouh, S. H., Kyhm, J. H., Lee, J., Kim, H. H., Yoo, T. H., Nam, S. H., Son, D. I., Ju, B. K., Park, M. C., Song, J. D., Choi, W. K., & Im, S. (2016). Ultrasensitive PbS quantum-dot-sensitized InGaZnO hybrid photoinverter for near-infrared detection and imaging with high photogain. *NPG Asia Materials, 8*(1), e233–e233. https://doi.org/10.1038/am.2015.137

Im, J. H., Lee, C. R., Lee, J. W., Park, S. W., & Park, N. G. (2011). 6.5% efficient perovskite quantum-dot-sensitized solar cell. *Nanoscale, 3*(10), 4088–4093. https://doi.org/10.1039/C1NR10867K

Ithurria, S., & Dubertret, B. (2008). Quasi 2D colloidal CdSe platelets with thicknesses controlled at the atomic level. *Journal of the American Chemical Society, 130*(49), 16504–16505. https://doi.org/10.1021/JA807724E

Jeong, Y. J., Yun, D. J., Noh, S. H., Park, C. E., & Jang, J. (2018). Surface modification of CdSe quantum-dot floating gates for advancing light-erasable organic field-effect transistor memories. *ACS Nano, 12*(8), 7701–7709. https://doi.org/10.1021/ACSNANO.8B01413

Kagan, C. R., Lifshitz, E., Sargent, E. H., & Talapin, D. V. (2016). Building devices from colloidal quantum dots. *Science, 353*, 6302. https://doi.org/10.1126/SCIENCE.AAC5523

Kaplan, A. E. K., Krajewska, C. J., Proppe, A. H., Sun, W., Sverko, T., Berkinsky, D. B., Utzat, H., & Bawendi, M. G. (2023). Hong–Ou–Mandel interference in colloidal CsPbBr3 perovskite nanocrystals. *Nature Photonics, 17*(9), 775–780. https://doi.org/10.1038/s41566-023-01225-w

Keuleyan, S., Lhuillier, E., Brajuskovic, V., & Guyot-Sionnest, P. (2011). Mid-infrared HgTe colloidal quantum dot photodetectors. *Nature Photonics, 5*(8), 489–493. https://doi.org/10.1038/nphoton.2011.142

Kim, Y., Che, F., Jo, J. W., Choi, J., García de Arquer, F. P., Voznyy, O., Sun, B., Kim, J., Choi, M. J., Quintero-Bermudez, R., Fan, F., Tan, C. S., Bladt, E., Walters, G., Proppe, A. H., Zou, C., Yuan, H., Bals, S., Hofkens, J., & Sargent, E. H. (2019). A facet-specific quantum dot passivation strategy for colloid management and efficient infrared photovoltaics. *Advanced Materials, 31*(17), 1805580. https://doi.org/10.1002/ADMA.201805580

Konstantatos, G., Badioli, M., Gaudreau, L., Osmond, J., Bernechea, M., De Arquer, F. P. G., Gatti, F., & Koppens, F. H. L. (2012). Hybrid graphene–quantum dot phototransistors with ultrahigh gain. *Nature Nanotechnology, 7*(6), 363–368. https://doi.org/10.1038/nnano.2012.60

Konstantatos, G., Howard, I., Fischer, A., Hoogland, S., Clifford, J., Klem, E., Levina, L., & Sargent, E. H. (2006). Ultrasensitive solution-cast quantum dot photodetectors. *Nature, 442*(7099), 180–183. https://doi.org/10.1038/nature04855

Kufer, D., Nikitskiy, I., Lasanta, T., Navickaite, G., Koppens, F. H. L., & Konstantatos, G. (2015). Hybrid 2D-0D MoS2 -PbS quantum dot photodetectors. *Advanced Materials, 27*(1), 176–180. https://doi.org/10.1002/ADMA.201402471

Li, H., Jiang, X., Ye, W., Zhang, H., Zhou, L., Zhang, F., She, D., Zhou, Y., & Han, S. T. (2019). Fully photon modulated heterostructure for neuromorphic computing. *Nano Energy, 65*, 104000. https://doi.org/10.1016/J.NANOEN.2019.104000

Lv, Z., Wang, Y., Chen, Z., Sun, L., Wang, J., Chen, M., Xu, Z., Liao, Q., Zhou, L., Chen, X., Li, J., Zhou, K., Zhou, Y., Zeng, Y. J., Han, S. T., & Roy, V. A. L. (2018). Phototunable biomemory based on light-mediated charge trap. *Advanced Science, 5*(9), 1800714. https://doi.org/10.1002/ADVS.201800714

Ma, J., Li, J., Zheng, W., & Fu, Y. (2023). In situ growth MoS2 quantum dots as promising interface materials for silicon solar cells. *Materials Science in Semiconductor Processing, 165*, 107653. https://doi.org/10.1016/J.MSSP.2023.107653

Mcdonald, S. A., Konstantatos, G., Zhang, S., Cyr, P. W., Klem, E. J. D., Levina, L., & Sargent, E. H. (2005). Solution-processed PbS quantum dot infrared photodetectors and photovoltaics. *Nature Materials, 4*(2), 138–142. https://doi.org/10.1038/nmat1299

Medintz, I. L., Uyeda, H. T., Goldman, E. R., & Mattoussi, H. (2005). Quantum dot bioconjugates for imaging, labelling and sensing. *Nature Materials, 4*(6), 435–446. https://doi.org/10.1038/nmat1390

Meinardi, F., Colombo, A., Velizhanin, K. A., Simonutti, R., Lorenzon, M., Beverina, L., Viswanatha, R., Klimov, V. I., & Brovelli, S. (2014). Large-area luminescent solar concentrators based on 'stokes-shift-engineered' nanocrystals in a mass-polymerized PMMA matrix. *Nature Photonics, 8*(5), 392–399. https://doi.org/10.1038/nphoton.2014.54

Meng, X., Chen, Y., Yang, F., Zhang, J., Shi, G., Zhang, Y., Tang, H., Chen, W., Liu, Y., Yuan, L., Li, S., Wang, K., Chen, Q., Liu, Z., & Ma, W. (2022). Perovskite bridging PbS quantum dot/polymer interface enables efficient solar cells. *Nano Research, 15*(7), 6121–6127. https://doi.org/10.1007/S12274-022-4195-8/METRICS

Mueller, T., & Malic, E. (2018). Exciton physics and device application of two-dimensional transition metal dichalcogenide semiconductors. *npj 2D Materials and Applications, 2*(1), 1–12. https://doi.org/10.1038/s41699-018-0074-2

Murray, C. B., Norris, D. J., & Bawendi, M. G. (1993). Synthesis and characterization of nearly monodisperse CdE (E = S, se, Te) semiconductor nanocrystallites. *Journal of the American Chemical Society, 115*, 8706–8715.

Nedelcu, G., Protesescu, L., Yakunin, S., Bodnarchuk, M. I., Grotevent, M. J., & Kovalenko, M. V. (2015). Fast anion-exchange in highly luminescent nanocrystals of cesium lead halide perovskites (CsPbX3, X = cl, Br, I). *Nano Letters, 15*(8), 5635–5640. https://doi.org/10.1021/ACS.NANOLETT.5B02404

Nozik, A. J. (2002). Quantum dot solar cells. *Physica E: Low-dimensional Systems and Nanostructures, 14*(1-2), 115–120. https://doi.org/10.1016/S1386-9477(02)00374-0

Peng, X., Manna, L., Yang, W., Wickham, J., Scher, E., Kadavanich, A., & Alivisatos, A. P. (2000). Shape control of CdSe nanocrystals. *Nature, 404*(6773), 59–61. https://doi.org/10.1038/35003535

Pradhan, S., Di Stasio, F., Bi, Y., Gupta, S., Christodoulou, S., Stavrinadis, A., & Konstantatos, G. (2018). High-efficiency colloidal quantum dot infrared light-emitting diodes via engineering at the supra-nanocrystalline level. *Nature Nanotechnology, 14*(1), 72–79. https://doi.org/10.1038/s41565-018-0312-y

Proppe, A. H., Berkinsky, D. B., Zhu, H., Šverko, T., Kaplan, A. E. K., Horowitz, J. R., Kim, T., Chung, H., Jun, S., & Bawendi, M. G. (2023). Highly stable and pure single-photon emission with 250 ps optical coherence times in InP colloidal quantum dots. *Nature Nanotechnology, 18*(9), 993–999. https://doi.org/10.1038/s41565-023-01432-0

Rainò, G., Becker, M. A., Bodnarchuk, M. I., Mahrt, R. F., Kovalenko, M. V., & Stöferle, T. (2018). Superfluorescence from lead halide perovskite quantum dot superlattices. *Nature, 563*(7733), 671–675. https://doi.org/10.1038/s41586-018-0683-0

Raja-Mogan, T., Ohtani, B., & Kowalska, E. (2020). Photonic crystals for plasmonic photocatalysis. *Catalysts, 10*(8), 827. https://doi.org/10.3390/CATAL10080827

Ramalingam, G., Kathirgamanathan, P., Ravi, G., Elangovan, T., Arjun Kumar, B., Manivannan, N., & Kaviyarasu, K. (n.d.). Materials research group (MRG), iThemba LABS-National Research Foundation (NRF).

Ramalingam, G., Kathirgamanathan, P., Ravi, G., Elangovan, T., Kumar, B. A., Manivannan, N., Kasinathan, K., Ramalingam, G., Kathirgamanathan, P., Ravi, G., Elangovan, T., Kumar, B. A., Manivannan, N., & Kasinathan, K. (2020). Quantum confinement effect of 2D nanomaterials. *Quantum Dots - Fundamental and Applications*. https://doi.org/10.5772/INTECHOPEN.90140

Rossetti, R., Nakahara, S., & Brus, L. E. (1983). Quantum size effects in the redox potentials, resonance Raman spectra, and electronic spectra of CdS crystallites in aqueous solution. *The Journal of Chemical Physics, 79*, 1086–1088. https://doi.org/10.1063/1.445834

Semonin, O. E., Luther, J. M., Choi, S., Chen, H. Y., Gao, J., Nozik, A. J., & Beard, M. C. (2011). Peak external photocurrent quantum efficiency exceeding 100% via MEG in a quantum dot solar cell. *Science, 334*(6062), 1530–1533. https://doi.org/10.1126/SCIENCE.1209845

Shen, K., Xu, H., Li, X., Guo, J., Sathasivam, S., Wang, M., Ren, A., Leong Choy, K., Parkin, I. P., Guo, Z., Wu, J., Shen, K., Xu, H., Li, X., Wu, J., Ren, A., Wang, M., Choy, K. L., Guo, J., & Guo, Z. (2020). Flexible and self-powered photodetector arrays based on all-inorganic CsPbBr3 quantum dots. *Advanced Materials, 32*(22), 2000004. https://doi.org/10.1002/ADMA.202000004

Shiau, S. Y., & Combescot, M. (2023). A fresh view on Frenkel excitons: Electron–hole pair exchange and many-body formalism. *Annals of Physics, 458*, 169431. https://doi.org/10.1016/J.AOP.2023.169431

Song, J., Wang, O., Shen, H., Lin, Q., Li, Z., Wang, L., Zhang, X., & Li, L. S. (2019). Over 30% external quantum efficiency light-emitting diodes by engineering quantum dot-assisted energy level match for hole transport layer. *Advanced Functional Materials, 29*(33), 1808377. https://doi.org/10.1002/ADFM.201808377

Srivastava, V., Kamysbayev, V., Hong, L., Dunietz, E., Klie, R. F., & Talapin, D. V. (2018). Colloidal chemistry in molten salts: Synthesis of luminescent In1- xGaxP and In1- xGaxAs quantum dots. *Journal of the American Chemical Society, 140*(38), 12144–12151. https://doi.org/10.1021/JACS.8B06971

Sun, Z., Yang, H., Ma, Z., Zhang, Z., Han, L., Wang, Z., Zhang, Y., Wang, X., & Wang, Q. (2023). AgAuSe quantum dots-based eco-friendly solar cells. *Solar, 7*(15), 2300353. https://doi.org/10.1002/SOLR.202300353

Tang, X., Ackerman, M. M., Shen, G., Guyot-Sionnest, P., Tang, X., Ackerman, M. M., Shen, G., & Guyot-Sionnest, P. (2019). Towards infrared electronic eyes: Flexible colloidal quantum dot photovoltaic detectors enhanced by resonant cavity. *Small, 15*(12), 1804920. https://doi.org/10.1002/SMLL.201804920

Terada, S., Ueda, H., Ono, T., & Saitow, K. I. (2022). Orange-red Si quantum dot LEDs from recycled rice husks. *ACS Sustainable Chemistry and Engineering, 10*(5), 1765–1776. https://doi.org/10.1021/ACSSUSCHEMENG.1C04985

Wang, Y., Lv, Z., Liao, Q., Shan, H., Chen, J., Zhou, Y., Zhou, L., Chen, X., Roy, V. A. L., Wang, Z., Xu, Z., Zeng, Y.-J., & Han, S.-T. (2018). Memory devices: Synergies of electrochemical metallization and valance change in all-inorganic perovskite quantum dots for resistive switching. *Advanced Materials, 30*(28), 1870207. https://doi.org/10.1002/ADMA.201870207

Yang, Z., Fan, J. Z., Proppe, A. H., De Arquer, F. P. G., Rossouw, D., Voznyy, O., Lan, X., Liu, M., Walters, G., Quintero-Bermudez, R., Sun, B., Hoogland, S., Botton, G. A., Kelley, S. O., & Sargent, E. H. (2017). Mixed-quantum-dot solar cells. *Nature Communications, 8*(1), 1–9. https://doi.org/10.1038/s41467-017-01362-1

Ye, S., Lv, J., Wu, D., Zhang, S., Peng, D., Lai, N., Yang, J., Wang, Y., Lin, F., Ke, S., & Wang, C. (2024). Microstructure and ferromagnetism of Mn0.05Ge0.95 quantum dots/graphene heterostructures for spintronic devices. *ACS Applied Nano Materials, 7*(14), 16542–16552. https://doi.org/10.1021/ACSANM.4C02559

Yuan, L., Michaels, H., Roy, R., Johansson, M., Öberg, V., Andruszkiewicz, A., Zhang, X., Freitag, M., & Johansson, E. M. J. (2020). Four-terminal tandem solar cell with dye-sensitized and PbS colloidal quantum-dot-based subcells. *ACS Applied Energy Materials, 3*(4), 3157–3161. https://doi.org/10.1021/ACSAEM.0C00030

Zhang, Y., Kan, Y., Gao, K., Gu, M., Shi, Y., Zhang, X., Xue, Y., Zhang, X., Liu, Z., Zhang, Y., Yuan, J., Ma, W., & Jen, A. K. Y. (2020). Hybrid quantum dot/organic heterojunction: A route to improve open-circuit voltage in PbS colloidal quantum dot solar cells. *ACS Energy Letters, 5*(7), 2335–2342. https://doi.org/10.1021/ACSENERGYLETT.0C01136

Zhang, Z., Sung, J., Toolan, D. T. W., Han, S., Pandya, R., Weir, M. P., Xiao, J., Dowland, S., Liu, M., Ryan, A. J., Jones, R. A. L., Huang, S., & Rao, A. (2022). Ultrafast exciton transport at early times in quantum dot solids. *Nature Materials, 21*(5), 533–539. https://doi.org/10.1038/s41563-022-01204-6

Zheng, J., Luo, C., Shabbir, B., Wang, C., Mao, W., Zhang, Y., Huang, Y., Dong, Y., Jasieniak, J. J., Pan, C., & Bao, Q. (2019). Flexible photodetectors based on reticulated SWNT/perovskite quantum dot heterostructures with ultrahigh durability. *Nanoscale, 11*(16), 8020–8026. https://doi.org/10.1039/C8NR08026G

Chapter 6
Nanocomposites: From Design to Multifunctionality in Modern Applications

M. Amarnath, Shreyash Mohite, Omkar Sadalkar, and Shital Palaskar

6.1 Introduction

The word "nano" correlates with a dimension of matter which uses the prefix 10^{-9} meter. Put simply, the size is at a level of nanometer (Nasrollahzadeh et al., 2019). Uniquely, at this size, materials behave unlike other substances. One important factor in scaling material size to the nanometer level is the generation of phase boundaries (Kumar & Krishnamoorti, 2010). These boundaries are very important since they substantially improve the characteristics of materials and thus their performance in a number of uses. Like pristine nanoparticles and nanomaterials, nanocomposites are materials that also exist at the nanoscale but have multiple dimensions such as 1D, 2D, and 3D shapes (Liu et al., 2018). These composites blend various nanoscale components and achieve enhanced properties that surpass the individual properties of the components. To understand the structure–property relationships of nanocomposites, the surface area-to-volume ratio of the reinforcing components is very important (Guler et al., 2022). The more pronounced interactions among the matrix and the reinforcing material nanocomposites are caused by increased surface area-to-volume ratio. The ratio significantly influences the mechanical, electrical, and thermal properties of the nanocomposites (Tjong, 2006). Thus, this particular ratio is a crucial variable in customizing material properties and optimizing the performance for desired applications. A single system encapsulating several properties makes multifunctional nanocomposites an important development in material science and engineering. These materials are designed to enhance performance with regard to sensing for environmental monitoring (Kausar & Ahmad, 2024), storing

M. Amarnath · S. Mohite · O. Sadalkar · S. Palaskar (✉)
Plasma Research Lab, The Bombay Textile Research Association, Mumbai, India
e-mail: pmebtra@btraindia.com

S. Kulkarni (ed.), *Next-Generation Nanomaterials for Sustainable Engineering*, Lecture Notes in Nanoscale Science and Technology 43,
https://doi.org/10.1007/978-3-032-05719-8_6

energy (Elalfy et al., 2024), biomedical (Baishya et al., 2024), and even addressing environmental problems (Chinnappan et al., 2017). By merging new features like electrical conductivity, thermal stability, mechanical strength, and chemical reactivity, nanocomposites make it possible to address complex problems between science and engineering (Ali et al., 2024). Different kinds of nanocomposites are the key to the subsequent development of the technology of the future, and this is how humanity can build more dependable, sustainable, and responsive systems. This book consist of various aspects of nanoscience and nanotechnologies. In this we will explore multifunctional nanocomposites extensively, concentrating on their different structural and functional characteristics as well as the fact that they facilitate the increase of performance criteria. This chapter explores the process of nanocomposite synthesis and application. This way, we talk about them in the context of the major aspects of energy storage implementation, implant materials used in the medical field, and environmental sustainability. Subsequently, we turn our attention to the basic causes of their performance: the structure–property relation, the ratio of the surface area to the volume, and the synergistic effects of the parts. Our expansive discussions include both our perspective and ideas to guide future study.

6.2 Definition and Overview of Multifunctional Nanocomposites

Nanocomposite materials acquire their multifunctionality through the integration of the suitable features from diverse constituent materials, so that a completely new material can always display a wide range of types of beneficial properties (Sharma et al., 2024). In this, there is a combination of different phases such as fibers (Chamanehpour et al., 2024), particles (Garavand et al., 2024), or layers (Saber et al., 2024), with each of these contributing different attributes that augment the properties and applications of the composite. The multifunctional property of these materials arises from the achievement of synergy by selecting components with complementary properties. The polymer matrix may be flexible and easy to process; inorganic fillers serve to add strength, thermal stability, or electrical conductivity (Afolabi & Ndou, 2024). This complementary interaction results in improved mechanical properties—high tensile strength, resistance, and very good elasticity that can be tuned to accommodate specified application yielding. Classification of these composites can be effectively placed under but not limited to polymers, ceramics, bio-derived, hybrid, conductive, thermal, magnetic, and so on. Figure 6.1 shows the role of nanofillers in producing nanocomposites and their applications.

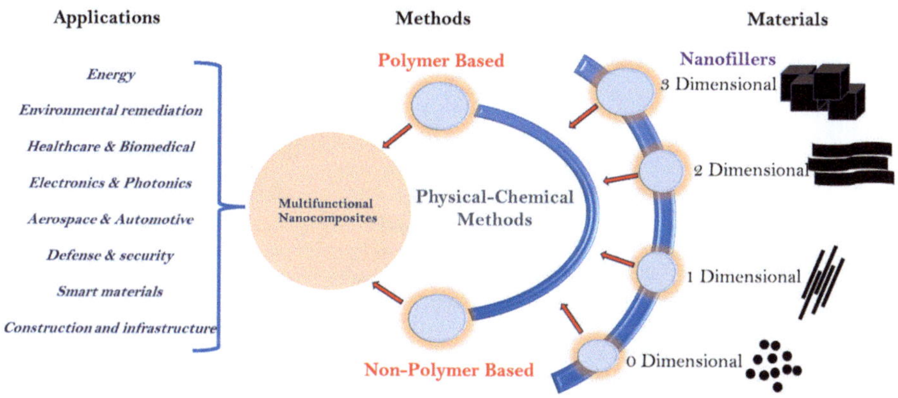

Fig. 6.1 Schematic representation of materials, methods, and application of multifunctional nanocomposites

6.3 Historical Development and Advancements

The technology for the production of multifunctional nanocomposites culminated into progressively encompassing understanding for nanomaterials. It was a finding of immense pertinence in today crafts. Need for such materials originated in generic need for materials possessing several functions a wish to judiciously combine multiple functionalities, from electroconductivity, thermal conductivity, and/or mechanical strength to stimuli-responsiveness all into one single material. The idea for composites (Fig. 6.2) dates back centuries, with examples being natural composites such as wood and bone, which combine differing material phases for structural integrity (Shalan et al., 2022). An important milestone in material science appeared in the twentieth century when the rise of synthetic composites, particularly polymer composites reinforced with fibers, provided upgraded mechanical properties for lightweight structures, which, in turn, led to their broad-based industrial application in the aerospace and automotive domains (Mahboubizadeh et al., 2024).

The late twentieth century observed a pattern swing between material science and other scientific areas with the development of nanoscience and nanotechnology, enabling the manipulation at the atomic/molecular regions. This led to the development of nanocomposites, where nanoscale reinforcements, such as nanoparticles, nanofibers, nanotubes were embedded into polymer, metal, and ceramic matrices (Pokropivny et al., 2007). These nanomaterials introduced exceptional properties, such as high surface area, improved strength, and enhanced electrical and thermal conductivity. The twenty-first century has seen rapid advancements in multifunctional nanocomposites, fuelled by innovations in nanomaterial synthesis, processing techniques, and a deeper understanding of nanoscale interactions (Gomes Souza et al., 2024). The incorporation of carbon nanomaterials, namely graphene and carbon nanotubes (CNTs), has been one of the most significant advancements. CNTs,

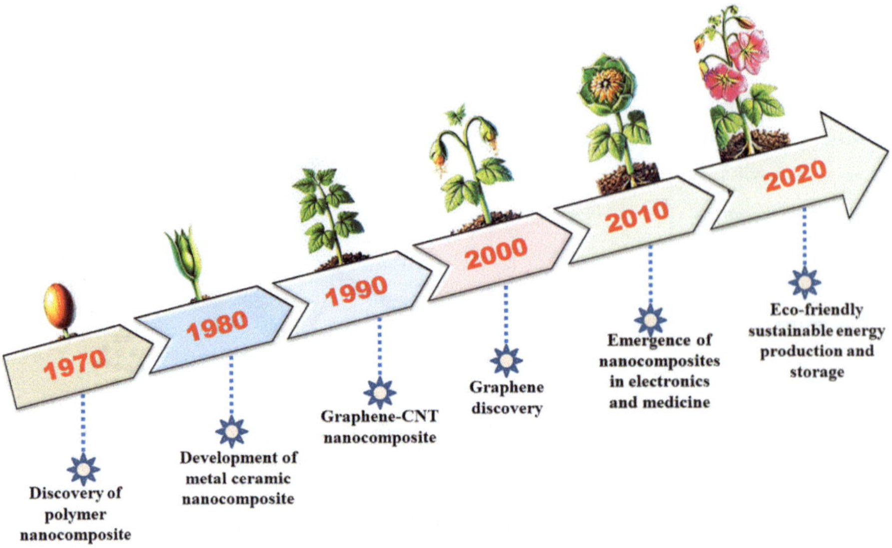

Fig. 6.2 History and milestones of nanocomposites

discovered in the early 1990s, and graphene, isolated in 2004, opened new possibilities for nanocomposites (Thakur & Thakur, 2018; Yeang et al., 2014).

One of the significant milestones in the advancement of novel multifunctional nanocomposites is the development of smart nanocomposites those that trigger instantaneous changes upon changes associated with temperature (presence or absence), pH, light, or magnetic fields. This has impacted innovations in self-healing materials, controlled drug delivery systems, and various sensors. The emergence of structural, thermal, and electrical functionality with the innovative features that involve smart behavior has already reached considerable heights with respect to multifunctionality.

6.4 Classification of Nanocomposites

Nanocomposites can be grouped according to the arrangement and integration of their constituent phases. The main process is the incorporation of one or more discrete nano-sized phase in a continuous macro phase that brings about synergy. This type of integration confers distinct physical and chemical characteristics that may, ideally, exceed the qualities of the constituent materials (Din et al., 2020). Typically, one of the components is in a much greater proportion than the others, forming a continuous matrix that surrounds the rest, which in this capacity is referred to as nanofillers or reinforcements (Njuguna et al., 2008). Classification of nanocomposites based on their structural and functional properties characterizes such nanocomposites and their unique multifunctional characteristics arising from the integration of these diverse phases (Singh & Agarwal, 2016). Nanocomposites are classified based on composition and structure, and four main functional groups were identified.

6.4.1 Polymer Nanocomposite

6.4.1.1 Multifunctional Polymer Fibers

In 1920, Hermann Staudinger published the first paper on polymers and polymerizations, coining the term macromolecules. He described that as big molecules constituted of many repeating units, which set the stage of modern polymer science (Mülhaupt, 2004). Multifunctional polymer nanocomposite cloths are fibers developed specially by adding nanomaterials or nanostructures into their composition to give rise to improved properties and specialized functionalities (Das et al., 2025). The composites increase and combine the strength and stiffness of reinforcing fibers because of smart properties of polymer matrices leading to materials with better tensile strength, fatigue resistance, and impact tolerance (Andrew et al., 2019). The lightweight and exceptional load-bearing capabilities of these composites give them value in sectors such as aerospace, marine, automotive, and offshore industries, where reducing weight matters without compromising the structural integrity (Phiri et al., 2024). More to this, being flexible allows them to incorporate functionalities such as thermal resistance, electrical conductivity, and corrosion resistance into the composite matrix or fiber structure by design (Ferreira et al., 2016). It is on account of these attributes that fiber-reinforced polymer composite materials are an essential material class in modern engineering applications by delivering significant performance improvements over conventional materials. Figure 6.3 shows the multifunctional characteristics of carbon fiber composite utilizing the electrochemical conversion technique.

Multifunctional basalt fiber polymer composite was prepared by Libing Liu et al. This study presents a novel approach for designing reinforced composites via electrostatically self-assembled multi-wall CNTs on basalt fibers. Melt mixing with polyarylene ether nitrile makes CNT network over the fiber as interconnected surface structure. This CNT networks impart excellent conductivity for enabling damage and temperature sensing. The structure enhanced the flexural strength and modulus of the composites by up to 8.9% and 12.9%, respectively, with conductivity of 1.56×10^{-2} S m^{-1} and this nanocomposite demonstrated excellent damage self-sensing (GF = 7.68) and temperature-sensitive behavior (TCR = 0.542%/°C). This method offers a novel multifunctional approach for integrating self-monitoring and temperature sensing in one material with only changing the filler addition (Liu et al., 2024).

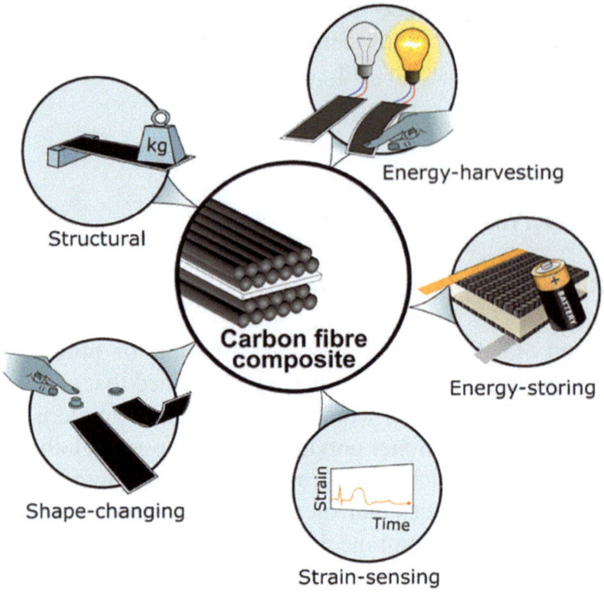

Fig. 6.3 Multifunctional application of carbon fiber by using electrochemical activation approach. Reproduced with permission obtained from Zenkert et al. (2024). Under the terms of the creative commons cc by license, Elsevier© 2024

6.4.2 Types and Synthesis Methods for Multifunctional Polymer Nanocomposites

The properties of polymer nanocomposites were controlled by various features begins with the size, shape, surface chemistry, and distribution of the nanofillers. These parameters are well defined and executed through various methods of synthesis and post-treatment (Norizan et al., 2022). Each method can be chosen based on the specific requirements of the final nanocomposite. A comprehensive pictorial representation of various methods for polymer nanocomposite preparation was shown in Fig. 6.4.

6.4.3 Melt Mixing and Intercalation

The melt mixing method incorporates nanomaterials directly into the polymer matrix during its melted state. This is the process where nano-reinforcements (nanoparticles, CNTs, graphene, clays, cellulose, and many more) are mixed with the polymer and melted above the polymer's softening point, so that uniform dispersion is obtained (Banerjee & Dutta, 2019; Chowdhury et al., 2022). This process

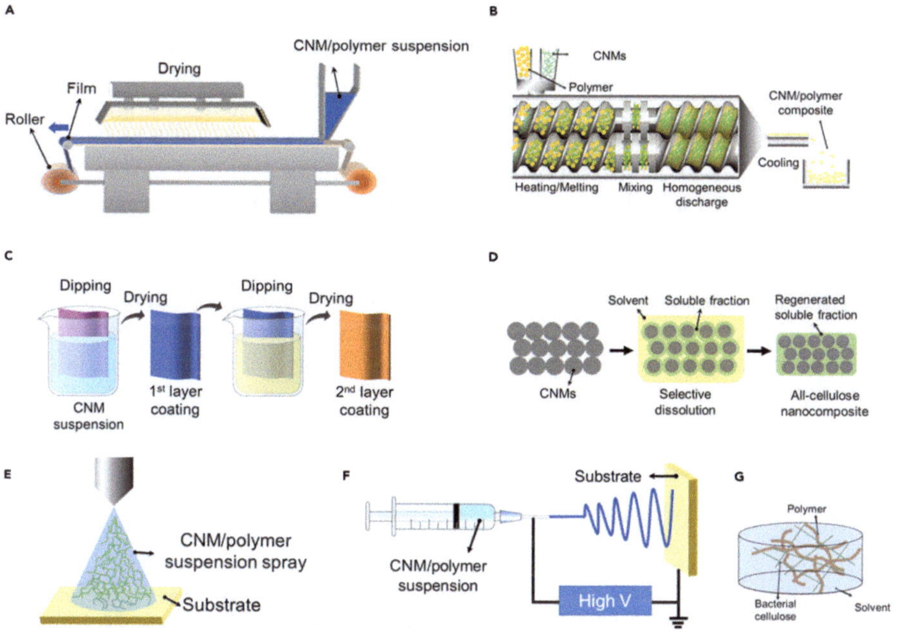

Fig. 6.4 Efficient preparation and manufacturing techniques for polymer nanocomposites reinforced with cellulose nanomaterials. (**a**) Solvent casting technique, (**b**) melt extrusion, (**c**) dip coating, (**d**) partial dissolution for cellulose nanocomposites, (**e**) spray coating, (**f**) spinning, (**g**) resin impregnation. Reproduced with permission obtained from Wang et al. (2023) copyright © Elsevier

is characterized by compatibility with standard processing techniques such as extrusion and injection molding, making it the easiest and most economical for industrial-scale production. This makes melt mixing an environmentally clean process for having no solvent usage, while at the same time eliminating the complications caused by limited nanofiller dispersion in solvent-based processes. Important processing parameters like temperature profile, screw speed, mixing time, and the chemistry of the nanofiller are vital to the mechanical and morphological properties of the resultant nanocomposites (Rane et al., 2018). While the reinforcement mechanisms do yield superior stiffness and barrier performance, this can however accompany reduced toughness due to the resultant increased cross-link density. It has also been noted that there is a shear-thinning behavior at high shear rates, probably due to the nanoparticles aligning themselves in the flow direction.

This study describes the use of clay (sodium montmorillonite and montmorillonite) polymer (eudragit RS) nanocomposites for drug delivery applications, which were prepared by the hot-melt extrusion method. This clay-polymer system presents interesting properties such as barrier effect in mixed composites, swelling index, mucoadhesion, and many others. Such properties influence drug release profiles, swelling rates, and drug uptakes. The clay nanocomposites allow for multifunctional opportunities in diverse applications. Solid nanocomposites would consist of poly (ethylene oxide) polymer intercalated with ammonium iodide (NH_4I) salt and nanoclay. In doing so, they create effective solid electrolyte films. The intercalation of these clays drastically affects the general morphology and structure of the augmented films (Liu et al., 2020). A separate work by Mou Saha et al. describes these nanocomposites that carry a higher degree of ion dissociation due to improved ion polymer and ion–clay interactions. This nanocomposite has shown a maximum conductivity of $\sim 1.13 \times 10^{-6}$ S cm^{-1}, comprised of 4 wt% clay at 10 wt% salt concentration. The room temperature value of these nanocomposites has a decisive role in enriching the direct current conductivity, which tends to be of high value in energy storage device research. These finding confirm the amorphic nature of clay layers supporting rapid ion diffusion into the polymer matrix (Saha et al., 2021).

6.4.4 Sol-Gel Method

This method is a straightforward approach to synthesize various nanocomposites, primarily employed to fabricate thin films. The overall synthesis can be compartmentalized into four main stages: mixing, casting, drying, and extended aging. The synthesis process is straightforward but provides vast possibilities of modifying and tailoring the nanocomposite‘s morphology, porosity, and thickness (Bokov et al., 2021). This method is advantageous due to its ability to produce uniform films at relatively low processing temperatures, which suits for a wide choice of applications, such as coatings, sensors, and optical device. Yaseen et al. prepared polyvinyl alcohol/CuO-SiO_2 nanocomposite for waste water treatment. The removal of pollutants from water reservoirs to ensure safe and secure potable water is a prominent area of research worldwide. Owing to this motivation these group prepared the nanocomposite and analyzed its performance for the degradation of methylene blue (MB) and nile blue (NB). Simultaneously, they examined the potential biological activities (antioxidant and antileishmanial). Photocatalytic property of the prepared nanocomposite has the potential to degrade (NB) 85% and (MB) 76% of the model dyes under UV irradiation (Yaseen et al., 2024). Similarly, polymer incorporated Fe_2O_3 and ZnO NPs exhibits interesting electrical and magnetic properties which will be useful for the real-world application. A precise sol-gel method with gel casting is used to synthesize $(ZnO)_x$ $(1 - x)$ Fe_2O_3 nanocomposites in (polystyrene-polyethylene glycol) support, enabling the arrangements of hydrogen bonds between the organic and inorganic components, which enhances the structural integrity and improves the material's mechanical and thermal abilities for potential multifunctional applications.

6.4.5 In-Situ Emulsion Polymerization

A well-dispersed nanofiller within the advanced polymer matrix is essential for achieving the desired multifunctional properties. In this case, in situ emulsion polymerization serves as a promising approach compared to other methods. In a conventional process, fillers are mixed inside the precursor solution consist of the required monomer, dispersing agent, emulsifier, and an initiator. Subsequently, the reaction of monomer polymerization proceeds by means of normal emulsion polymerization toward the final nanocomposite materials. Based on this synthesis method, Oguzkan et al. studied the tribological properties of polyamide 6/GO nanocomposites. They were able to graft polyamide 6 onto GO, and with the addition of GO, the tensile, flexural, and impact properties bulked up significantly (Senturk & Palabiyik, 2023). Similarly, in the pursuit of achieving synergy between polyaniline and phosphorus, nitrogen dual co-doped graphene, this combination has demonstrated significant potential for various multifunctional applications. Ravinder Singh et al. prepared the aforementioned 15% nanocomposite sensor, which exhibited high sensing performance and achieved a specific capacitance of 750.41 F g^{-1} at 5 A g^{-1}, along with outstanding rate capability and cycling. The dopants were incorporated to give multifunctional properties to the composite material and thereby improve the performance in gas sensing and supercapacitor applications to a large extent (Singh et al., 2024). Wastewater treatment using hydrogel was developed by Qian et al. with a unique multifunctional nanocomposite of ZIF-8@IL-MXene/Poly (N-isopropylacrylamide). This unique hydrogel can efficiently absorb ions, dyes and phenol components. Swelling makes the hydrogels to affect directly the adsorption and desorption process. Owing to the growth of these ZIF-8 over the composite wall decreases the swelling effect due to the formation of a metal-organic framework which occupies the hydrophilic groups and their equilibrium swelling ratios got decreased and it was rectified by using the chelation of ZIF-8. The multifunctionality shows not only absorption and also the reduction of 4-NP and Cu^{2+}. With these useful impacts in-situ polymerization is regarded as one of the prominent synthesis methods for polymer nanocomposite preparation (Wang et al., 2024).

6.5 Carbon Nanocomposite

Carbon nanocomposites are advanced materials that combine carbon with other nanomaterials act as a host matrix, typically polymers, metals, or ceramics (Malaki et al., 2019). Pristine and engineered carbon-based nanocomposites are in incessant demand worldwide due to their distinct and intrinsic properties. Allotropes of carbon have been utilized and are almost irreplaceable in every field. The main advantage of using carbon is that it can be synthesized in many forms (carbon aerogels, carbon nanotubes, graphene, graphene oxide, carbon nanofibers, and fullerenes). Generally, the presence of free mobile electrons in the sp^2 hybridized sheets of carbon makes it more reliable for conductivity and bond formation with other materials in composite configurations. Delocalization of π-electrons along the planar carbon structure allows charge transfer to occur with much ease, thereby improving the electrical as well as thermal conductivity of the material. Moreover, there are covalent bonds within the layers that are very strong and also provide weak van der Waals interaction both between layers as well as with other components in composite materials which make a multifunctional property (Amarnath et al., 2020a, 2020b; Amarnath & Gurunathan, 2021a, 2021b; Idumah & Hassan, 2016; Mallakpour & Khadem, 2016). Different types of carbon-based nanocomposites have been prepared and characterized step by step, and this section focuses on the most unique amongst them.

6.5.1 Carbon Nanotube Nanocomposite

Carbon nanotubes (CNTs) are allotropes of carbon with a cylindrical structure, composed of one or more layers of graphene rolled into seamless tubes. They are categorized into single-wall and multi-wall structure, depending on the number of concentric graphene cylinders. CNTs exhibit extraordinary mechanical properties (tensile strength and Young's modulus) due to the strong sp^2 carbon–carbon bonds. Their electrical properties range from metallic to semiconducting, depending on their chirality and diameter. The multifunctional property of CNT lies in the increment of mechanical, electrical, thermal, sensing, and energy storage (Varshney, 2014). Reis et al. assessed the role of CNT addition in concrete cement to improve physical and mechanical properties. Concrete is a generally used as a construction material. Its strength, durability, and ability to be molded into various shapes make it suitable for widespread applications, from residential buildings and infrastructure to bridges and highways. Here, the authors explored the addition of CNT at 0.05% and 0.10% by weight relative to the cement to CNT ratio to determine the optimal results. The pre-dispersion of CNTs in isopropanol prior to their addition into cement introduces a novel approach, yielding more effective results compared to traditional paste and mortar methods (Reis et al., 2024). Similarly, Han Du et al. developed a pressure sensor based on a softer elastomer matrix reinforced with graphene and CNTs. This hybrid elastomer nanocomposite was studied under bending, where the increased volume fractions of graphene and CNTs enhance electrical conductivity and sensing performance. The composite model demonstrates a significant decrease in sensor resistance as bending pressure increases from 0 to 1 KPa, making the sensor highly conductive and ideal for pressure sensing. Likewise, dielectric elastomers, like conducting elastomers, are highly responsive to electromechanical stimuli. These materials are generally flexible and voltage-sensitive, undergoing significant deformation when subjected to a high-electric field during actuation (Du et al., 2023). Marco Alfonso et al. prepared MWCNT@SiO_2

distributed in liquid silicone rubber which shows great potential for mechatronic applications due to their efficient electromechanical conversion. These nanocomposites, sandwiched between stretchable, conductive carbon-black electrodes, exhibited a dielectric constant increase from 2.6 to 5.9 at 1 kHz with low dielectric loss. They demonstrated excellent elastic properties in tensile tests, with Young's modulus ranging from 0.024 to 0.084 MPa, and maintained mechanical integrity with minimal impact from the electrodes. Additionally, the actuation strain exceeded 125% compared to the pure polymer, highlighting their scalability for electroactive soft material development (Alfonso et al., 2024). The application of CNTs in next-generation microelectronic devices requires high thermal transport performance. To address this, Zhang et al. developed CNT-reinforced polydimethylsiloxane (PDMS) composites, leveraging the superior thermal properties of CNTs to enhance heat dissipation in these devices. In this work, the authors address the issue of intrinsic thermal resistance among individual tubes of CNTs, which bounds the thermal conduction of CNT-reinforcements, by developing an welding strategy to form graphitic structure welded CNT networks. The resulting nanocomposite, with 4.75 wt% GS, achieves thermal conductivity of 5.58 W/m·K, representing a 410% improvements. Molecular dynamics simulations further demonstrate that GS welding reduces phonon scattering and interfacial thermal resistance, enhancing heat transfer efficiency (Zhang et al., 2020). By manipulating the versatility of CNTs, it has applications across various fields, from electronics to biology with multifunctional capabilities.

6.5.2 Carbon Quantum Dot Nanocomposite

Carbon quantum dots (CQDs) are very novel materials with a size of less than 10 nm, possessing strong fluorescent properties and size-dependent photoluminescence. This further enriches the particular properties of high photostability, biocompatibility, and low toxicity of CQDs which are highly encouraging for broad applications in electronics, sensing, and biomedical imaging (Yang et al., 2023). Functionalization of CQDs with polymer matrices or other nanomaterials creates synergistic effects of CQDs and the matrix, thereby improving functionality. A multifunctional application of CQDs was worked on by Shen et al. Here, the authors developed an electronic skin with high electrical conductivity alongside antibacterial, ultraviolet absorption, and fluorescence emission properties. The combination of CQDs enhances the performance of the skin by making it suitable for employing into environments for protecting against microbial contamination and UV radiation, while at the same time allowing for monitoring in real time through its fluorescence properties. A new transparent electronic skin was fabricated based on a top-down design strategy improved by hydrogen bond reconstruction. This e-skin demonstrated remarkable tensile strengths, with a tensile strength of 4.94 MPa and an all-out elongation of 263.42%, alongside a Young's modulus comparable to that of human skin. The incorporation of functionalized carbon quantum dots derived from discarded goat hair imparted excellent anti freezing and moisturizing properties. E-Skin effectively senses human motion and physiological signals, achieving a signal-to-noise ratio of 25.1 dB, positioning it as a promising candidate for real-time monitoring and innovative applications in human-computer interaction (Shen et al., 2024).

CQD-epoxy nanocomposites have been synthesized by Bifulco et al., which achieved multifunctionality: fire safety, hydrophobicity, and flame-triggered shape memory action. The authors describe a waste-to-wealth approach of synthesizing CQDs by the hydrothermal method using humic acid. The addition of CQDs and hexadecyltrimethoxysilane improved the photoluminescence, hydrophobicity, and fire resistance of the nanocomposites, with a substantial increase in the ignition temperature and delay in structural degradation under reduced flame exposure. In addition, these multifunctional nanocomposites could be an identifier for polypropylene packaging and a new way to defeat counterfeiting (Bifulco et al., 2024).

A self-healing hydrogel, modified with bio-derivable carbon quantum dots, was prepared by Shengli Lu et al. This hydrogel addresses the clinical challenge of rapid healing of wounds infected with multidrug-resistant bacteria, particularly methicillin-resistant Staphylococcus aureus (MRSA). Bio-derived carbon quantum dots have antibacterial properties and can scavenge excessive ROS, thus relieving oxidative stress and inflammation, which hinder cell migration and wound regeneration. CQDs promote migration and angiogenesis of human umbilical vein endothelial cells in a condition of downregulated pro-inflammatory cytokine production in macrophages. In vivo study results confirmed that this multifunctional hydrogel is efficient in eliminating MRSA, reducing inflammation, and exceedingly increasing wound healing, hence being a promising candidate for the treatment of infected skin wounds (Lu et al., 2024).

6.5.3 Graphene Nanocomposite

Graphene, an allotrope of carbon, possesses a two-dimensional single-layer honeycomb structure. Scientists are quite keen on the potentialities projected by graphene in its pure state (Muthuvinayagam et al., 2023). But when incorporated into polymers and other host matrices, it enables multifunctional applications that become our field of interest. For example, reduced simultaneous motion and other phenomena before electrical conductivity would be significantly reduced with less other filler addition, such as carbon black, due to its unique structure in the possibility to form a three-dimensional network with the polymer. Just as electrical conductivity is enhanced when using graphene, when added to strengthen the structure, the mechanical fracture toughness can also be enhanced by inhibiting crack propagation. Salimiyan et al. synthesized a multifunctional hydrogel composed of chitosan/thiol functionalized graphene nanocomposite. They adopted a thiol-ene click reaction which proceeds through a free-radical mechanism whereby a radical initiator like heat, light, or chemical initiator generates radicals that now instigate the addition of thiol to alkenes. The click reaction is a highly efficient, selective, and highly versatile chemical reaction; fast, simple reactions, which yield more product with very few by-products. The graphene hydrogel produced as a result certainly exhibits multifunctional behavior by being flexible, self-healing, conducting, strain-feeling, degrading rapidly, and remaining non-drying (Salimiyan et al., 2023).

Incorporation of mixed metal oxides with graphene derivatives is also a good candidate for various applications. A quick method for preparing Cu on GO layers is illustrated in Fig. 6.5. It demonstrates a direct approach to depositing metal oxides onto GO layers using an inorganic salt precursor through a molecular level mixing technique. These types of mixed nanocomposites can be synthesized easily, making it a cost-effective approach. Sahoo et al. synthesized perovskite oxide Dy_2NiMnO_6 and reduced graphene oxide nanocomposite by sol–gel and solvothermal method. This nanocomposite was developed for catalytic reactions (4-NP reduction), supercapacitor applications (specific capacitance of 260 F g^{-1} at a current density of 1 A g^{-1}), and electromagnetic shielding. The combination of perovskite metal oxides with carbon derivatives usually results in higher dielectric and magnetic losses owing to the double attenuation mechanism toward enhancing electromagnetic shielding. Graphene alone cannot achieve effective shielding because its high magnetic loss does not match the required impedance value. And so, here the authors apply a concept of nanocomposite which is beneficial for multifunctionality (Sahoo et al., 2024).

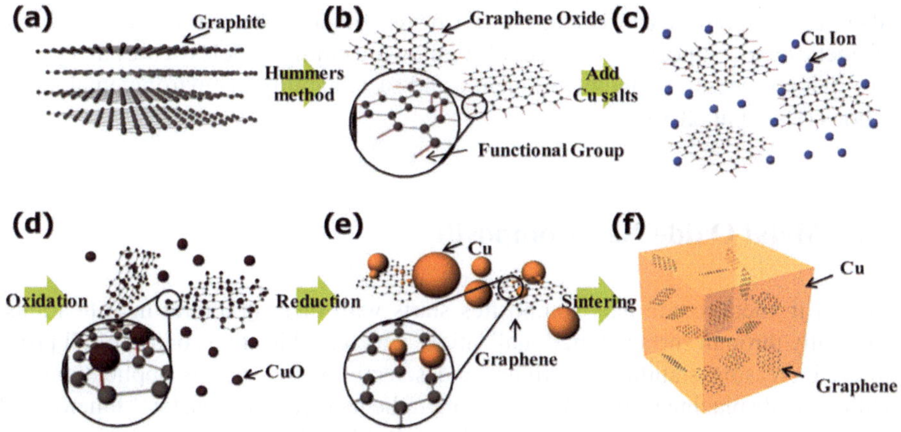

Fig. 6.5 Schematic representation of the preparation method for rGO/Cu nanocomposite by molecular level mixing method (**a**) graphite precursor, (**b**) GO with functional groups, (**c**) Mixing of Cu salt in GO starting solution, (**d**) Oxidation of Cu ions to Cu-oxide over the GO layers. (**e**) Reduction of Cu-oxide and GO, (**f**) Final rGO/Cu nanocomposite. Reproduced with permission obtained from Hwang et al. (2013). Copyright © 2024 Wiley

6.5.4 Magnetic Carbon Nanocomposite

This class of nanocomposites namely magnetic carbon nanocomposites consists of magnetic nanoparticles (Fe_3O_4, Fe_2O_3, and γ-Fe_2O_3) or mixed with Fe_2O_4 (Co, Ni, Mn, Nd, Sm, Cu, Zn) which is dispersed within a non-magnetic carbon matrix. These composites constitute composite materials exhibiting a unique assortment of structural, functional, and magnetic characteristics for the endeavor. They have found broad applications in medicine, environmental remediation, energy storage catalysis, sensing, and electromagnetic shielding. It establishes that Fe_3O_4 is conventionally used to prepare magnetic nanocomposite especially here where carbon dot-coated Fe_3O_4 was used for cancer treatment. It indicated the use of the magnetic particle for hyperthermia and negative contrast agents in magnetic resonance imaging. Carbon, graphene, and Fe_2O_3 nanoparticles are all nanocomposites integrated into a collective electrical network to improve electron transport in their application as an anode material in lithium-ion batteries. Carbon wrapping on top of each Fe_2O_3 nanoparticle prevents direct contact with the electrode–electrolyte, which contributes to extremely high reversible specific capacity of the nanocomposite material. The applications of these magnetite nanocomposites are not limited solely to the fields of biomedicine and energy, but are highly relevant to the environmental remediation for water treatment. Dai et al. found simple and environmental-friendly way of preparing core-shell Fe_3O_4@C@MnO_2 nanocomposite. The whole process consisted of two steps of synthesis of green nanocluster by solvothermal method and decoration of the MnO_2 nanosheets via self-sacrificing redox reaction. The obtained nanocomposites presented high specific surface area, which could function as adsorbents to absorb U(VI) and Eu(III) from effluent. The advantage of using a magnetic nanocomposite is its reusability, allowing multiple cycles of the adsorption-desorption process, enhancing efficiency and sustainability in environmental applications (Dai et al. 2019).

6.6 Metal Oxide Nanocomposite

Interest in metal oxide nanocomposites starts with their structural modifications, especially through surface functionalization, which enables fine control of all properties, increasing stability, reactivity, and selectivity for various applications: In addition to displacing bi-metallic and tri-metallic nanocomposites, low-dimensional nanofillers such as CNT, graphene, BNTs, BNSs, MXene, and transition metal dichalcogenides are effective fillers in metal nanocomposites; they outdo the performance problems of traditional metal composites (Liu et al., 2022); aluminum, copper, titanium, and magnesium played a key role in composite preparation (selection) according to end application. Chemical and physical approaches served as the former for synthesizing these composites; newly, numerous trends in preparation and analysis have recently been on the rise in metal nanocomposite research (Casati &

Vedani, 2014). In such studies, mixed oxide nanocomposites will prove crucial, as they generate more than two-oxide composites through nucleation-to-growth processes. Their structural, morphological, and compositional distribution characterizations by electronic microscopy and/or spectroscopy provide optimization of their functional properties. These innovations continue fueling interest in such mixed metal oxide nanocomposites for materials science research (Al-Hossainy et al., 2019). A mixed metal oxide nanocomposite of CuO, ZnO, and TiO_2 embedded into an electrically conducting polypyrrole polymer matrix has been fabricated by Ankita Yadav et al.; minimal indicator for photocatalytic, antibacterial, anticorrosive, electrical, and magnetic properties establishes the realization of multifunctionality for advancement applications. The structural defects, mixed oxidation states, and huge surface area provide these compositions with catalytic properties, thereby making them effective for catalytic applications such as pollution control, CO_2 reduction, and energy production. The hybrid dinuclear medium-free and spin-crossover complexes offer several catalytic perks, including oxidation, hydrogenation, and splitting of the water. Presence of mixed valency of metallic ions with good conductance solicits for better energy storage and transportation with a great density and stability. Ferromagnetic or antiferromagnetic characteristics arise thereof due to the presence of unpaired electrons in the metal ions (Yadav et al., 2021).

Multifunctional metal oxides, free of noble metals are sustainable alternatives in catalysis. These materials exhibit versatile catalytic characteristics, high thermal stability, and tunable electronic characteristics. Unlike noble metals, metal oxides such as TiO_2, Fe_2O_3, CuO, and MnO_2 are abundantly available sources with cost-effectiveness, permitting scalable deployment into industrial practice. Their multifunctionality is achieved with the redox participations, adsorption–desorption cycles, and even with some exhibiting photocatalytic activities at the blindness of visible or ultraviolet light. Moreover, the advances in nano-structuring and doping techniques have further developed surface area, catalytic activity, and selectivity out of metal oxides, which makes them an attractive class in sustainable, green catalytic applications, rather than on expensive noble metal counterparts. In this regard, the core–shell magnetic nanocomposite developed by Alina Tirsoaga et al. has convincingly demonstrated bulk-free catalysis. The authors designed a base-free catalytic process for the selective oxidation of 5-hydroxymethylfurfural into 2,5-furandicarboxylic acid: quite a leap forward in mass catalyst preparation. Manganese, cobalt, and iron oxides served as the core, while niobium-involved mesoporous-siliceous matrix served as the shell. These metal oxides have the role of co-catalysts for niobium centers: an odd feature relationship tied via hydroxyl species to the decomposition of t-BuOOH. This synergism itself boosts catalytic activity and selection toward biomass conversion processes. The core–shell structure enhances the durability and reusability of the catalyst, while it also offers extended control over reaction conditions, thus paving a way for radical advancements toward eco-friendliness in catalytic systems for industrial purposes (Tirsoaga et al., 2020).

6.7 Applications in Modern Technology

Multifunctional nanocomposites are shaking up modern technology with their enhanced performance and functionality of nanomaterials in their different applications. They improve the conductivity and flexibility of electronic materials, allowing better and more durable devices to be made such as flexible electronics and semiconductors. They allow detection at high sensitivities, hence increasing precision of chemical and biological sensing and enabling environmental monitoring and medical diagnostics. For storage, nanocomposites enhance storage batteries and supercapacitors through their performance by contributing to improved electrochemical characteristics. They also find essential applications in coatings providing corrosion resistance to self-healing in automotive and aerospace. Nanocomposites play a key role in various environmental applications, including water purification and air filtration, as well as contribute to energy efficiency. In medicine, nanocomposites allow targeted drug delivery and enhanced disease diagnostics so as to improve the thermal management applications in electronic goods and electronics themselves. To summarize, nanocomposites forge ahead in many fields, offering versatile solutions to modern technological challenges. Table 6.1 gives novel nanocomposites and their functions in multifunctional applications.

6.8 Challenges and Future Perspectives

Multifunctional nanocomposites emerged as a broad theme in material science with the ability to deliver several properties and end user functions at the same time. Nonetheless, the obstacles in developing such sophisticated materials still remain high. Existing trends in some of the nanocomposite material preparation approaches and their applications have been presented in this chapter in brief. While a fundamental problem is realizing homogeneous dispersion of nanomaterials in the composite matrix particularly polymer precursors, which is vital for realizing their maximum properties (Shalan et al., 2022).

Nanomaterials of metal and oxide tend to cluster together due to the powerful van der Waals forces due to its nanoregime. Aggregation not only impairs function but also makes scaling up manufacturing processes challenging, making cost-efficient, large-scale synthesis problematic. Furthermore, the nanomaterial–matrix interface also needs to be designed carefully so that strong interfacial bonding is achieved for effective load transfer and composite mechanical property enhancement. All this can be accomplished by refining surface functionalization or coating processes to enhance compatibility and the regulation of distribution and orientation of the nanomaterials in the matrix. This engineering is essential for optimally exploiting the performance advantages offered by the nanomaterials, for example, enhanced thermal, electric, or mechanical characteristics, and also to secure long-term stability and reliability of the composite material (Shrestha et al., 2020).

Table 6.1 List of various nanocomposite materials having multifunctional applications

S. No.	Nanocomposite material	Synthesis technique	Potential application	References
1.	Graphite nanosheet/electrospun polyamide 66 nanocomposites	Electrospinning	ECG, strain, temperature measurement	Lu et al. (2020)
2.	MXene embedded hydrogel	Chemical	Pressure, strain, human motion, wireless transmission, photothermal	Wang et al. (2022)
3.	PVDF/PEO blend with Al_2O_3, SnO_2, TiO_2, ZnO	Casting	Nanodielectrics, UV shielding, optical bandgap tuners, porous membranes	Sengwa et al. (2020)
4.	MXene@CeO_2	Chemical	Anti-inflammatory, antibacterial and conductive	Zheng, Wang, et al. (2021)
5.	rGO/MWCNT/Fe_3O_4/ZnO	Green and solvothermal	Photocatalyst, cytotoxicity, antibacterial	Rabiee et al. (2021)
6.	CeO_2–x@C-rGO	Chemical	Energy storage	Li et al. (2022)
7.	MnO_2/CNT	Electrodeposition	Flexible supercapacitor	Zhou et al. (2021)
8.	Au-PVA/CMC	Green and casting	Electronics and optoelectronics	Morsi et al. (2021)
9.	Cu_2Te/CoTe	Microwave irradiation	Optoelectronics	Supriya et al. (2024)
10.	ZnO-unsaturated polyester resin	Hand layup method	UV shielding	Noor et al. (2024)
11.	PVDF/PAN/ZrO_2	Solution casting	Flexible electronics	Vilvanatha Prabu et al. (2024)
12.	PVA/Nd_2O_3	Solution casting	Optoelectronics	Atta et al. (2024)
13.	Si_3N_4/CeO_2	Casting	Nanodielectrics	Kareem et al. (2024)
14.	PMMA-PEO-SiC–$BaTiO_3$	Casting	Pressure sensor	Jaafar et al. (2024)
15.	$CoMoO_4$@*h*BN	Hydrothermal	Supercapacitor	Zhao et al. (2024)
16.	$ZnFe_2O_4$/graphene	Co-precipitation	Photocatalysis, energy storage	Israr et al. (2020)
17.	Cellulose nanocrystal/ konjac glucomannan/ MXene	Chemical	EMI shielding, solar-thermal and electrothermal conversion	Cao et al. (2022)
18.	RGO/$Ti_3C_2T_x$	Chemical	Energy storage, emi shielding, electrothermal, human motion sensing	Zheng, Nie, et al. (2021)
19.	Spinel $MnCo_2O_4$	Chemical	Supercapacitor Batteries Oxygen reduction reaction, hydrogen evolution reaction	Gonçalves et al. (2021)
20.	MXene/Ag nanowire	Chemical	Energy storage, shape memory, and emi shielding	He et al. (2022)

Resistance toward various environmental factors such as moisture, heat, or chemicals is another challenge. These factors might lead to product degradation or performance impairment in a specific time frame. In such scenarios, composites may be made more resistant to these agents by the application of protective coatings, crosslinking techniques, or addition of stabilizing compounds. The material should be designed to provide inherent resistance to environmental aggressors, and encapsulation strategies. Such approaches would enhance longevity and uniformity of composites in practical applications.

In the years to come, advancement in surface modification of nanomaterials, self-assembly routes, and hybrid materials can overcome these limitations with better dispersion, adhesion, and stability. Besides, the advent of smart features, for instance, self-healing, environmental responsiveness, and recyclability, will evolve multifunctional nanocomposites on the lines of sustenance and the circular economy. Near future discovery will be grippingly fantastic, as multipurpose nanocomposites will elevate technologies and applications in the industries, concentrating on electronics and energy storage, to biomedicine and environmental cleansing.

Bright prospects for nanocomposites with multiple functions are within reach because of innovations in how they are synthesized and processed. Development of these techniques, such as surface functionalization, hybrid nanostructure synthesis, and 3D printing, may make easier the integration of multiple functions. These developments can expedite advance in applications. Aside from the research, the development of eco-friendly and sustainable nanocomposite materials would become paramount, as environmental stresses require materials that would perform well yet remain recyclable or biodegradable. Henceforth, multifunctional nanocomposites are influenced center-stage in next-generation technologies wherein smarter, greener, and more responsive material systems are set into motion.

6.9 Conclusion

To summarize, the hurdles (separation, incompatibility of materials, and scale of production) in the processes of multifunctional nanocomposites synthesis and fabrication have a direct consequence on performance, cost efficiency, and the degree of universal applicability. This area is braced for disruptive change. It is anticipated that advancements in synthesis, surface engineering, and well fabricating techniques will eliminate existing barriers and nanocomposites with specific properties will be achieved. There is reason to believe that the future of nanocomposites is indeed there, as the market thirsts for sustainable and economical materials with multifunctionality. Along with increased efforts in research and development, nanostructures are expected to profoundly impact technological advances in electronics, energy, healthcare, and environmental management. After all, it enables the creation of more intelligent and efficient systems.

References

Afolabi, O. A., & Ndou, N. (2024). Synergy of hybrid fillers for emerging composite and nanocomposite materials—A review. *Polymers, 16*, 1907.

Alfonso, M., He, D., Molinié, P., et al. (2024). Improved electromechanical coupling in silicone-based elastomer thin membranes filled by silica-coated multiwalled carbon nanotubes. *ACS Applied Engineering Materials, 2*, 376–387.

Al-Hossainy, A. F., Ibrahim, A., & Zoromba, M. S. (2019). Synthesis and characterization of mixed metal oxide nanoparticles derived from co–Cr layered double hydroxides and their thin films. *Journal of Materials Science: Materials in Electronics, 30*, 11627–11642.

Ali, Z., Yaqoob, S., Yu, J., & D'Amore, A. (2024). Critical review on the characterization, preparation, and enhanced mechanical, thermal, and electrical properties of carbon nanotubes and their hybrid filler polymer composites for various applications. *Composites Part C: Open Access, 13*, 100434.

Amarnath, M., & Gurunathan, K. (2021a). Highly selective CO_2 gas sensor using stabilized NiO-In_2O_3 nanospheres coated reduced graphene oxide sensing electrodes at room temperature. *Journal of Alloys and Compounds, 857*, 157584.

Amarnath, M., & Gurunathan, K. (2021b). Selective ammonia sensing response of vanadium doped cerium oxide nanorods wrapped reduced graphene oxide electrodes at room temperature. *Sensors & Actuators, B: Chemical, 336*, 129679.

Amarnath, M., Heiner, A., & Gurunathan, K. (2020a). Surface bound nanostructures of ternary r-GO/Mn_3O_4/V_2O_5 system for room temperature selectivity of hydrogen gas. *Ceramics International, 46*, 7336–7345.

Amarnath, M., Heiner, A., & Gurunathan, K. (2020b). Highly sensitive room temperature liquefied petroleum gas sensor based on CoSnO2 nanoislands deposited graphene layers. *Synthetic Metals, 270*, 116607.

Andrew, J. J., Srinivasan, S. M., Arockiarajan, A., & Dhakal, H. N. (2019). Parameters influencing the impact response of fiber-reinforced polymer matrix composite materials: A critical review. *Composite Structures, 224*, 111007.

Atta, A., Alshammari, K., Alanazi, A. H., et al. (2024). Fabrication, surface characterization and optical behavior of flexible PVA/Nd_2O_3 polymer nanocomposites materials for optoelectronics applications. *Inorganic Chemistry Communications, 170*, 113249.

Baishya, G., Parasar, B., Limboo, M., et al. (2024). Advancements in nanocomposite hydrogels: A comprehensive review of biomedical applications. *Discover Materials, 4*, 40.

Banerjee, J., & Dutta, K. (2019). Melt-mixed carbon nanotubes/polymer nanocomposites. *Polymer Composites, 40*, 4473–4488.

Bifulco, A., Imparato, C., Climaco, I., et al. (2024). Multifunctional fire-resistant and flame-triggered shape memory epoxy nanocomposites containing carbon dots. *Chemical Engineering Journal, 484*, 149327.

Bokov, D., Turki Jalil, A., Chupradit, S., et al. (2021). Nanomaterial by sol-gel method: Synthesis and application. *Advances in Materials Science and Engineering, 2021*, 5102014.

Cao, Y., Zeng, Z., Huang, D., et al. (2022). Multifunctional phase change composites based on biomass/MXene-derived hybrid scaffolds for excellent electromagnetic interference shielding and superior solar/electro-thermal energy storage. *Nano Research, 15*, 8524–8535.

Casati, R., & Vedani, M. (2014). Metal matrix composites reinforced by nano-particles—A review. *Meta, 4*, 65–83.

Chamanehpour, E., Thouti, S., Rubahn, H., et al. (2024). Smart nanofibers: Synthesis, properties, and scopes in future advanced technologies. *Advanced Materials Technologies, 9*, 2301392.

Chinnappan, A., Baskar, C., Baskar, S., et al. (2017). An overview of electrospun nanofibers and their application in energy storage, sensors and wearable/flexible electronics. *Journal of Materials Chemistry C, 5*, 12657–12673.

Chowdhury, M. I. S., Autul, Y. S., Rahman, S., & Hoque, M. E. (2022). Polymer nanocomposites for automotive applications. In *Advanced polymer nanocomposites* (pp. 267–317). Elsevier.

Dai, S., Wang, N., Qi, C., et al. (2019). Preparation of core-shell structure Fe_3O_4@C@ MnO_2 nanoparticles for efficient elimination of U (VI) and Eu (III) ions. *Science of the Total Environment, 685*, 986–996.

Das, N., Panda, S., Das, D. K., et al. (2025). Polymer nanocomposites and nanostructured polymers. In *Exploring nanomaterial synthesis, characterization, and applications* (pp. 213–240). IGI Global.

Din, S. H., Shah, M. A., Sheikh, N. A., & Butt, M. M. (2020). Nano-composites and their applications: A review. *Characterization and Application of Nanomaterials, 3*, 40–48.

Du, H., Mazzeo, A. D., Shan, J. W., et al. (2023). Electrical response, elastic property, and pressure sensing under bending of hybrid graphene/CNT/elastomer nanocomposites. *Composite Structures, 311*, 116838.

Elalfy, D. A., Gouda, E., Kotb, M. F., et al. (2024). Comprehensive review of energy storage systems technologies, objectives, challenges, and future trends. *Energy Strategy Reviews, 54*, 101482.

Ferreira, A. D. B. L., Nóvoa, P. R. O., & Marques, A. T. (2016). Multifunctional material systems: A state-of-the-art review. *Composite Structures, 151*, 3–35.

Garavand, F., Khodaei, D., Mahmud, N., et al. (2024). Recent progress in using zein nanoparticles-loaded nanocomposites for food packaging applications. *Critical Reviews in Food Science and Nutrition, 64*, 3639–3659.

Gomes Souza, F., Bhansali, S., Pal, K., et al. (2024). A 30-year review on nanocomposites: Comprehensive bibliometric insights into microstructural, electrical, and mechanical properties assisted by artificial intelligence. *Materials, 17*, 1088.

Gonçalves, J. M., Silva, M. N. T., Naik, K. K., et al. (2021). Multifunctional spinel $MnCo_2O_4$ based materials for energy storage and conversion: A review on emerging trends, recent developments and future perspectives. *Journal of Materials Chemistry A, 9*, 3095–3124.

Guler, S. H., Guler, O., & Dikici, B. (2022). Structure–property relationships in polymer nanocomposites. In *Handbook of carbon nanotubes* (pp. 659–685). Springer.

He, Y., Shao, Y., Xiao, Y., et al. (2022). Multifunctional phase change composites based on elastic MXene/silver nanowire sponges for excellent thermal/solar/electric energy storage, shape memory, and adjustable electromagnetic interference shielding functions. *ACS Applied Materials & Interfaces, 14*, 6057–6070.

Hwang, J., Yoon, T., Jin, S. H., et al. (2013). Enhanced mechanical properties of graphene/copper nanocomposites using a molecular-level mixing process. *Advanced Materials, 25*, 6724–6729.

Idumah, C. I., & Hassan, A. (2016). Emerging trends in graphene carbon based polymer nanocomposites and applications. *Reviews in Chemical Engineering, 32*, 223–264.

Israr, M., Iqbal, J., Arshad, A., et al. (2020). A unique $ZnFe_2O_4$/graphene nanoplatelets nanocomposite for electrochemical energy storage and efficient visible light driven catalysis for the degradation of organic noxious in wastewater. *Journal of Physics and Chemistry of Solids, 140*, 109333.

Jaafar, H. K., Hashim, A., & Rabee, B. H. (2024). Fabrication and unraveling the morphological, structural, and dielectric features of PMMA-PEO-SiC–$BaTiO_3$ promising quaternary nanocomposites for multifunctional nanoelectronics applications. *Journal of Materials Science: Materials in Electronics, 35*, 128.

Kareem, A., Hashim, A., & Hassan, H. B. (2024). Ameliorating and tailoring the morphological, structural, and dielectric features of Si_3N_4/CeO_2 futuristic nanocomposites doped PEO for nanoelectronic and nanodielectric applications. *Journal of Materials Science: Materials in Electronics, 35*, 461.

Kausar, A., & Ahmad, I. (2024). Highpoints of carbon nanotube nanocomposite sensors—A review. E-prime-advances in electrical engineering. *Electronics and Energy, 7*, 100419.

Kumar, S. K., & Krishnamoorti, R. (2010). Nanocomposites: Structure, phase behavior, and properties. *Annual Review of Chemical and Biomolecular Engineering, 1*, 37–58.

Li, Y., Zhang, X., Zhang, Q., et al. (2022). Layer-by-layer assembly of CeO_2–x@ C-rGO nanocomposites and CNTs as a multifunctional separator coating for highly stable lithium–sulfur batteries. *ACS Applied Materials & Interfaces, 14*, 18634–18645.

Liu, J., Hui, D., & Lau, D. (2022). Two-dimensional nanomaterial-based polymer composites: Fundamentals and applications. *Nanotechnology Reviews, 11*, 770–792.

Liu, L., Xiang, D., Ma, J., et al. (2024). Multifunctional basalt fiber reinforced polymer composites with in-situ damage self-sensing and temperature-sensitive behavior. *Composites Communications, 50*, 101990.

Liu, P., Cottrill, A. L., Kozawa, D., et al. (2018). Emerging trends in 2D nanotechnology that are redefining our understanding of "nanocomposites". *Nano Today, 21*, 18–40.

Liu, X., Lu, X., Su, Y., et al. (2020). Clay-polymer nanocomposites prepared by reactive melt extrusion for sustained drug release. *Pharmaceutics, 12*, 51.

Lu, L., Yang, B., & Liu, J. (2020). Flexible multifunctional graphite nanosheet/electrospun-polyamide 66 nanocomposite sensor for ECG, strain, temperature and gas measurements. *Chemical Engineering Journal, 400*, 125928.

Lu, S., Chen, Z., Tu, H., et al. (2024). Multifunctional carbon quantum dots decorated self-healing hydrogel for highly effective treatment of superbug infected wounds. *Chemical Engineering Journal, 480*, 148218.

Mahboubizadeh, S., Sadeq, A., Arzaqi, Z., et al. (2024). Advancements in fiber-reinforced polymer (FRP) composites: An extensive review. *Discover Materials, 4*, 22.

Malaki, M., Xu, W., Kasar, A. K., et al. (2019). Advanced metal matrix nanocomposites. *Metals, 9*, 330.

Mallakpour, S., & Khadem, E. (2016). Carbon nanotube–metal oxide nanocomposites: Fabrication, properties and applications. *Chemical Engineering Journal, 302*, 344–367.

Morsi, M. A., Asnag, G. M., Rajeh, A., & Awwad, N. S. (2021). Nd: YAG nanosecond laser induced growth of au nanoparticles within CMC/PVA matrix: Multifunctional nanocomposites with tunable optical and electrical properties. *Composites Communications, 24*, 100662.

Mülhaupt, R. (2004). Hermann Staudinger and the origin of macromolecular chemistry. *Angewandte Chemie International Edition, 43*, 1054–1063.

Muthuvinayagam, M., Ashok Kumar, S. S., Ramesh, K., & Ramesh, S. (2023). Introduction of graphene: The "mother" of all carbon allotropes. In *Graphene: Fabrication, properties and applications* (pp. 5–20). Springer.

Nasrollahzadeh, M., Sajadi, S. M., Sajjadi, M., & Issaabadi, Z. (2019). An introduction to nanotechnology. In *Interface science and technology* (pp. 1–27). Elsevier.

Njuguna, J., Pielichowski, K., & Desai, S. (2008). Nanofiller-reinforced polymer nanocomposites. *Polymers for Advanced Technologies, 19*, 947–959.

Noor, H., Zafar, A., Raza, A., et al. (2024). Advancement in optical and dielectric properties of unsaturated polyester resin/zinc oxide nanocomposite: Synthesis to application in electronics. *Journal of Materials Science: Materials in Electronics, 35*, 1598.

Norizan, M. N., Shazleen, S. S., Alias, A. H., et al. (2022). Nanocellulose-based nanocomposites for sustainable applications: A review. *Nanomaterials, 12*, 3483.

Phiri, R., Rangappa, S. M., Siengchin, S., et al. (2024). Advances in lightweight composite structures and manufacturing technologies: A comprehensive review. *Heliyon, 10*, e39661.

Pokropivny, V., Lohmus, R., Hussainova, I., et al. (2007). *Introduction to nanomaterials and nanotechnology*. Tartu University Press Tartu.

Rabiee, N., Bagherzadeh, M., Ghadiri, A. M., et al. (2021). Multifunctional 3D hierarchical bioactive green carbon-based nanocomposites. *ACS Sustainable Chemistry & Engineering, 9*, 8706–8720.

Rane, A. V., Kanny, K., Abitha, V. K., & Thomas, S. (2018). Methods for synthesis of nanoparticles and fabrication of nanocomposites. In *Synthesis of inorganic nanomaterials* (pp. 121–139). Elsevier.

Reis, E. D., Resende, H. F., Christoforo, A. L., et al. (2024). Assessment of physical and mechanical properties of concrete with carbon nanotubes pre-dispersed in cement. *Journal of Building Engineering, 89*, 109255.

Saber, O., Osama, A., Shaalan, N. M., & Osama, M. (2024). Engineering the integration of titanium and nickel into zinc oxide nanocomposites through nanolayered structures and nanohybrids to design effective photocatalysts for purifying water from industrial pollutants. *Catalysts, 14*, 340.

Saha, M., Ray, R., Choudhury, A. R., et al. (2021). Impact of exfoliation/intercalation of nano-clay on structure, morphology and electrical properties of poly (ethylene oxide) based solid nanocomposite electrolytes. *Journal of Polymer Research, 28*, 1–21.

Sahoo, B. B., Sahoo, P. K., Rao Bhaviripudi, V., et al. (2024). Multifunctional Dy_2NiMnO_6/reduced graphene oxide nanocomposites and their catalytic, electromagnetic shielding, and electrochemical properties. *ACS Omega, 9*, 4600–4612.

Salimiyan, N., Gholami, M., & Sedghi, R. (2023). Preparation of degradable, biocompatible, conductive and multifunctional chitosan/thiol-functionalized graphene nanocomposite hydrogel via click chemistry for human motion sensing. *Chemical Engineering Journal, 471*, 144648.

Sengwa, R. J., Dhatarwal, P., & Choudhary, S. (2020). A comparative study of different metal oxide nanoparticles dispersed PVDF/PEO blend matrix-based advanced multifunctional nanodielectrics for flexible electronic devices. *Materials Today Communications, 25*, 101380.

Senturk, O., & Palabiyik, M. (2023). Exploring the tribological properties of PA6/GO nanocomposites produced by in situ polymerization. *Journal of Materials Science, 58*, 10318–10339.

Shalan, A. E., Makhlouf, A. S. H., & Lanceros-Méndez, S. (2022). Nanocomposites materials and their applications: Current and future trends. In *Advances in nanocomposite materials for environmental and energy harvesting applications* (pp. 3–14). Springer.

Sharma, S., Agarwal, A., Chauhan, P. S., et al. (2024). Recent advancements in materials: Nanocomposites. In *Biomedical implants* (pp. 24–38). CRC Press.

Shen, J., Yang, Y., Zhang, J., et al. (2024). Carbon quantum dot-functionalized dermis-derived transparent electronic skin for multimodal human motion signal monitoring and construction of self-powered triboelectric nanogenerator. *ACS Applied Materials & Interfaces, 16*, 46771–46788.

Shrestha, S., Wang, B., & Dutta, P. (2020). Nanoparticle processing: Understanding and controlling aggregation. *Advances in Colloid and Interface Science, 279*, 102162.

Singh, N. B., & Agarwal, S. (2016). Nanocomposites: An overview. *Emerging Materials Research, 5*, 5–43.

Singh, R., Agrohiya, S., Rawal, I., et al. (2024). Multifunctional porous polyaniline/phosphorus-nitrogen co-doped graphene nanocomposite for efficient room temperature ammonia sensing and high-performance supercapacitor applications. *Applied Surface Science, 665*, 160368.

Supriya, S., Das, S., Samal, S. K., et al. (2024). Rapid microwave-assisted synthesis and characterization of a novel CuCoTe nanocomposite material for optoelectronic and dielectric applications. *Nanoscale, 16*, 7566–7581.

Thakur, V. K., & Thakur, M. K. (2018). *Chemical functionalization of carbon nanomaterials*. CRC Press.

Tirsoaga, A., El Fergani, M., Nuns, N., et al. (2020). Multifunctional nanocomposites with non-precious metals and magnetic core for 5-HMF oxidation to FDCA. *Applied Catalysis. B, Environmental, 278*, 119309.

Tjong, S. C. (2006). Structural and mechanical properties of polymer nanocomposites. *Materials Science and Engineering: R: Reports, 53*, 73–197.

Varshney, K. (2014). Carbon nanotubes: A review on synthesis, properties and applications. *International Journal of Engineering Research and General Science, 2*, 660–677.

Vilvanatha Prabu, A., Vijayaraghavan, G. V., Ahamed, M. B., & Suganya, R. (2024). Enhanced dielectric, thermal, energy storage, quality factor, impedance and conductivity properties of novel PVDF/pan/ZrO_2 polymer nanocomposites for flexible electronics applications. *Polymers for Advanced Technologies, 36*, 9.

Wang, H., Xiang, J., Wen, X., et al. (2022). Multifunctional skin-inspired resilient MXene-embedded nanocomposite hydrogels for wireless wearable electronics. *Composites. Part A, Applied Science and Manufacturing, 155*, 106835.

Wang, L., Kelly, P. V., Ozveren, N., et al. (2023). Multifunctional polymer composite coatings and adhesives by incorporating cellulose nanomaterials. *Matter, 6*, 344–372.

Wang, Q., Qiao, J., Xiong, Y., et al. (2024). A novel ZIF-8@ IL-MXene/poly (N-isopropylacrylamide) nanocomposite hydrogel toward multifunctional adsorption. *Environmental Research, 242*, 117568.

Yadav, A., Kumar, H., Sharma, R., & Kumari, R. (2021). Influence of polyaniline on the photocatalytic properties of metal nanocomposites: A review. *Colloid and Interface Science Communications, 40*, 100339.

Yang, H.-L., Bai, L.-F., Geng, Z.-R., et al. (2023). Carbon quantum dots: Preparation, optical properties, and biomedical applications. *Materials Today Advances, 18*, 100376.

Yaseen, M., Khan, A., Humayun, M., et al. (2024). Fabrication and characterization of CuO–SiO_2/PVA polymer nanocomposite for effective wastewater treatment and prospective biological applications. *Green Chemistry Letters and Reviews, 17*, 2321251.

Yeang, Q. W., Sulong, A. B., & Tan, S. H. (2014). Carbon nanotubes and graphene. *Nano Convergence, 1*, 15.

Zenkert, D., Harnden, R., Asp, L. E., et al. (2024). Multifunctional carbon fibre composites using electrochemistry. *Composites Part B: Engineering, 273*, 111240.

Zhang, D., Sun, J., Lee, L. J., & Castro, J. M. (2020). Overview of ultrasonic assisted manufacturing multifunctional carbon nanotube nanopaper based polymer nanocomposites. *Engineered Science, 10*, 35–50.

Zhao, Y., Ghanem, M. A., Reddy, G. R., & Joo, S. W. (2024). Probing structural and electrochemical properties of a 2D $CoMoO_4$@ hBN nanocomposite as pseudocapacitive electrode for high-performance supercapacitors. *Ceramics International, 50*, 55330–55341.

Zheng, H., Wang, S., Cheng, F., et al. (2021). Bioactive anti-inflammatory, antibacterial, conductive multifunctional scaffold based on MXene@ CeO_2 nanocomposites for infection-impaired skin multimodal therapy. *Chemical Engineering Journal, 424*, 130148.

Zheng, X., Nie, W., Hu, Q., et al. (2021). Multifunctional RGO/Ti_3C_2Tx MXene fabrics for electrochemical energy storage, electromagnetic interference shielding, electrothermal and human motion detection. *Materials and Design, 200*, 109442.

Zhou, Y., Cheng, X., Tynan, B., et al. (2021). High-performance hierarchical MnO_2/CNT electrode for multifunctional supercapacitors. *Carbon, 184*, 504–513.

Chapter 7
Economic and Social Implications of Nanomaterials

Vishal Vasistha and A. V. Raghavendra Rao

7.1 Introduction

The scientific community is intrigued with nanomaterials with their amazing features at the nanoscale that differ substantially from their bulk counterparts. Due to their outstanding properties such as enhanced strength, electrical conductivity, chemical reactivity, and thermal stability, nanomaterials are distinguished. The nanomaterials are those which have at least one-dimension fewer than 100 nm. These features make possible a wide variety of applications, including the medicine and electronics fields as well as the energy and environmental protection fields. These materials are, however, used in various industries, hence the need to do the in-depth research into the economic and social ramifications of these materials (Baig et al., 2021).

The potential of nanomaterials is impossible to overstate when it comes to economic potential. Enabled are manufacturing of highly efficient and cost-effective products which has led to the revolution in the different industries. In the field of electronics, nanomaterials have made possible the reducing the size of gadgets and the development of more powerful batteries. In the world of healthcare, this has proved to be a key element in developing advanced diagnostic instruments, targeted drug delivery tools and treatments for vast number of diseases. This makes the aerospace and automotive industries to gain from the light weight and high strength

V. Vasistha (✉)
Department of Mechanical Engineering, School of Engineering and Technology, Monad University, Hapur, Uttar Pradesh, India
e-mail: vishal.vasistha@monad.edu.in

A. V. R. Rao
Department of Chemical Engineering, B V Raju Institute of Technology, Narsapur, Medak, Telangana, India

S. Kulkarni (ed.), *Next-Generation Nanomaterials for Sustainable Engineering*, Lecture Notes in Nanoscale Science and Technology 43,
https://doi.org/10.1007/978-3-032-05719-8_7

outputs of nanomaterials and further makes the fuel efficiency better and decreases the impact that the industry creates to the environment (Chatterjee & Dhibar, 2023).

From an economic perspective, nanomaterials have the power to alter innovation and productivity by a massive amount. Incorporating nanotechnology in their products allows businesses to be able to provide better performance and efficiency in their products resulting in competitive advantage in the international market. In addition, broad implementation of nanomaterials could lead to the creation of employment, specifically in R&D, and manufacturing activities. Nations that are likely to make tremendous investments in nanotechnology may become the leading nations technologically and thus economically (Charitidis et al., 2014).

Although these promising promises hold, several barriers to nanomaterial commercialization remain. For smaller- and medium-sized businesses (SMEs), the costs inherent in research and development and production can be prohibitive. Although nanoparticle development requires a unique set of processes and equipment, smaller companies are hindered by the fact that their technologies do not produce results as quickly or on the same scale as larger companies. And this could lead to market consolidation, which may mean the nanomaterials industry becomes dominated by a few large firms creating little innovation and less competition (Hobson, 2009).

Nanomaterials also face the constantly changing regulatory landscape as an economic consideration. Because they have new qualities, their use is still in the process of being regulated through regulatory frameworks. This uncertainty incurs to businesses because compliances standards may change subsequently incurring expenses. However, the long-term effects of nanomaterials have not been fully understood yet on human health and the environment, and in case the product was found out to be damaging in a later date, the corporations can be subjected to legal responsibility (Sajid et al., 2015).

Nanomaterials have big consequences for modern society in addition to their economic impact. Nanotechnology can also help to improve the quality of life in such areas as health care advancement, better protection of the environment, and facilitation of transition to the renewable energy sources. More accurately and efficiently the nanomaterials application applies in medicine. For example, nanoparticles can be specifically designed to target cancer cells so that drug can be delivered directly to the tumor site and the risk of side effects minimized. Such a focused approach helps make treatments more effective and also makes healthcare more cost effective due to the shorter stays in hospitals as well as the time it requires to recuperate (Prasad et al., 2018).

Nanomaterials are being used to improve pollution control and resource conservation strategies in the environmental protection field. Nanotechnology enabled water filtration systems, for example, can remove impurities better than traditional methods in order to achieve access to clean water in places where it is sometimes not available. Similarly, nanomaterials are introduced into renewable energy technologies, such as solar cells, to enhance the efficiency of energy conversion. It is being used as contribution toward shifting in the direction of sustainable energy in the entire world (Das et al., 2015).

However, hazards are also associated with the societal benefits that nanomaterials afford. The possible toxicity of nanoparticles is still object of concern, and nano materials environmental and health concerns are being investigated. However, the nanoparticles, which are of extremely small size, have the capacity to enter the biological systems through inhalation, ingestion, and contact with skin and may create heal issues. However, workers who are involved in the production and handling of nanomaterials may be at greatest risk, and strict occupational safety procedures need to be in place. To dispose of these materials, nanomaterials have also raised concerns about the long-term effects on ecosystems. Although able to remain and accumulate when released into the environment, nanoparticles pose a threat to wildlife, and could find their way into the distribution chain (Handy & Shaw, 2007).

We also need to consider the social equity of nanomaterials. Despite its promise to enhance healthcare, energy access, and environmental protection, nanotechnology might not equitably spread its benefits. Overpriced products and difficulties in accessing nanotechnology could exacerbate already existing inequities in countries with low and middle incomes. For example, new treatments in medicine based on nanotechnology could be available only to more affluent people. If this occurs it would create a bigger gap between those who can afford the most cutting-edge therapies and those who cannot (Bruns, 2004).

Another important aspect that should be considered before nanomaterials have a potential ethical consequence that we have to deal with. Just like any technology that promises to change the world, the development and deployment of nanomaterials bring privacy, security, and informed consent issues. However, the use of nanosensors in the healthcare industry and other industries, for instance, may raise concerns of data privacy and misuse of personal information. There is a want for fully open and inclusive processes of decision making around the development and regulation of nanomaterials (Singh et al., 2024).

7.2 Applications of Nanomaterials

Nanomaterials have rapidly turned out to be a fundamental component in different enterprises because of their unique properties. The use of nanomaterials in such diverse fields as medical, energy, environmental research and electronics is made possible by these properties. Based on their nanoscale size, large surface area, and tunability, these materials can interact with systems that are not possible with bulk materials. We will explore in the next subchapter innumerable application of nanoparticles and will explore transformative potential of these materials in various industrial fields (Findik, 2021). Figure 7.1 shows the applications of nanomaterials. The extensive, multi-industry applications of nanomaterials are shown in detail in Table 7.1 as well.

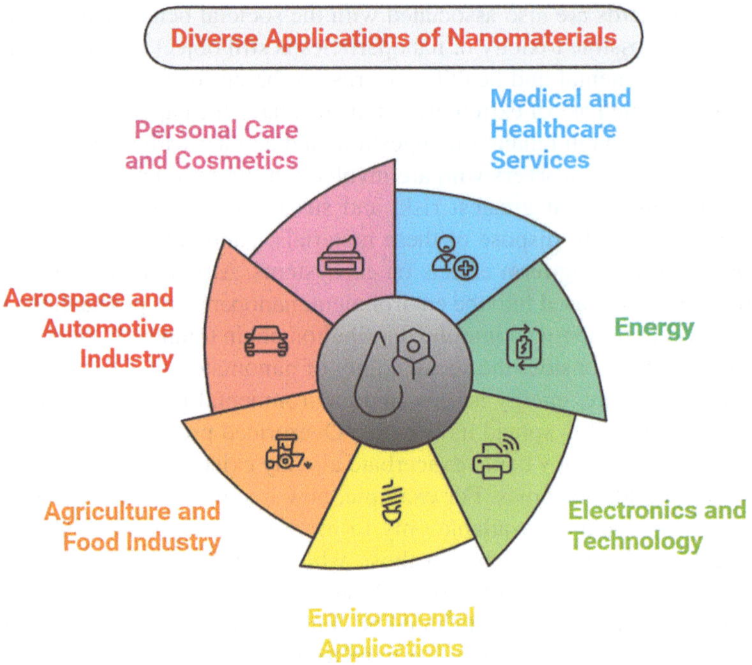

Fig. 7.1 Diverse applications of nanomaterials

7.2.1 Medical and Healthcare Services

The revolutionary effects of nanomaterials in healthcare have specifically left a big impact on drug delivery, diagnostics, and regenerative medicine. Nanotechnology has promising applications in medicine through the development of targeted medication delivery systems. Nanoparticles can be genetically modified to deliver medications directly where they are needed to particular cells, tissues, or organs. Such would decrease the chance of undesirable side effects and enhance the impact of treatment. Anticancer drugs are being transported in the form of liposomes and polymeric nanoparticles for example. This tailored strategy decreases the effect on healthy cells which in turn lowers the toxicity which is common with chemotherapy administered (Verma et al., 2011).

Another important area of nanomaterials is diagnostic. Utilization of quantum dots and gold nanoparticles is being used to improve imaging techniques used for early illness detection. One of a kind optical property make them usable as contrast agents for imaging modalities as computed tomography (CT) and magnetic resonance imaging (MRI). The ability of these nanoparticles to bind to specific biomarkers of diseases enhances their detection in early stages (Welch et al., 2021).

In regenerative medicine, the field of tissue engineering and wound healing utilize nanomaterials. For example, biodegradable polymers fabricated into nanofibers

Table 7.1 Extensive, multi-industry applications of nanomaterials

Industry	Application	Specific nanomaterials	Unique benefit	Key reference
Medical and healthcare	Drug delivery	Liposomes, polymeric nanoparticles	Targeted drug delivery with reduced side effects	Verma et al. (2011)
	Diagnostics	Quantum dots, gold nanoparticles	Early disease detection with enhanced imaging capabilities	Welch et al. (2021)
	Regenerative medicine	Biodegradable polymer nanofibers, silver nanoparticles	Tissue engineering and infection-resistant wound healing	Fathi-Achachelouei et al. (2019)
Energy	Solar cells	Quantum dots, carbon nanotubes	Enhanced light absorption and charge transfer	Hussein (2015)
	Batteries	Graphene, silicon nanoparticles	Increased capacity, faster charge/discharge rates	Pomerantseva et al. (2019)
	Fuel cells	Platinum nanoparticles	Higher efficiency, reduced catalyst material	Abdalla et al. (2018)
Electronics	Transistors	Silicon nanowires, carbon nanotubes	Miniaturization for faster, compact devices	Lah et al. (2018)
	Flexible electronics	Graphene	Used in transparent displays and wearable electronics	Triana et al. (2022)
	Data storage	Magnetic nanoparticles	Enhanced storage density	Zhang et al. (2013)
Environmental applications	Water purification	Carbon nanotubes, graphene oxide	Efficient heavy metal and pathogen removal	Singh et al. (2022)
	Air filtration	Titanium dioxide nanoparticles	Degradation of pollutants via photocatalysis	David and Niculescu (2021)
	Soil remediation	Iron nanoparticles	Effective breakdown of pesticides and contaminants	Bakshi and Abhilash (2020)
Agriculture and food industry	Crop protection	Silver nanoparticles, nano-fertilizers	Controlled release for sustained nutrient supply	Fu et al. (2020)
	Food packaging	Nanocomposites, silver nanoparticles	Antibacterial packaging for prolonged freshness	Alfadul and Elneshwy (2010)
	Smart packaging	Nanosensors	Real-time monitoring of food freshness	Dodero et al. (2021)

(continued)

Table 7.1 (continued)

Industry	Application	Specific nanomaterials	Unique benefit	Key reference
Aerospace and automotive	Lightweight components	Carbon nanotubes, graphene	Improved strength-to-weight ratio for fuel efficiency	Mohanavel et al. (2021)
	Protective coatings	Nanoparticles-based coatings	Enhanced resistance to temperature and corrosion	Anirudh et al. (2023)
	Emission control	Nanocatalysts	Lower harmful emissions from vehicles	Basha et al. (2022)
Personal care and cosmetics	Skincare formulations	Liposomes, nanoemulsions	Better delivery of vitamins and antioxidants	Souto et al. (2020)

are utilized as scaffolds for tissue regeneration by promoting the proliferation and differentiation of biological cells. A very common practice nowadays is to incorporate silver nanoparticles into the wound dressings in which silver nanoparticles are known for their antibacterial qualities for both preventing infection and increasing the rate of wound healing (Fathi-Achachelouei et al., 2019).

7.2.2 *Energy*

While developing all the other sectors, nanotechnology has also been a key driver of significant energy sector, especially in the areas of both energy storage and renewable energy. As an example, integrating nanoparticles has been really good for solar cells. Conventional silicon-based solar cells are being combined with nanoparticles such as quantum dots or carbon nanotubes to make them more efficient. Enhanced cell ability to absorb light and carry charge is done in order to achieve this. Furthermore, perovskite solar cells made up with nanostructured materials are being developed as an alternative of conventional solar cells that are very efficient as well as cost effective (Hussein, 2015).

The field of energy storage has a lot of interesting developments going on and both batteries and supercapacitors are made using nanomaterials. Significant improvement of lithium ion batteries (used in almost all consumer electronic devices and electric cars) has also helped by the incorporation of nanomaterials such as graphene and silicon nanoparticles. These materials boost battery capacity and battery lifetime and charge/discharge rates, all at times making them suitable to more demanding applications such as electric automobiles. The increased surface area of nanoparticles benefits supercapacitors from faster charge and discharge cycles. (Pomerantseva et al., 2019).

In addition, nanomaterials are being used to enhance the fuel cells, which are other type of sustainable energy source. Using nanoparticles of platinum as catalysts in fuel cells leads to a significant increase in the efficiency of these cells while decreasing the amount of costly catalyst material needed (Abdalla et al., 2018).

7.2.3 Electronics and Technology

Nanomaterials have had a significant impact on the electronics industry, as they have enabled the production of small sized, swift, economically efficient devices. The domain of semiconductors has already emerged silicon nanowires and carbon nanotubes for nanoscale transistors. This is among the most important uses of nanotechnology. These nanotransistors are very needed to continue the process of miniaturizing electronic devices. Nanomaterials make it possible for cramming more transistors onto a single chip is in accordance with Moore's Law (Lah et al., 2018).

Nanomaterials are used in the development of flexible and transparent electronics and used in electrical transistors, nanomaterials. Graphene is a nanomaterial that is formed of a single sheet of carbon atoms that is a very flexible material, and an exceptional conductor of electricity. It is a subject of investigation that could be used in transparent conductive films, wearable electronics, and flexible displays. But quantum dots are also being taken into display technologies, making panels that are more energy efficient and brighter. These will be characterized by screens on televisions, cellphones, and other electronic gadgets (Triana et al., 2022).

Another area of nanomaterials application is to improve is data storage. Magnetic nanoparticles are also being used in hard drives to increase the density of the stored materials. By changing the magnetic characteristics of these nanoparticles, researchers have been able to store more data in less space. As a result, it has opened the door to the development of the following generation of high capacity storage devices (Zhang et al., 2013).

7.2.4 Environmental Applications

Nanotechnology application has the potential to be a solution to many of the most pressing environmental problems including water purifying, air filtration, and pollution control. One of the most important applications of nanomaterials in environmental research is the treatment of water. Carbon nanotubes, graphene oxide, and silver nanoparticles form the basis of sophisticated filtration systems that can purify water. Heavy metals, bacteria, and viruses are included in these contaminants. These nanoparticles are capable of trapping more contaminants than conventional filtration methods owing to the large surface area (Singh et al., 2022).

Nanomaterials are being incorporated into filters used in air filtering to capture particulate matter, gasses, and volatile organic compounds (VOCs). Photocatalytic air purifiers, which use titanium dioxide nanoparticles, can able to degrade harmful pollutants when exposed to UV light, improving both indoor and outdoor air quality (David & Niculescu, 2021).

On the other hand, the process of soil revitalization and pollution control is carried out using nanomaterials. For example, iron nanoparticles are processed to treat contaminated soils. The decomposition of pesticides, heavy metals, and chlorinated chemicals are among the contaminants these nanoparticles can decompose. This technique named as nano remediation is an effective method for cleaning up of hazardous waste, which is cost effective and environment friendly (Bakshi & Abhilash, 2020).

7.2.5 Applications in Agriculture and Food Industry

The agricultural sector is now investigating nanomaterials with a focus on increasing crop productivity and protecting the plants from diseases and pest. Reducing the amount of chemicals, they need and minimizing the way those chemicals impact the environment, nanoparticles have potential to make it so nutrients or insecticides are delivered more efficiently. For instance, nano fertilizers release the nutrients in a controlled and a slow fashion. After separating the nutrients from a solution over a long period of time, plants can get the essential nutrients it needs, leading to larger yields and less wasted nutrients (Fu et al., 2020).

In food sector, nanotechnology is used to make food product packaging and food safety better. By including nanomaterials, including silver nanoparticles, to the packaging material, one may prolong the shelf life of the food products. It prevents the forming of bacteria. Furthermore, nanocomposites are used in manufacturing lightweight, robust, and biodegradable packaging material. These materials are not only good to the environment, but also useful for keeping food quality as well (Alfadul & Elneshwy, 2010).

Additionally, nanotechnology is being utilized toward the development of intelligent packaging that is capable of monitoring the quality of the food and providing real time real valve feedback regarding to the freshness of the food. The container can change its color and give signal that the food is going to be bad (Dodero et al., 2021).

7.2.6 Aerospace and Automotive Industry

The unique properties of nanomaterials are being used by aerospace and automotive industries to fabricate more durable, lighter, and fuel-efficient vehicles. Carbon nanotubes and graphene, for example, are being added into composite materials to increase the strength and longevity of components of cars, airplanes, and similar structures. These nanomaterials decrease the overall weight of cars and provide higher fuel efficiency and lesser emissions (Mohanavel et al., 2021).

For the development of protective coatings in severe temperatures, corrosion, and wear, nanomaterials are used by the aerospace industry. These coatings also extend the life time of important airplane and spacecraft components, thus reducing maintenance costs and improving safety. On similar lines, in the automobile sector the applications of nanoparticles are being explored in making lightweight, high strength body panel and engine components that increase the overall performance and safety of the vehicles (Anirudh et al., 2023).

Furthermore, gasoline additives that increase the efficiency of combustion are being investigate as possible application for nanotechnology, leading to engines that are both cleaner and more efficient. Furthermore, the nano catalysts are being used in the catalytic converters to minimize the amounts of poisonous emissions generated by the vehicles in the battle against air pollution (Basha et al., 2022).

7.2.7 Personal Care and Cosmetics

Nanomaterials have a specific application in the cosmetics and personal care business. They were used to improve the efficacy in hair care, sunscreen, and skincare formulations. Titanium dioxide and zinc oxide nanoparticles are commonly used in sunscreens, as an example, to provide broad spectrum protection from ultraviolet (UV) radiation. As compared to traditional sunscreens whose residue frequently remains on the skin, sunscreens that use nanoparticles are transparent, making them to look better (Salvioni et al., 2021).

In the cosmetics industry, nanomaterials such as liposomes and nano emulsions, are used to improve the delivery of active substances, as vitamins or antioxidants, into the skin. Here, such nanocarriers enable the substances in the compounds to be able to enter the skin more deeply than a topical alone, thus improving the magnitude of wrinkle diminution, dryness reduction, and other skin issues (Souto et al., 2020).

7.3 Effect of Nanomaterials

7.3.1 Ethical Impact

With the progression of nanotechnology and its importation into many industries, ethical issues of nanomaterials are held up to debate. How these nanomaterials are produced, deployed, and governed has both societal, health, environmental, and ethical ramifications associated with the expansion of nanotechnology. As the ethical issues of nanomaterials, we need to consider their long-term effects, equitable distribution of benefits, and avoid downsides (Allhoff & Lin, 2009). Figure 7.2 illustrates ethical impact in different areas. Table 7.2 displays ethical consideration

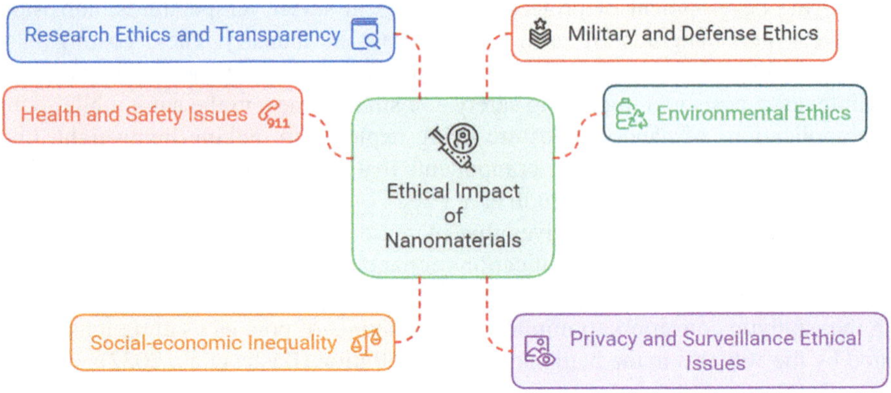

Fig. 7.2 Ethical impact of nanometerials

Table 7.2 Ethical considerations and key issues in nanomaterial development and deployment

Category	Ethical concerns	Key considerations
Health and safety	Nanoparticles may cause toxicity and long-term health risks; regulations lag behind innovations	Enforce safety guidelines, rigorous testing, and responsible commercialization
Environmental ethics	Potential ecological harm due to nanoparticle accumulation in ecosystems	Promote sustainable design, recycling, and careful disposal practices
Socio-economic inequality	Wealthier nations and corporations may disproportionately benefit, widening economic gaps	Support equitable access through inclusive policies and support for small-scale initiatives
Privacy and surveillance	Nanotechnology enables advanced, potentially intrusive surveillance devices	Regulate to protect privacy, ensure transparency, and respect autonomy
Research ethics and transparency	Concerns over lack of transparency, accountability, and unclear long-term risks	Uphold ethical research standards, engage the public, and provide clear information
Military and defense ethics	Enhanced weaponry and surveillance pose risks of misuse and human rights violations	Establish international guidelines to prevent escalation and misuse in conflict

and key issues in nanomaterial development and deployment integrates into several areas.

7.3.2 *Health and Safety Issues*

The intense ethical concern is the questions of whether nanomaterials are dangerous for human and environmental health. This is because nanomaterials are themselves nanoscale and highly chemically reactive materials that are unpredictable in biological systems leading to concern for toxicity and long-term exposure. Nanoparticles easily pass through cell membranes and can cause inflammation, oxidative stress, damage DNA, and so on. The question raised is whether nanomaterial workers and customers are adequately protected or not (Handy & Shaw, 2007).

Safety regulation of nanomaterial uses and its manufacture leads to another ethical issue. Safety rules for nanomaterial handling, disposal, and testing are being developed by many regulatory agencies, the USA and Europe for example. Yet, in this industry, regulations tend to lag behind innovation. Because of regulatory delays, nanoparticles in consumer products where risk to consumers is not studied. Before novel nanomaterials are released into the world and used by all, they should be tested by corporations, governments, and researcher (Bernard, 2017).

7.3.3 Environmental Ethics

Another issue is negative impacts of nanoparticles to the environment. On the one hand, nanotechnology has potential applications in different industries, on the other hand, it also haunts the ecosystems. Still it is unknown about the environmental toxicity of nanomaterials in soil, air, and water. They can change water quality by an accumulation of antibacterial silver nanoparticles used in them (Purohit et al., 2017).

The challenge is the balance of nanomaterial benefits and potential environmental impacts. Companies and governments want to develop and use nanomaterials, but they must do so without harming ecosystems. The ethical growth should look beyond short-term advantage and consider a strategy of minimizing ecological damage.

The disposal of nanomaterials already has ethical issues. End of life management of nanomaterials contained in consumer products is germane as nanomaterials become more widespread. Nanoparticles released into the environment from the disposal of water containing these particles may have unintended effects such as nanoparticles collecting (aggregating) in organisms. However, for the reasons that are ethical it is the work of the manufacturers and the policymakers to provide solutions on how they can re-use and dispose of the nanomaterials so as to reduce the environmental air pollution (Besha et al., 2020).

7.3.4 Social-Economic Inequality

Nanotechnology can create new jobs, new technology, and new sectors, to boost the economy. However, the benefits are not equally distributed and the inequality of such benefits is but a question of ethics. Also, that, in most of the developing countries, there is a shortage of money, infrastructure, and technological ability to compete with the nanomaterials, which adds the difference between countries which are able to invest in the nanotechnology and the poor nations which are not (Miller & Scrinis, 2011).

This inequality poses ethical considerations on fairness and equity. The regions and sociodemographic categories see an even distribution of benefits from nanomaterials. Small businesses and emerging nations fear the concentration of the market (in which a few giant corporations rule the market). For ethically valid reasons, access to nanotechnology needs to be both egalitarian and inclusive and not addition to inequities (Cozzens & Wetmore, 2010).

If the nanotechnology research and development costs are not the level playing field, poorer governments and industries will be left behind. Maybe nanomaterials can only be used by the wealthy for new medical treatments or cleaner energy solutions. One of the major ethical problem of nanotechnology is fair access, which will result in need for global cooperation and acceptable investment of considerable funds on the research and development (Babatunde et al., 2019).

7.3.5 Privacy and Surveillance Ethical Issues

The employment of nanotechnology creates privacy and surveillance ethical issues. The case of nanomaterials can use nanoscale surveillance sensors and tracking devices. The logistical and medical as well as security have valid uses of these technologies but such constant surveillance and control are unethical.

You can put nanosensors into everything, which leads to mass surveillance, without even consenting. This steals privacy and autonomy because you might not even know that you're being tracked, that your data is being collected and scored. Nanotechnology surveillance is an area which should be very cautiously examined for ethical reasons, and privacy rights must be protected (Khan, 2012).

In behavior that constitutes clandestine surveillance, for example, espionage or invasive monitoring, such devices raise ethical issues at a societal and policy level. Nanotechnology guidelines and restrictions on the use of the technology, for example, should be dully precise and restrictive to inhibit from being used unwontedly for monitoring and to protect human rights (Rajendran et al., 2015).

7.3.6 Research Ethics and Transparency

Nanomaterial research raises questions of openness, accountability, and responsible innovation. Nanotechnology research has to be undertaken by individual scientists, corporations, and governments in an ethical and transparent way. Society norms and expectations must be met with the development of nanomaterials, therefore public engagement and conversation on the benefits and risks of the development is crucial.

There has been a need for the study of nanomaterials effects in the long run. The study should cover immediate advantages and unforeseen risks of extensive use.

Researchers should be concerned not on short-term gains but on safety and long term sustainability (Malakar et al., 2022).

Nanomaterial dangers also have to be ethically disclosed to the public by industries or researchers. Honest communication about the benefits and risks of nanotechnology, customers, workers, and legislators will be able to make well-informed decisions. In nanomaterial development and use decision-making, stakeholders are also asked to participate in public accountability (Shrader-Frechette, 2007).

7.3.7 Military and Defense Ethics

Military and defense can use nanotechnology in the development of advanced weapons, lines of surveillance, protective gear, etc. These are ethical cases about uncovering the militarization of technology and the potential for exacerbating such conflicts or causing unintentional injury with nanotechnology (Abed & Jawad, 2022).

The use of nanomaterials for the creation of more efficient chemical or biological weapons could lead to real threats to global security and to human rights. As a result of using the technology for both civilian and military purposes, ethical problem occurs in nanotechnology. To prevent nanomaterials from violating ethical norms such as nonmaleficence (that is, not doing harm), and fairness, international agreements and laws are needed (Ratner & Ratner, 2004).

7.4 Conclusion

Nanomaterials have brought drastic changes in the industries and continue to offer capabilities to revolutionize economies as well as society. The properties of nanomaterials have let to advancements in medicine, electronics, energy, environmental protection innovation, cost reduction, and sustainability. On general economic grounds, nanotechnology has boosted efficiency, carried competition advantages, and employment opportunities in researching, innovating as well as manufacturing fields. Those countries that continue to subsidize research in nanotechnology stand in advantage point to lead technologically and economically. Still, threats like costly R&D initiatives, legislation barriers, and high chances of market saturation might slow down fair development for small firms and development countries.

Nanomaterials contributed greatly on the quality of life improvement. Examples include applications in targeted drug delivery and diagnostics, renewable energy and water purification, to mention but a few, all of which address serious global needs. However, risks to human health and the environment cannot be undermined. This raises questions: how toxic are the nanoparticles and how they were able to find themselves within ecosystems; and, to what extent can nanoparticles affect the human health? There is occupational danger that workers exposed to such materials might encounter, however impacts of these materials might have on ecosystems in the long run are not well understood.

From the ethical perspectives, concerning the distribution of the impact of nanotechnology, socio economical disadvantages seem imminent. The access to high quality nanotechnology goods may shift to an oligopolistic group of high-income countries, thereby escalating the technological divide between developed and developing nations. Some other concerns relating to privacy and surveillance, products such as nanosensors, tracking devices also require protection of human rights. In addition, the prospect of nanotechnology weapons like officer equipment and superior defense system increase global security risks.

Realization of the opportunities of nanomaterials coupled with aggravation of their risks requires intersectoral and international cooperation. It is necessary to talk about openness and responsibility as the main strategies to establish transparency; equal participation in the policy-making process; and effective regulation. The continued fair advancement of nanotechnology will not only unlock its improved potential for economical and societal progression, but also the safety of human being's health and ethical standards together with the environment.

References

Abdalla, A. M., Hossain, S., Azad, A. T., Petra, P. M. I., Begum, F., Eriksson, S. G., & Azad, A. K. (2018). Nanomaterials for solid oxide fuel cells: A review. *Renewable and Sustainable Energy Reviews, 82*, 353–368.

Abed, M. S., & Jawad, Z. A. (2022). Nanotechnology for defence applications. In *Nanotechnology for electronic applications* (pp. 187–205). Springer.

Alfadul, S. M., & Elneshwy, A. A. (2010). Use of nanotechnology in food processing, packaging and safety–review. *African Journal of Food, Agriculture, Nutrition and Development, 10*(6), 58068.

Allhoff, F., & Lin, P. (Eds.). (2009). *Nanotechnology & society: Current and emerging ethical issues*. Springer.

Anirudh, S., Krishnamurthy, S., Kandasubramanian, B., & Kumar, P. (2023). Probing into atomically thin layered nano-materials protective coating for aerospace and strategic defence application–a review. *Journal of Alloys and Compounds, 968*, 172203.

Babatunde, D. E., Denwigwe, I. H., Babatunde, O. M., Gbadamosi, S. L., Babalola, I. P., & Agboola, O. (2019). Environmental and societal impact of nanotechnology. *IEEE Access, 8*, 4640–4667.

Baig, N., Kammakakam, I., & Falath, W. (2021). Nanomaterials: A review of synthesis methods, properties, recent progress, and challenges. *Materials Advances, 2*(6), 1821–1871.

Bakshi, M., & Abhilash, P. C. (2020). Nanotechnology for soil remediation: Revitalizing the tarnished resource. In *Nano-materials as photocatalysts for degradation of environmental pollutants* (pp. 345–370). Elsevier.

Basha, J. S., Al Balushi, M., Soudagar, M. E. M., Safaei, M. R., Mujtaba, M. A., Khan, T. Y., & Elfasakhany, A. (2022). Applications of nano-additives in internal combustion engines: A critical review. *Journal of Thermal Analysis and Calorimetry, 147*(17), 9383–9403.

Bernard, D. (2017). Ethics and nanomaterials industrial production. In *Metrology and standardization of nanotechnology: Protocols and industrial innovations* (pp. 485–504). Wiley.

Besha, A. T., Liu, Y., Bekele, D. N., Dong, Z., Naidu, R., & Gebremariam, G. N. (2020). Sustainability and environmental ethics for the application of engineered nanoparticles. *Environmental Science & Policy, 103*, 85–98.

Bruns, B. (2004). Applying nanotechnology to the challenges of global poverty: Strategies for accessible abundance. In *1st conference on advanced Nanotechonology: Research, applications, and policy*.

Charitidis, C. A., Georgiou, P., Koklioti, M. A., Trompeta, A. F., & Markakis, V. (2014). Manufacturing nanomaterials: From research to industry. *Manufacturing Review, 1*, 11.

Chatterjee, P., & Dhibar, S. (2023). Nanomaterial marvels: Pioneering applications and cutting-edge advancements in drug delivery. *Nano and Medical Materials, 3*(2), 220.

Cozzens, S. E., & Wetmore, J. (Eds.). (2010). *Nanotechnology and the challenges of equity, equality and development* (Vol. 2). Springer Science & Business Media.

Das, S., Sen, B., & Debnath, N. (2015). Recent trends in nanomaterials applications in environmental monitoring and remediation. *Environmental Science and Pollution Research, 22*, 18333–18344.

David, E., & Niculescu, V. C. (2021). Volatile organic compounds (VOCs) as environmental pollutants: Occurrence and mitigation using nanomaterials. *International Journal of Environmental Research and Public Health, 18*(24), 13147.

Dodero, A., Escher, A., Bertucci, S., Castellano, M., & Lova, P. (2021). Intelligent packaging for real-time monitoring of food-quality: Current and future developments. *Applied Sciences, 11*(8), 3532.

Fathi-Achachelouei, M., Knopf-Marques, H., Ribeiro da Silva, C. E., Barthès, J., Bat, E., Tezcaner, A., & Vrana, N. E. (2019). Use of nanoparticles in tissue engineering and regenerative medicine. *Frontiers in Bioengineering and Biotechnology, 7*, 113.

Findik, F. (2021). Nanomaterials and their applications. *Periodicals of Engineering and Natural Sciences, 9*(3), 62–75.

Fu, L., Wang, Z., Dhankher, O. P., & Xing, B. (2020). Nanotechnology as a new sustainable approach for controlling crop diseases and increasing agricultural production. *Journal of Experimental Botany, 71*(2), 507–519.

Handy, R. D., & Shaw, B. J. (2007). Toxic effects of nanoparticles and nanomaterials: Implications for public health, risk assessment and the public perception of nanotechnology. *Health, Risk & Society, 9*(2), 125–144.

Hobson, D. W. (2009). Commercialization of nanotechnology. *Wiley Interdisciplinary Reviews: Nanomedicine and Nanobiotechnology, 1*(2), 189–202.

Hussein, A. K. (2015). Applications of nanotechnology in renewable energies—A comprehensive overview and understanding. *Renewable and Sustainable Energy Reviews, 42*, 460–476.

Khan, A. S. (2012). Nanotechnology: An introduction to applications and ethical and social issues. In *Nanotechnology: Ethical and social implications*. CRC Press.

Lah, N. A., Zubir, M. N., Mahendran, A., & Samykano, L. (2018). Engineered nanomaterial in electronics and electrical industries. In *Handbook of nanomaterials for industrial applications* (pp. 324–364). Elsevier.

Malakar, Y., Lacey, J., & Bertsch, P. M. (2022). Towards responsible science and technology: How nanotechnology research and development is shaping risk governance practices in Australia. *Humanities and Social Sciences Communications, 9*(1), 1–14.

Miller, G., & Scrinis, G. (2011). Nanotechnology and the extension and transformation of inequity. In *Nanotechnology and the challenges of equity, equality and development* (pp. 109–126). Springer.

Mohanavel, V., Ali, K. A., Ranganathan, K., Jeffrey, J. A., Ravikumar, M. M., & Rajkumar, S. (2021). The roles and applications of additive manufacturing in the aerospace and automobile sector. *Materials Today: Proceedings, 47*, 405–409.

Pomerantseva, E., Bonaccorso, F., Feng, X., Cui, Y., & Gogotsi, Y. (2019). Energy storage: The future enabled by nanomaterials. *Science, 366*(6468), 8285.

Prasad, M., Lambe, U. P., Brar, B., Shah, I., Manimegalai, J., Ranjan, K., & Prasad, G. (2018). Nanotherapeutics: An insight into healthcare and multi-dimensional applications in medical sector of the modern world. *Biomedicine & Pharmacotherapy, 97*, 1521–1537.

Purohit, R., Mittal, A., Dalela, S., Warudkar, V., Purohit, K., & Purohit, S. (2017). Social, environmental and ethical impacts of nanotechnology. *Materials Today: Proceedings, 4*(4), 5461–5467.

Rajendran, J., Karri, R., Wendt, J. B., Potkonjak, M., McDonald, N., Rose, G. S., & Wysocki, B. (2015). Nano meets security: Exploring nanoelectronic devices for security applications. *Proceedings of the IEEE, 103*(5), 829–849.

Ratner, D., & Ratner, M. A. (2004). *Nanotechnology and homeland security: New weapons for new wars*. Prentice Hall Professional.

Sajid, M., Ilyas, M., Basheer, C., Tariq, M., Daud, M., Baig, N., & Shehzad, F. (2015). Impact of nanoparticles on human and environment: Review of toxicity factors, exposures, control strategies, and future prospects. *Environmental Science and Pollution Research, 22*, 4122–4143.

Salvioni, L., Morelli, L., Ochoa, E., Labra, M., Fiandra, L., Palugan, L., & Colombo, M. (2021). The emerging role of nanotechnology in skincare. *Advances in Colloid and Interface Science, 293*, 102437.

Shrader-Frechette, K. (2007). Nanotoxicology and ethical conditions for informed consent. *NanoEthics, 1*, 47–56.

Singh, K. K., Singh, A., & Rai, S. (2022). A study on nanomaterials for water purification. *Materials Today: Proceedings, 51*, 1157–1163.

Singh, P., Gupta, P., Verma, V., Yadav, N., & Pandey, N. K. (2024). Nanotechnology: Social acceptance, cultural impact and privacy. In *Nanotechnology in societal development* (pp. 65–93). Springer.

Souto, E. B., Fernandes, A. R., Martins-Gomes, C., Coutinho, T. E., Durazzo, A., Lucarini, M., & Santini, A. (2020). Nanomaterials for skin delivery of cosmeceuticals and pharmaceuticals. *Applied Sciences, 10*(5), 1594.

Triana, M. A., Hsiang, E. L., Zhang, C., Dong, Y., & Wu, S. T. (2022). Luminescent nanomaterials for energy-efficient display and healthcare. *ACS Energy Letters, 7*(3), 1001–1020.

Verma, S., Domb, A. J., & Kumar, N. (2011). Nanomaterials for regenerative medicine. *Nanomedicine, 6*(1), 157–181.

Welch, E. C., Powell, J. M., Clevinger, T. B., Fairman, A. E., & Shukla, A. (2021). Advances in biosensors and diagnostic technologies using nanostructures and nanomaterials. *Advanced Functional Materials, 31*(44), 2104126.

Zhang, Q., Uchaker, E., Candelaria, S. L., & Cao, G. (2013). Nanomaterials for energy conversion and storage. *Chemical Society Reviews, 42*(7), 3127–3171.

Part III
The Future of Sustainable Nanomaterials

Chapter 8
Carbon Nanotubes: Strengthening the Future

P. Saranraj, Shrikaant Kulkarni, K. Gayathri, B. Lokeshwari, and E. Tharani

8.1 Introduction

Carbon nanoparticles (CNTs) have become materials in the technology due to their unique mechanical, electrical, and thermal properties combinations. These cylindrical nanostructures are carbon atoms arranged in a hexagonally packed pattern as essentially rolled-up graphene sheets. It is a single layer of carbon atoms arranged in a honeycomb configuration. This unique structure also CNTs with exceptional strength, elasticity, and electrical conductivity, making them ideal candidates for various applications. The tubes are made up of reinforced composite material with tensile strengths several times that of steel. CNT is a promising candidate for electric and thermal conduction (Tiwari et al., 2016).

CNTs, thermal conductivity as high as diamond—that carbon-carbon bonds, are among the strongest known in materials science. This property makes them useful for electronics because heating is a critical problem for energy storage and conversion systems. CNTs can have various electronic properties depending on the configuration, from metallic conductors to semiconductors. Such tuning of CNT microstructure, synthesis method, and macro-phenomena features boosts CNTs in many state-of-the-art electronics applications like transistors, sensors, and flexible electronics (Sadegh & Shahryari Ghoshekandi, 2015). An important factor is that many electronic devices that save energy and have high electrical and thermal

P. Saranraj (✉) · K. Gayathri · B. Lokeshwari · E. Tharani
PG and Research Department of Microbiology, Sacred Heart College (Autonomous), Tirupattur, Tamil Nadu, India

S. Kulkarni
Faculty of Engineering and Technology, Sanjivani University, Kopargaon, Maharashtra, India

S. Kulkarni (ed.), *Next-Generation Nanomaterials for Sustainable Engineering*, Lecture Notes in Nanoscale Science and Technology 43,
https://doi.org/10.1007/978-3-032-05719-8_8

conductivity combined with unique molecular structures are made CNTs potentials in supercapacitors and batteries. CNTs have been utilised in pharmaceutical applications, diagnostic imaging, and tissue engineering for medical purposes. CNTs can serve as scaffolds for tissue engineering and cell and tissue growth that can be game changers in regenerative medicine (Rauti et al., 2019).

There are many intrinsic and extrinsic challenges with the widespread applications of CNTs. We provide a mechanism for the production of high-quality CNTs in large quantities. The production methods of carbon nanotubes (CNTs), such as chemical vapour deposition (CVD), arc discharge, and laser ablation, a very intricate process that is heavily reliant on managing temperature and pressure, as well as the chemical environment in which the CNTs are created, to develop the numerous characteristics of the CNTs (Anzar et al., 2020). CNTs have found many applications, and their purity can be essential to their application properties. The turmoil of synthetic methods is considerable for the safe handling and environmental implications of CNTs. In addition, CNTs are toxic to the environment and humans, especially in inhalation, and are as harmful to humans as lead, leading to respiratory diseases and other diseases. Such strategies aim to improve the quality of CNTs for mass production and the environmental and health hazards associated with the applications of CNTs. The CNTs are expected to be essential components for many advanced technologies, from next-generation electronics and renewable energy systems to medical devices and structural materials. Some fields include aerospace, automotive, electronics, and medicine (Dhinesh Babu et al., 2020).

8.2 Synthesis of Carbon Nanotubes

The production of CNTs is the most significant approach for the advanced processing and applications of CNTs, and it has a substantial role in controlling their structural, electric, mechanical, and thermal characteristics. Carbon nanotubes, which are composed of carbon molecules in the configuration of hexagons, are synthesised in several ways that decompose carbon materials in controlled situations. Chemical vapour deposition (CVD), arc discharge, and laser ablation are three popular CNT growth methods, all of which have pros and cons in yield, purity, and scale (Radhamani et al., 2018). The primary technique used for CNT fabrication is chemical vapour deposition (CVD), as this method can generate high-quality CNTs under relatively low-cost and more controlled environments. In the chemical vapour deposition (CVD) method, a hydrocarbon gas such as methane, ethylene, or acetylene is introduced to a catalyst reaction chamber (Kumar & Ando, 2010).

Gas decomposes at the needed temperature of 600 °C–1200 °C, and carbon atoms are deposited to form nanotube catalysts. In carbon vapour deposition (CVD), carbon species are diffused on the catalyst particles, which can generate single-walled carbon nanotubes (SWCNTs) or multi-walled carbon nanotubes (MWCNTs) depending on the process circumstances (Nessim, 2010). CVD has many advantages over other CNT synthesis methods, like forming aligned, spaced, and sized CNTs that allow exclusive CNTs to be manufactured. On the contrary, catalyst deactivation, residual additives, and post-synthesis purification resolution lower its efficiency. One of the first methods developed for synthesising CNTs is the arc discharge method, which is based on conducting a high-temperature plasma arc between two graphite electrodes in an inert gas atmosphere (Hou et al., 2022).

Amorphous carbon and fullerenes also are produced and must be purified. Synthesis techniques such as plasma-enhanced CVD, flame, and electrochemical deposition are being utilised to improve the performance quality, yield, and functionalisation of CNTs (Bell et al., 2006). As a result of this outcome, plasma-enhanced CVD lays the foundation for the growth of CNTs based on the substrate supporting low temperature, thus making the synthesis more advantageous on temperature-sensitive materials, polymers, and flexible substrates. CNTs have various approaches for separating them post-CNT synthesis that are frequently very inefficient. Impurities from amorphous carbon, residual catalyst particles, and defects in CNT structure harm properties that may be detrimental (Hofmann et al., 2003).

8.3 Mechanisms of Carbon Nanotubes

The mechanisms of carbon nanotubes (CNTs) underpinned their special properties. Carbon nanotubes are unique cylindrical nanostructures consisting of sp^2-hybridised carbon atoms arranged in a hexagonal lattice that exhibits mechanical and electrical, surface, and thermal properties (Wei et al., 2015). Quantum effects and atomic-scale interaction strength combined with nanoscale electrical conductivity and thermal stability. The hexagonally structured carbon-carbon bonds in the CNTs are responsible for the dominant mechanical properties shown by the remaining, such as two-piece strength and ductility. CNTs have a much higher tensile strength than steel for a much lower density of material (Oncel & Yurum, 2006).

The CNTs regulate their geographical characteristics, and it is well known that they can either be metallic or semiconducting, depending on their scalability. Graphene sheets used in the formation of CNT have a band structure influenced by the electronic density of CNT, showing less resistance and scattering. CNTs have a bandgap that can be opened in transistor applications, paving the way for high-performance, low-power electronic devices. Because of their superb thermal conductivity, CNTs make ideal candidates for microelectronics thermal management (Yang et al., 2022). CNT-polymer composite materials can also achieve increased thermal stability, which is necessary for space exploration applications and high-temperature utilisation. CNTs would rather agglomerate due to their superhydrophobicity to increase surface modification procedures. Due to their exceptional surface sensitivity and electrical conductivity, CNT-based devices were utilised to detect minute amounts of gases, biomolecules, and environmental parameters (Zhang et al., 2022).

CNTs are used in applications such as high-resolution displays, vacuum electronic devices, and energy-efficient lighting systems. Despite their incredible properties, challenges remain for the scale-up, integration, and commercialisation of CNTs. Their growth, structural growth, and defect formation mechanisms are yet to concur fully, and accurate adjustments made, as response regulation of CNT chirality and length is necessary for their practical applications (Liu et al., 2017). The CNT's interactions with biological systems are essential for the application of these nanomaterials in the medical and environmental fields. The CNT mechanisms were the future for the growth and diversity of nanotechnology in their innovations in several industries (He et al., 2019). Common methods for synthesising carbon nanotubes, including arc discharge, electrolysis, sonochemical, and hydrothermal processes, chemical vapour deposition, and solid-state/liquid pyrolysis, are illustrated in Fig. 8.1.

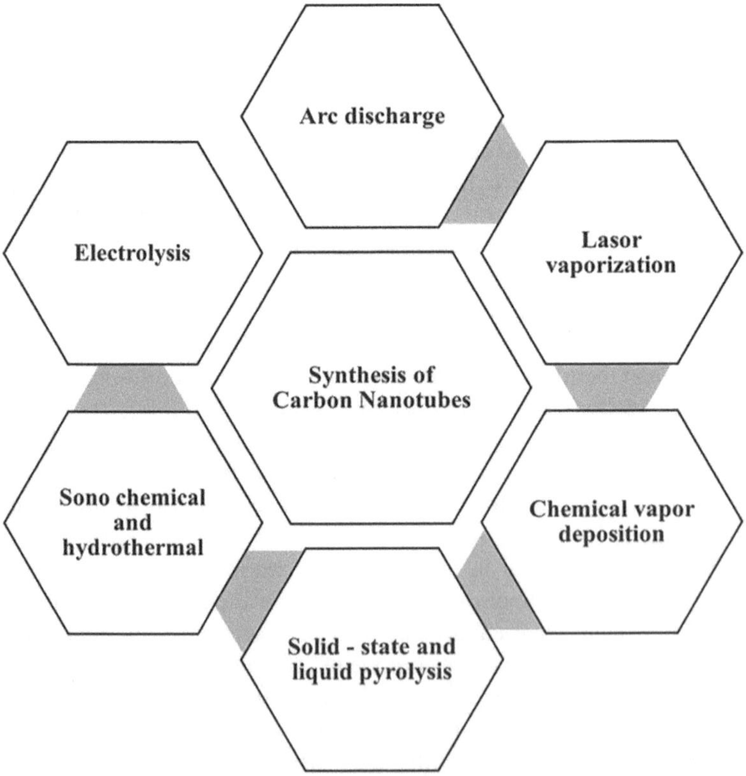

Fig. 8.1 Structural and functional mechanisms of carbon nanotubes

8.3.1 Load Transfer Mechanism

CNTs have excellent tensile strength and potential to strengthen polymer-, metal-, and ceramic-matrix composites: CNTs and surrounding matrix through interfacial interactions and nanotube dispersion. The dominant load transfer mechanism in CNT-reinforced composites is different and is attributed to van der Waals forces, mechanical interlocking, and chemical bonding at the CNT-polymer interface. CNTs have smooth graphitic surfaces that have low van der Waals interaction with the matrix stemming from inefficient load transfer and uneven stress transfer to CNTs. Covalent functionalisation refers to the process of attaching chemical groups to the CNT surface, which results in the formation of strong chemical bonds with the matrix, improving the load transfer efficiency (Chiang & Sankaran, 2009).

Non-covalent functionalisation involves the stacking interactions, surfactant adsorption, or polymer wrapping of CNTs for their maintenance during functionality and is used to improve the interfacial interactions. Again, the dispersion of the CNTs can change the strength of the composite and help decide the efficiency of load transfer (Sánchez Bastardo et al., 2021). Homogeneous dispersion allows for more uniform stress distribution and stops the agglomeration of CNT, which can bring about high concentrations and weak zones in the material. Such structure of CNTs, as shown, has a considerable impact on load transfer efficiency and allows uniform load transfer along their axial direction. CNTs could happen by lowering or increasing the mechanical properties, which rely on the configuration and fracture tough surrealism. The advanced aerospace, automotive, and structural applications of CNT-based composite materials and strain range stress optimisation (Korányi et al., 2022).

8.3.2 Grain Refinement or Texture Hardening

CNTs act as nucleating particles for grain refinement in metallic or ceramic materials, and a mixture of CNTs and other metals influences grain shape in the microstructural growth of the final products. Dislocations are defects in the inner network of the crystal structure of the material that cause plastic deformation in material with fine grains, thus contributing towards their properties. This material is characterized by a high aspect ratio and a substantial specific surface area, coupled with superior mechanical performance (Sun et al., 2023). CNTs demonstrate heterogeneous nucleation sites in the solidification or sintering process and give rise to the formation of refined grains. MMNCs stand for metal matrix nanocomposites, and ceramic composites incorporate a second complementary phase to achieve greater hardness, wear resistance, and tensile strength. This type and size of the second phase contribute to the grain refinement effects. The texture hardening is caused by the anisotropic mechanical behaviours induced by the arrangement and confinement of grains by CNTs (Wang et al., 2019).

A well-bonded interface ensures efficient load transfer from the matrix to CNTs, resulting in porosity and reduced grain refinement efficiency. Spark plasma sintering, mechanical alloying, and severe plastic deformation are some of the different processing techniques to optimise grain refinement and texture hardening in the CNT/metal matrix composite (Kondoh et al., 2009). This facilitates controlling microstructural features towards strengthening mechanisms without sacrificing material integrity using such advanced processing methods. The mechanical properties of CNT-based composites are enhanced through grain refinement and texture hardening. These mechanisms improve strength by restricting grain size, blocking dislocation motion, and providing anisotropic strengthening. These mechanisms are responsible for the yield strength, hardness, and wear resistance of CNT-reinforced composites, which are the characteristics that have led to their use in advanced engineering applications (Zhou et al., 2016).

8.3.3 Second-Phase Hardening

The second-phase hardening is a dominant mechanism in CNT-reinforced materials, which can enhance the interface between the CNT and the matrix. The composite material is of two phases: the material of the primary phase and the CNTs, which are embedded in the matrix of primary materials for property enhancement. Such an increase can be ascribed to second-phase hardening processes, which allow CNTs to provide barriers to dislocation motion, stress dispersion, and load transfer (Chen et al., 2015). This second dispersed phase of CNT will contribute to the enhanced toughness and hardness with complex dislocation, grain boundaries, and microstructural interaction in the corresponding CNT-reinforced matrix composites with metal, ceramic, or polymer matrices. Load transfer strengthening from the softer matrix is effectively transferred to the stiffer, stronger CNTs, increasing the overall integrity of the composite. The hardening occurring in the second stage is related to the dispersion of the composition, bonding between the interface and the quantity of the CNTs in the matrix (Boesl et al., 2014).

Different functionalisation methods, such as covalent and non-covalent interactions and surface modifications, improve CNT-matrix interfacial adhesion and promote transfer, allowing maximum strengthening effect. The described hardening stage enhances the wear resistance of CNT-reinforced materials and their thermal stability. Incorporating CNTs as a second phase reduces friction and prevents overloss of material, leading to better resistance during mechanical loading (Choi et al., 2011). Thermal performance optimisation of CNTs in the heat applications is mainly due to retardation of coarsening of grain. Dislocation hardening is responsible for the increased mechanical properties in CNT-reinforced materials. The Combat dispersal, interfacial strength, and engineering still functioning in the direction of these are effective next-generation composite materials with high strength, high hardness, long-term performance characteristics, and broad application level in aeronautical and automotive and civil engineering in the CNT nanostructured materials (Fan et al., 2017).

8.3.4 Strain Hardening

Strain hardening or work hardening during deformation are strengthened mechanisms that improve mechanical properties in CNT-reinforced materials through plastic deformation and raising dislocation density. In classical materials, work hardening is mainly produced through dislocation motion at the interactions with other dislocations and grain boundaries, and phase particles with increased deformation are produced (Akbarpour & Pouresmaeil, 2018). CNT mediations enhance a dislocation mechanism, and several dislocation densities are present, which bring out strain hardening phenomena. The interface interaction between CNTs and polymer matrix is enhanced by CNTs dispersion and alignment in the matrix. The uniform distribution allows CNTs to function as a highly effective barrier along with the entire fracture cross-sectional area. At the same time, the strong bond between the matrix and CNTs achieves more suitable stress transfer and dislocation interaction (Yoon et al., 2013).

Grain refinements directly ensure better mechanical properties through reduced dislocation motion, as the CNTs are likely to pin the grains, restricting grain growth. This provides better mechanical properties up to yield strain hardening and grain refinement, which ensures UMS in metal matrix nanocomposites (Yoo et al., 2013) such strain-induced texture evolution by CNTs facilitate anisotropic mechanical properties for directional strength. The strain hardening in CNT-based composites is utilised in various processing techniques, including high-pressure torsion, mechanical alloying, and severe plastic deformation (SPD). In short, they facilitate the dispersion of CNTs and their microstructures and improve interfacial bonding to develop better reinforcement efficiency. Since CNT-reinforced composites inhibit crack propagation and delay failure from cyclic loading, they have improved fatigue resistance. The effect of the strain hardening mechanism on the mechanical properties of products reinforced by CNT is highly employed for their mechanical properties with increasing dislocation density, the suppression of plastic deformation and microstructure strengthening (Suárez et al., 2014).

8.4 Functionalization of Carbon Nanotubes

CNT functionalisation is vital for enhancing interfacing with the proposed matrices, such as drug delivery, sensors, energy storage, and environmental remediation matrices, by increasing the compatibility, dispersion, and energy of interactions between the CNTs and matrix. Functionalisation involves modifying CNT surfaces chemically or physically. To enhance their solubility, processability, and material compatibility with different materials (Patil et al., 2021). Covalent functionalisation refers to adding functional groups onto the CNT surface, enhancing its reactivity and compatibility with the materials. It includes oxidation, amination, or esterification processes that introduce carboxyl (–COOH), hydroxyl (–OH), or amine ($-NH_2$) functional groups to CNT walls, enhancing their solvability in aqueous and organic solutions. At the same time, the covalent modification may destroy the intrinsic π-conjugated structure of CNTs, which may change the electrical and mechanical properties. Functionalisation strategy and application in biological and sensor applications and structural integrity of CNTs. CNTs and polymers during composite formation had better mechanical performance, effective force transfer, and improved thermal and electrical conductance (Nam et al., 2012).

Functionalised CNTs form polymer nanocomposites in the polymer matrix, and CNTs interact with each other as sieving structures, increasing the distribution of stress, resisting fracture, and improving durability. CNTs and their composites have been investigated as promising reinforcements for improved grain refinement, dislocation pinning and second-phase strengthening mechanisms (Sharma et al., 2020). CNTs, when functionalised, have also been developed as carriers for drug delivery, biosensing, and tissue engineering for biomedicine. They can target diseased cells, enhance therapeutic efficiencies, and reduce side effects. Functionalized CNTs are essential for energy storage applications because they enhance ion transport, stabilise electrodes, and increase ionic or electrical capacities to develop high-performance supercapacitors and lithium-ion batteries. In environmental remediation, functionalised CNTs are efficient catalysts and adsorbents for removing heavy metals, organic pollutants, and greenhouse gases from the water and air. The development of CNT functionalization strategies remains crucial as they enter the future of nanomaterials technologies (Kwon et al., 2010).

8.5 Biomedical Applications of Carbon Nanotubes

Carbon nanotubes (CNTs) have become nanomaterials for various biomedical uses due to their mechanical, electrical, and chemical properties. The material's high surface area and biocompatibility make it suitable for diverse biomedical applications, including drug delivery, biosensing, medical imaging, tissue engineering, and cancer therapy. Some CNTs in biomedicine include drug delivery systems, they act as nanocarriers to transport therapeutic molecules to specific cells or tissues (Chen, Hu, & Smith, 2014). They can efficiently penetrate cell membranes, and the surface can be modified to bind drug molecules and precise drug transport with excellent targeting and side effects. Functionalised CNTs can also be designed to deliver anticancer drugs, antibiotics, or gene therapy agents to target sites in the body, increasing their solubility to minimise toxicity and maximise treatment response.

CNTs have been widely used for anticancer therapy, including photothermal and photodynamic therapy (Guo et al., 2018).

This localised thermal ablation has the advantage of sparing adjacent healthy tissues, indicating that CNT-based therapy is a possible novel strategy in oncology. CNTs had has also been used in biosensors for the rapid and ultra-sensitive detection of biomolecules, including glucose, proteins, and DNA. Their high electrical conductivity and ease of functionalisation with specific recognition elements allow the real-time monitoring of biological processes to realise an early diagnosis and precision medicine (Sun & Chen, 2009). CNTs have been functionalised and integrated into biomaterials for enhanced cell adhesion, proliferation, and differentiation, improving scaffold performance in regenerative medicine. Functionalised CNTs can be conjugated with imaging probes, enhancing imaging of biological structures for disease detection and monitoring. CNTs also help to prevent bacterial biofilm formation and bacterial infections by disrupting the bacterial membranes and making them suitable for antimicrobial applications. The translation of CNT-based biomedical applications requires biocompatibility, toxicity, and long-term effects (Choi et al., 2010). Carbon nanotubes exhibit diverse biomedical applications, including drug delivery, cancer treatment, biosensors, bio-imaging, and tissue engineering, as illustrated in Fig. 8.2.

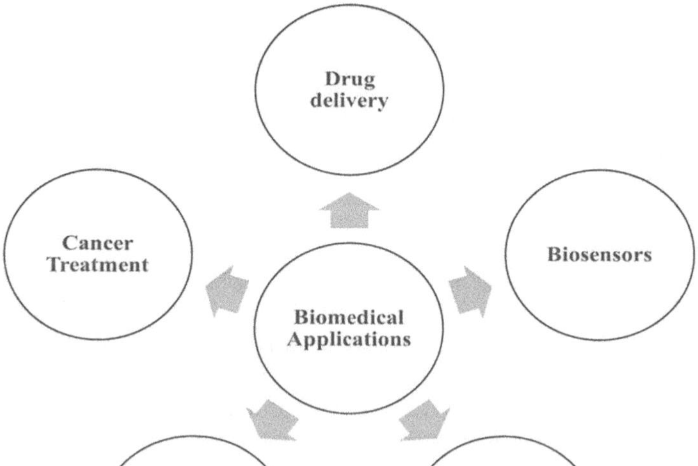

Fig. 8.2 Biomedical applications of carbon nanotubes

8.5.1 Drug Delivery

Due to their biomedical application, CNTs are used for pharmaceutical drug delivery and treating diseases like cancer, neurological diseases, and infectious disorders. Due to their unique physicochemical properties, such as high surface area, excellent mechanical properties, high aspect ratio, and good biocompatibility, CNTs are ideal drug molecule carriers. Functionalisation is crucial for drug delivery in CNTs to be biocompatible, non-toxic, and stable in the physiological state (Merchan Merchan et al., 2010). CNTs, as such, are functionalised with drug molecules, peptides, nucleic acids, or targeting ligands on CNT surfaces via the covalent and non-covalent functionalisation methods, their solubility and bioavailability. CNTs have the most extensive use in drug delivery as cancer therapies. Chemotherapeutic agents were encapsulated in the CNTs, and direct delivery to the tumour cells reduced systemic toxicity and increased treatment efficiency (Sajid et al., 2016).

The ability of CNTs to strongly absorb NIR lights and generate local heat that kills tumour cells has already been utilised in photothermal and photodynamic therapy. CNTs have also been applied for gene delivery and as nanocarriers to transport DNA, RNA, or gene editing tools such as CRISPR/Cas9 into cells. They allow the efficient transport of genetic material across membranes, enabling specific genomic editing, thus becoming interesting targets for gene and genetic disease therapy (De Menezes et al., 2019). CNTs can treat neurological diseases stemming from their ability to breach the blood–brain barrier (BBB) and improve neuroprotective drug delivery into the brain for neurodegenerative conditions such as Alzheimer's and Parkinson's. CNT-based drug delivery systems would revolutionise nanomedicine, offering site-specific effective individual therapy and empowering the future of biomedical science (He & Swager, 2016).

8.5.2 Biosensors

High-sensitive, selective biosensors based on carbon nanotubes (CNTs) are used for the real-time detection of biomolecules, pathogens, and disease markers with very high accuracy. CNTs are one of the best transducers in the field of biosensing devices due to their electrical, mechanical, and structural characteristics, which improve their properties compared to conventional biosensors (Meng et al., 2009). The surface area-to-volume ratio, excellent electrical conductivity, and capacity for biomolecule functionalisation (enzymes, antibodies, nucleic acids, etc.) allow them to conduct effective signal transduction to improve sensitivity and reduce the detection limit in biosensing applications. CNT-based biosensors function by recognising biological processes and converting these signalling events into quantifiable electrical, optical, or electrochemical signals with applications for early disease diagnosis, personalised medicine and crime scene investigation. Their application in glucose biosensors for diabetes diagnosis, enabling enzymatic detection of glucose concentrations in blood or interstitial fluids at very sensitive levels, is one of their key roles in CNT-based biosensing technologies (Panchakarla & Govindaraj, 2008).

CNTs functionalised with glucose oxidase enzymes can enhance the electron transfer to the electrodes and provide an accurate quantification of glucose concentration to receive a successful diabetes diagnosis. Biosensors based on CNT have gained some importance in cancer diagnostics by identifying the tumour biomolecules, such as proteins, nucleic acids, and exosomes present in biological fluids, which help in the early diagnosis and prognosis of the patients (Umasankar & Chen, 2008). Their ability to detect low-level markers in physiologic fluids makes them a non-invasive and efficient diagnostic tool. The CNT biosensors can also be applied to screen viral or bacterial antibodies or antigens in virus detection. CNT-based sensors detect neuron neurotransmitters, e.g., dopamine and serotonin, and neuronal nervous tissue. CNT-based biosensors have the potential to revolutionise disease detection, health tracking, and personalised medicine, hence boosting future biomedical research and nanotechnology-based medical approaches (Packhaeuser et al., 2004).

8.5.3 Bio-Imaging

Due to their high sensitivity for the ultra-high-resolution visualisation of cellular structures, tissues, and organs at molecular levels in biomedical applications, carbon nanotubes (CNTs) offer the potential for bio-imaging (Hamman, 2010). Due to their relevant properties, the optical, electrical, and mechanical properties of CNTs make them a suitable candidate to enhance imaging technologies of fluorescent, MRI, CT, and photoacoustic imaging. Functionalisation of CNTs can be carried out using various biomolecules, dyes, and contrast agents for targeted imaging of specific cells, tissues, or disease biomarkers. The large surface area with high modification of respective functional groups accounts for the loading of relevant imaging agents (e.g., fluorophores, gadolinium-based compounds) that could enhance the imaging signals, such as by increasing resolvable contrast (Bianco et al., 2005). CNTs offer some specificity in targeting the corresponding biomarkers for various diseases, such as cancer. The proposed mechanism reflects the characteristic of utilising NIR-regional absorption and emission of photons by CNTs to achieve deep tissue penetration with minimal autofluorescence and light scattering (Akhter et al., 2018), making it suitable for fluorescence cellular imaging. This property takes full advantage of living bodies based on their ability to monitor cell activity, drug delivery, and disease processes.

In addition, CNT-based imaging agents are used in MRI as contrast enhancements, increasing the signal intensity and resolution. One possibility is the use of gadolinium-functionalized CNTs as paramagnetic contrast agents that considerably enhance the imaging of the target tissues, thus providing an excellent approach for MRI of neurological diseases and tumours and, therefore, also applicable to imaging of cardiovascular diseases. In addition to fluorescence and MRI, CNTs have shown massive potential in photoacoustic imaging. This hybrid imaging modality combines the spatial resolution from ultrasound with molecular specificity from optical imaging. Additionally, CNTs have strong absorption in the NIR range, which enables them to act as suitable contrast agents in photoacoustic imaging, making it possible to non-invasively visualise structures with high-resolution images of blood vessels, tumours, and other tissues without the use of ionising radiation. The other application of CNTs is in CT imaging, and their high atomic number enables considerable X-ray attenuation, which permits better tumour identification.

CNTs also have been playing more and more critical roles in the development of multimodal imaging; they can be functionalised for multiple imaging modalities like fluorescence, MRI, and photoacoustic imaging to provide complementary information that may provide a more integrated view of biological events and disease progression (Chen, Jia, et al., 2014). CNTs have significant functionalisation challenges to overcome to enable their clinical use regarding toxicity, biocompatibility, and long-term stability in biological systems. Nevertheless, more studies are warranted to counter such issues: the safety of CNTs, functionalisation approaches, and body clearance strategies must be improved. The success of bio-imaging will tremendously impact the further development of precision medicine and promote the refinements of modern biomedical imaging techniques with the unremitting development of CNT-based imaging techniques (Mirsalari et al., 2021).

8.5.4 Tissue Engineering

CNT-composite-based hydrogel is used to develop electroactive tissue engineering scaffolds. CNTs have been incorporated into the cells for tissue engineering applications due to their unique mechanical properties, electrical conductivity, and large surface area. The specific properties are close to the structural features of the extracellular matrix (ECM) and offer physical and biochemical signals, which are needed to sustain relevant cellular processes such as adhesion, migration, and proliferation (Yoosefian & Jahani, 2019). It is well-acknowledged that CNT-based scaffolds promote multifunctional tissue restoration through the induction of functional structures, including neural, bone, and cardiac tissues. As a neural tissue engineering approach, CNTs have shown great potential for promoting the growth and regeneration of neural tissue. These cells become electrically conductive, assist in transmitting signals between neurons, and play a critical role in repairing neuronal signalling after injury (Bhalla et al., 2016).

CNTs serve as a scaffold for neural cell attachment and proliferation and induce axonal outgrowth and synaptogenesis. The beneficial effects of CNTs on proliferation and mineralisation are vital in bone tissue engineering, which is involved in bone repair and regeneration. Additionally, CNTs can modify bonding with conductive hydrogels or 3D-printed scaffolds for heart patches to restore cardiac function and promote damage repair in the myocardium (Iguchi et al., 2007). CNTs show unique mechanical and electrical characteristics that can be used to pack bioactive molecules that promote tissue regeneration and speed healing, including growth factors, cytokines, and nucleic acids. There are still some difficulties concerning the biocompatibility, toxicity, and degradation of CNTs in these fields. Several features, such as functionalisation techniques, biosystems of scaffold formations, and the need for more toxicity assessments, would shape the CNT's footprint in tissue engineering. CNT-based scaffolds are an innovative and versatile approach for creating tissue substitutes with more optimal characteristics to induce tissue regeneration and the application of regenerative medicine, implementing more efficient therapeutic strategies to address injuries and degenerative illnesses (Timur et al., 2007).

8.5.5 Cancer Treatment

One important application is that CNTs have been utilised for active targeting delivery of anticancer drugs, genes, and other therapeutic molecules to the tumour cell to improve the therapeutic effect and reduce the side effects of cancer treatment. As a drug carrier for chemotherapeutic agents, CNTs are also widely used in cancer therapy. CNTs can be combined with targeting ligands, such as antibodies or peptides, which can bind to the cancer cell receptors. CNTs can transport various anticancer agents in the form of small molecules, nucleotides, and proteins and are, therefore, a versatile delivery platform for personalised cancer therapy (Singh et al., 2016). CNTs are also being applied in photothermal therapy for cancer treatment. CNTs quickly take up near-infrared (NIR) light and convert the light energy to heat, resulting in selective heating and destruction of the tumour cells when irradiated with NIR light. CNTs have high light-absorption characteristics and are amenable to functionalisation by targeting tumour-specific molecules, which makes them usable for treatment in this manner. CNTS PDT is a procedure that uses reactive oxygen species ROBS to ablate the tumour cells (Sireesha et al., 2018).

CNTs could additionally be used as a nanocarrier of therapeutics in combination therapies that permit a simultaneous delivery of chemotherapeutic drugs and also improve the influence of photothermal or PDT in an improved anti-tumour reaction (Sánchez Bastardo et al., 2021). Localised treatments have also been used as a contrast agent in diagnostic techniques for cancer imaging, such as magnetic resonance imaging (MRI), computed tomography (CT), and photoacoustic imaging. In addition, CNTs can be functionalised with imaging agents, allowing the real-time tracking of tumour progression and treatment efficacy monitoring. It also improves accurate cancer diagnosis, aids in early detection, and improves patient outcomes. Although CNTs have great promise for cancer therapy in the clinics, their application still faces many challenges, such as toxicity and long-term biocompatibility. CNTs promote their functionalization, which minimises potential toxicity while retaining their therapeutic potential (Yun et al., 2007).

8.6 Future Perspectives

With the unique properties of CNTs, which can be utilised in many fields, especially in biomedical and nanotechnologies studies, suggesting new focus areas such as optimising CNT synthesis and functionalisation and addressing scalability and implementation barriers, the potentials of CNTs are unprecedented. CNTs can have applications in drug delivery, cancer therapy, tissue engineering, and diagnostics. These nanoparticles may be functionalised with specific ligands or drugs to create targeted and personalised therapies with minimum side effects and maximum therapeutic efficacy (Kamil Reza et al., 2014). CNTs can be combined with other medicines, including gene therapy and immunotherapy; therefore, they are potent vectors for treating complex disorders (currently cancer, neurodegenerative disorders and genetic disorders). CNT-based biosensors and imaging agents possess great potential for early disease detection at the molecular level and as scaffolds with highly functional properties to facilitate the regeneration of damaged tissues and organs (Goutam et al., 2011; Zhang et al., 2011). CNT-based scaffolds produce more sophisticated biological structures, including but not limited to complex tissue like neural, cardiac, and bone. CNTs are expected to change the traditional processes for fabricating lightweight, efficient devices. In the newly emerging field of quantum computing, CNTs could serve as the basis for such an architecture, with their unique properties, such as their potential for electron manipulation and nanoscale transportation of electricity, leading to inherently faster and more efficient computers (Ghrera et al., 2018). Due to the high surface area of these materials, they can efficiently adsorb pollution, which can lead to environmental cleanup and sustainability, thus providing a solution to global issues such as pollution and water scarcity (Savalia & Chatterjee, 2017).

8.7 Conclusion

Carbon nanotubes (CNTs) represent a groundbreaking advancement in materials science, offering unparalleled strength, conductivity, and versatility. Their remarkable mechanical properties—exceptional tensile strength and lightweight structure make them ideal for reinforcing composites in aerospace, automotive, and construction industries. Their electrical and thermal conductivity also opens new possibilities in electronics, energy storage, and medical technologies. Despite challenges in large-scale production and cost-effectiveness, ongoing research and innovation continue to optimise their applications. As scientists and engineers unlock their full potential, carbon nanotubes are poised to revolutionise multiple industries, driving progress towards a stronger, more sustainable future. By overcoming current limitations, CNTs could redefine material performance, paving the way for next-generation technologies that enhance durability, efficiency, and environmental sustainability. The future, strengthened by carbon nanotubes, holds immense promise ushering in an era of smarter, lighter, and more resilient materials.

References

Akbarpour, M. R., & Pouresmaeil, A. (2018). The influence of CNTs on the microstructure and strength of Al-CNT composites produced by flake powder metallurgy and hot-pressing method. *Diamond and Related Materials, 88*, 6–11.

Akhter, M. H., Rizwanullah, M., Ahmad, J., Ahsan, M. J., Mujtaba, M. A., & Amin, S. (2018). Nanocarriers in advanced drug targeting: Setting novel paradigm in cancer therapeutics. *Artificial Cells, Nanomedicine, and Biotechnology, 46*(5), 873–884.

Anzar, N., Hasan, R., Tyagi, M., Yadav, N., & Narang, J. (2020). Carbon nanotube - A review on synthesis, properties and plethora of applications in biomedical science. *Sensors International, 1*, 100003.

Bell, M. S., Teo, K. B. K., Lacerda, R. G., Milne, W. I., Hash, D. B., & Meyyappan, M. (2006). Carbon nanotubes by plasma-enhanced chemical vapour deposition. *Pure and Applied Chemistry, 78*(6), 1117–1125.

Bhalla, N., Jolly, P., Formisano, N., & Estrela, P. (2016). Introduction to biosensors. *Essays in Biochemistry, 60*(1), 1–8.

Bianco, A., Kostarelos, K., & Prato, M. (2005). Applications of carbon nanotubes in drug delivery. *Current Opinion in Chemical Biology, 9*(6), 674–679.

Boesl, B., Lahiri, D., Behdad, S., & Agarwal, A. (2014). Direct observation of carbon nanotube induced strengthening in aluminum composite *via* tensile tests. *Carbon, 69*, 79–85.

Chen, B., Jia, L., Li, S., Imai, H., Takahashi, M., & Kondoh, K. (2014). In situ synthesized Al_4C_3 Nanorods with excellent strengthening effect in aluminum matrix composites. *Advanced Engineering Materials, 26*, 972–975.

Chen, B., Li, S., Imai, H., Umeda, J., Takahashi, M., & Kondoh, K. (2015). Inter-Wall bridging induced peeling of multi-walled carbon nanotubes during tensile failure in aluminum matrix composites. *Micron, 69*, 1–5.

Chen, S., Hu, S., & Smith, E. F. (2014). Aqueous cationic, anionic and non-ionic multi-walled carbon nanotubes, functionalised with minimal framework damage, for biomedical application. *Biomaterials, 35*(17), 4729–4738.

Chiang, W. H., & Sankaran, R. M. (2009). Linking catalyst composition to chirality distributions of as-grown single-walled carbon nanotubes by tuning Ni(x) Fe(1-x) nanoparticles. *Nature Materials, 8*(11), 882–886.

Choi, H. J., Shin, J. H., & Bae, D. H. (2011). Grain size effect on the strengthening behavior of aluminum-based composites containing multi-walled carbon nanotubes. *Composites Science and Technology, 71*, 1699–1705.

Choi, H. J., Shin, J. H., Min, B. H., & Bae, D. H. (2010). Deformation behavior of Al-Si alloy based nanocomposites reinforced with carbon nanotubes. *Composites Part A: Applied Science and Manufacturing, 41*, 327–329.

De Menezes, B. R. C., Rodrigues, K. F., Fonseca, B. C. D. S., Ribas, R. G., Montanheiro, T. L. D. A., & Thim, G. P. (2019). Recent advances in using carbon nanotubes as smart biomaterials. *Journal of Materials Chemistry B, 7*(9), 1343–1360.

Dhinesh Babu, N. R., Mahadevi, N., & Assein, D. (2020). Electronic applications of multi-walled carbon nanotubes in polymers. A short review. *Materials Today Proceedings, 33*, 382–386.

Fan, G., Huang, H., Tan, Z., Xiong, D., Guo, Q., Naito, M., Li, Z., & Zhang, D. (2017). Grain refinement and superplastic behavior of carbon nanotube reinforced aluminum alloy composite processed by cold rolling. *Materials Science and Engineering, 708*, 537–543.

Ghrera, A. S., Pandey, C. M., & Malhotra, B. D. (2018). Multiwalled carbon nanotube modified microfluidic-based biosensor chip for nucleic acid detection. *Sensors & Actuators, B: Chemical, 266*, 329–336.

Goutam, P. J., Singh, D. K., Giri, P. K., & Iyer, P. K. (2011). Enhancing the photostability of poly(3-hexylthiophene) by preparing composites with multiwalled carbon nanotubes. *The Journal of Physical Chemistry. B, 115*(5), 919–924.

Guo, B., Chen, B., Zhang, X., Cen, X., Wang, X., Song, M., Ni, S., Yi, J., Shen, T., & Du, Y. (2018). Exploring the size effects of Al_4C_3 on the mechanical properties and thermal behaviors of Al-based composites reinforced by SiC and carbon nanotubes. *Carbon, 135*, 224–235.
Hamman, J. H. (2010). Chitosan-based polyelectrolyte complexes as potential carrier materials in drug delivery systems. *Marine Drugs, 8*(4), 1305–1322.
He, M., & Swager, T. M. (2016). Covalent functionalisation of carbon nanomaterials with iodonium salts. *Chemistry of Materials, 28*(23), 8542–8549.
He, M., Zhang, S., Wu, Q., Xue, H., Xin, B., Wang, D., & Zhang, J. (2019). Designing catalysts for chirality-selective synthesis of single-walled carbon nanotubes: Past success future opport. *Advanced Materials, 31*(9), 1800805.
Hofmann, S., Ducati, C., Robertson, J., & Kleinsorge, B. (2003). Low-temperature growth of carbon nanotubes by plasma-enhanced chemical vapour deposition. *Applied Physics Letters, 83*, 135–137.
Hou, P. X., Zhang, F., Zhang, L., Liu, C., & Cheng, M. (2022). Synthesis of carbon nanotubes by floating catalyst chemical vapor deposition and their applications. *Advanced Functional Materials, 32*(2), 108541.
Iguchi, S., Kudo, H., & Saito, T. (2007). A flexible and wearable biosensor for tear glucose measurement. *Biomedical Microdevices, 9*(4), 603–609.
Kamil Reza, K., Srivastava, S., Yadav, S. K., & Biradar, A. M. (2014). Biofunctionalized carbon nanotubes platform for biomedical applications. *Materials Letters, 126*, 126–130.
Kondoh, K., Threrujirapapong, T., Imai, H., Umeda, J., & Fugetsu, B. (2009). Characteristics of powder metallurgy pure titanium matrix composite reinforced with multi-wall carbon nanotubes. *Composites Science and Technology, 69*, 1077–1081.
Korányi, T. I., Németh, M., Beck, A., & Horváth, A. (2022). Recent advances in methane pyrolysis: Turquoise hydrogen with solid carbon production. *Energies, 15*(17), 6342.
Kumar, M., & Ando, Y. (2010). Chemical vapour deposition of carbon nanotubes: A review on growth mechanism and mass production. *Journal of Nanoscience and Nanotechnology, 10*(6), 3739–3758.
Kwon, H., Park, D. H., Silvain, J. F., & Kawasaki, A. I. (2010). Investigation of carbon nanotube reinforced aluminum matrix composite materials. *Composites Science and Technology, 70*, 546–550.
Liu, B., Wu, F., Gui, H., Zheng, M., & Zhou, C. (2017). Chirality controlled synthesis and applications of single-wall carbon nanotubes. *ACS Nano, 11*(1), 31–53.
Meng, L., Fu, C., & Lu, Q. (2009). Advanced technology for functionalising carbon nanotubes. *Progress in Natural Science, 19*(7), 801–810.
Merchan Merchan, W., Saveliev, A. V., Kennedy, L., & Jimenez, W. C. (2010). Combustion synthesis of carbon nanotubes and related nanostructures. *Progress in Energy and Combustion Science, 36*(6), 696–727.
Mirsalari, H., Maleki, A., Raissi, H., & Soltanabadi, A. (2021). Investigation of the pristine and functionalised carbon nanotubes as a delivery system for the anticancer drug dacarbazine: Drug encapsulation. *Journal of Pharmaceutical Sciences, 110*(5), 2005–2016.
Nam, H. D., Cha, S. I., Lim, B. K., Park, H. M., Han, D. S., & Hong, S. H. (2012). Synergistic strengthening by load transfer mechanism and grain refinement of CNT/Al-cu composites. *Carbon, 50*, 2417–2423.
Nessim, G. D. (2010). Properties, synthesis, and growth mechanisms of carbon nanotubes with special focus on thermal chemical vapour deposition. *Nanoscale, 2*(8), 1306–1323.
Oncel, C., & Yurum, Y. (2006). Carbon nanotube synthesis *via* the catalytic CVD method: A review on the effect of reaction parameters. *Fullerenes, Nanotubes, and Carbon Nanostructures, 14*, 17–37.
Packhaeuser, C. B., Schnieders, J., Oster, C. G., & Kissel, T. (2004). *In situ* forming parenteral drug delivery systems: An overview. *European Journal of Pharmaceutics and Biopharmaceutics, 58*(2), 445–455.

Panchakarla, L. S., & Govindaraj, A. (2008). Covalent and non-covalent functionalisation and solubilisation of double-walled carbon nanotubes in non-polar and aqueous media. *Journal of Chemical Sciences, 120*(6), 607–611.

Patil, A., Nartu, M. Y., Ozdemiz, F., Banerjee, R., Gupta, R. K., & Borkar, T. (2021). Strengthening effects of multi-walled carbon nanotubes reinforced nickel matrix nanocomposites. *Journal of Alloys and Compounds, 876*, 159981.

Radhamani, A. V., Lau, H. C., & Ramakrishna, S. (2018). CNT-reinforced metal and steel nanocomposites: A comprehensive assessment of progress and future directions. *Composites. Part A, Applied Science and Manufacturing, 114*, 170–187.

Rauti, R., Musto, M., Bosi, S., Prato, M., & Ballerini, L. (2019). Properties and behaviour of carbon nanomaterials when interfacing neuronal cells. *Carbon, 143*, 430–446.

Sadegh, H., & Shahryari Ghoshekandi, R. (2015). Functionalization of carbon nanotubes and its application in nanomedicine. *Nano, 2*(4), 231–248.

Sajid, M. I., Jamshaid, U., Jamshaid, T., Zafar, N., Fessi, H., & Elaissari, A. (2016). Carbon nanotubes from synthesis to *in vivo* biomedical applications. *International Journal of Pharmaceutics, 501*(1–2), 278–299.

Sánchez Bastardo, N., Schlögl, R., & Ruland, H. (2021). Methane pyrolysis for zero-emission hydrogen production: A potential bridge technology from fossil fuels to a renewable and sustainable hydrogen economy. *Industrial and Engineering Chemistry Research, 60*(32), 11855–11881.

Savalia, R., & Chatterjee, S. (2017). Sensitive detection of brucine an anti-metastatic drug for hepatocellular carcinoma at carbon nanotubes - Nafion composite based biosensor. *Biosensors & Bioelectronics, 98*, 371–377.

Sharma, A., Fujii, H., & Paul, J. (2020). Influence of reinforcement incorporation approach on mechanical and Tribological properties of AA6061-CNT nanocomposite fabricated *via* FSP. *Journal of Manufacturing Processes, 59*, 604–620.

Singh, A., Choudhary, M., Kaur, S., Singh, S. P., & Arora, K. (2016). Molecular functionalisation of carbon nanomaterials for immuno-diagnosis of cancer. *Materials Today Proceedings, 3*, 157–161.

Sireesha, M., Jagadeesh Babu, V., Kranthi Kiran, A. S., & Ramakrishna, S. (2018). A review on carbon nanotubes in biosensor devices and their applications in medicine. *Nano, 4*(2), 36–57.

Suárez, S., Ramos-Moore, E., Lechthaler, B., & Mucklich, F. (2014). Grain growth analysis of multiwalled carbon nanotube-reinforced bulk nickel composites. *Carbon, 70*, 173–178.

Sun, E., Zhai, S., Kim, D., Gigantino, M., Haribal, V., Dewey, O. S., Williams, S. M., Wan, G., Nelson, A., & Marin-Quiros, S. (2023). A semi-continuous process for co-production of CO_2 - Free hydrogen and carbon nanotubes *via* methane pyrolysis. *International Journal of Nanotechnology, 4*(4), 101338.

Sun, Y., & Chen, Q. (2009). Diameter dependent strength of carbon nanotube reinforced composite. *Applied Physics Letters, 95*, 021901.

Timur, S., Anik, U., Odaci, D., & Gorton, L. (2007). Development of a microbial biosensor based on carbon nanotube (CNT) modified electrodes. *Electrochemistry Communications, 9*(7), 1810–1815.

Tiwari, S. K., Kumar, V., Huczko, A., Oraon, R., de Adhikari, A., & Nayak, G. C. (2016). Magical allotropes of carbon: Prospects and applications. *Critical Reviews in Solid State and Materials Sciences, 41*(4), 257–317.

Umasankar, Y., & Chen, S. M. (2008). Recent trends in the application of carbon nanotubes-polymer composite modified electrodes for biosensors: A review. *Analytical Letters, 41*(2), 210–243.

Wang, I. W., Kutteri, D. A., Gao, B., Tian, H., & Hu, J. (2019). Methane pyrolysis for carbon nanotubes and cox - Free H_2 over transition metal catalysts. *Energy & Fuels, 33*(1), 197–205.

Wei, D., Peng, L., Li, M., Mao, H., Niu, T., Han, C., Chen, W., & Wee, A. T. (2015). Low-temperature critical growth of high-quality nitrogen-doped graphene on dielectrics by plasma-enhanced chemical vapour deposition. *ACS Nano, 9*(1), 164–171.

Yang, F., Zhao, H., Li, R., Liu, Q., Zhang, X., Bai, X., Wang, R., & Li, Y. (2022). Growth modes of single-walled carbon nanotubes on catalysts. *Science Advances, 8*(41), 794.

Yoo, S. J., Han, S. H., & Kim, W. J. (2013). A combination of Ball milling and high-ration differential speed rolling for synthesizing carbon/copper composites. *Carbon, 61*, 487–500.

Yoon, E. Y., Lee, D. J., Park, B., Akbarpour, M., Farvizi, M., & Kim, H. S. (2013). Grain refinement and tensile strength of carbon nanotube reinforced cu matrix nanocomposites processed by high-pressure torsion. *Metals and Materials International, 19*, 927.

Yoosefian, M., & Jahani, M. (2019). A molecular study on drug delivery system based on carbon nanotube for the novel norepinephrine prodrug Droxidopa. *Journal of Molecular Liquids, 284*, 258–264.

Yun, Y., Dong, Z., Shanov, V., Heineman, W. R., Halsall, H. B., Bhattacharya, A., Conforti, L., Narayan, R. K., Ball, W. S., & Schulz, M. J. (2007). Nanotube electrodes and biosensors. *Nano Today, 2*(6), 30–37.

Zhang, H., Jiang, H., & Sun, F. (2011). Rapid diagnosis of multidrug resistance in cancer by electrochemical sensor based on carbon nanotubes-drug supramolecular nanocomposites. *Biosensors & Bioelectronics, 26*(7), 3361–3366.

Zhang, X., Qian, L., Yao, X., Zhang, L., Wu, Q., Li, D., Ma, C., Zhao, N., Xin, L., & Liu, C. (2022). Solid supported ruthenium catalyst for growing single-walled carbon nanotubes with narrow chirality distribution. *Carbon, 193*, 35–41.

Zhou, W., Yamamoto, G., Fan, Y., Kwon, H., Hashida, T., & Kawasaki, A. (2016). *In situ* characterization of interfacial shear strength in multi-walled carbon nanotube reinforced aluminum matrix composites. *Carbon, 106*, 37–47.

Chapter 9
Environmental Impacts and Safety of Nanomaterials

John Elisa Kumar, Munmee Goswami, Pooja Paul, and Tsungom Mulai

9.1 Introduction

Nanomaterials (NMs) have found an essential place in the field of medicine, wastewater treatment, pharmaceutical, and many more. NMs have been used since the introduction of the archaic synthetic pigment "Egyptian blue" and can be dated back to 4500 years. Michael Faraday was the first to study nanoparticles (NPs). However, it was not until Richard P. Feynman stepped in and explained the concept of "nanotechnology" during his seminal talk, "There's Plenty of Room at the Bottom" (Martinez et al., 2020). Moving forward, nanotechnology has gained immense interest in the emerging fields of agriculture, electronics, medicine, and many more, leading to greater manufacture of NPs. They are defined as materials having a size between 1 and 100 nm in any of the three dimension (Mekuye & Abera, 2023). In the past few years, NMs have served as an important component in reducing global climate change, global warming owing to their smaller size than their bulk counterparts (Apoorva et al., 2023; Chausali et al., 2023; Subramanian et al., 2020). One of the prominent examples of NMs use in the field of medicine can be seen in diagnosing and controlling the global epidemic Covid-19 (Ayan et al., 2023; Iqbal et al., 2022; Papi et al., 2023; Patel et al., 2022; Skwarek et al., 2021). Its use in the field of electronics is the most progressive since everybody uses electronic gadgets such as mobiles, laptops, fridges, smart watches, television, and many more, NMs, for example, graphene can be used in biosensors, photovoltaic devices (Mhetre et al., 2021), and so on. NMs in electronics enhance the devices

J. E. Kumar
Department of Chemical Engineering, Institute of Chemical Technology Mumbai, Mumbai, India

M. Goswami · P. Paul · T. Mulai (✉)
Department of Chemistry, North Eastern Hill University, Shillong, Meghalaya, India

S. Kulkarni (ed.), *Next-Generation Nanomaterials for Sustainable Engineering*, Lecture Notes in Nanoscale Science and Technology 43,
https://doi.org/10.1007/978-3-032-05719-8_9

density, enlarge memory chip capacity, and reduce size, weight, and power consumption. In the field of cosmetics, NMs like, titanium oxide (TiO_2), zinc oxide (ZnO) can be seen using as UV-filters (ultraviolet) (Gupta et al., 2022), silver, and fullerenes for anti-aging and brightening (Fytianos et al., 2020; Raj et al., 2012). NMs are also utilized to treat organic pollutants or industrial waste specially dyes run-off (Kumar et al., 2021). NMs are also used in the food processing industry due to competition between different industries to produce genuine, expedient, fresh, and flavorful food products (Ameta et al., 2020). Metallic NMs such as Ag, Cu, Fe, etc. have been incorporated in food packaging due to their antibacterial and antifungal properties. Carbon NMs such as graphene-based is used for prolonging the shelf life of food products. Organic NMs such as chitosan, agar, starch, gelatine, and many more, are known to have been used to preserve food (Ozcakir, 2023). Metal NPs such as Cu, Au, Ti, Zn, and Ag are emerging NPs used in the food packing industry (Onyeaka et al., 2022).

Despite its vast and positive impact in various fields of technology, they are also credited with some disadvantages (Soliman, 2023). For example, in medicine, it is known to cause cytotoxic effect such as mesothelioma by carbon nanotubes, gold NPs of 1.4 nm display toxicity (Wolfram et al., 2015). It is also possible for NPs from food packing to enter the cell membrane, affecting human organs and leading to deposition in the kidney, brain, liver, and testicles, causing inflammatory and other immune responses such as neurotoxicity, genotoxicity etc. (Ashfaq et al., 2022; Naseer et al., 2018). NMs present in the environment agglomerates leading to the change in their properties and cause negative impact inside and outside the environment, such as affecting the biogeochemical cycles (Ray et al., 2009). In this chapter, the issues relating with the environmental impact of NMs and their safety will be addressed. Along with that, their positive and negative effects in the fields of food, textile, medicine, electronics, etc., will be addressed. This chapter aims to explore the potential benefits and drawbacks of nanoparticles (NMs) on human health and ecosystems. In addition, it will examine strategies for mitigating their negative effects.

9.2 Properties and Preparation of Nanoparticles

NPs show numerous applications in different industries, such as in the environment (bioremediation, sensors, catalysts, diagnostics, tissue engineering, antimicrobials), medicine (drug delivery, diagnostics, tissue engineering, antimicrobials), agriculture pesticides and herbicides; fertilizers and plant growth; food safety; water purification), food (food processing; preservation; packaging; food fortification; sensors), defense (sensors; protective coatings; weapons; manufacturing; energy storage), electronics and automotive (display technologies/storage devices; data storage; chemical processing/catalysis; Separation and purification) industries due to its specific properties: (1) small size; (2) high surface area; (3) surface chemistry;

(4) biocompatibility; (5) target ability; (6) stability; and (7) tunable optical properties (Fig. 9.1) (Altammar, 2023; Tiple et al., 2020).

Nanoparticles are synthesized using two approaches, that is, top-down and bottom-up. The different methods used for nanoparticle synthesis are discussed briefly below (Fig. 9.2) (Abid et al., 2022).

9.2.1 Top-Down

The top-down method works well for long-range ordered structures because it breaks down bulk materials into their atomic components.

9.2.1.1 Sputtering

Sputtering is when high-energy plasma or gas particles, usually Ar^+ ions, assault a solid object, causing microparticles to be expelled off its surface. The films produced have compositions comparable to the target material and include fewer impurities. The ejected atoms are deposited on a substrate, creating a homogenous coating. Metals, semiconductors, and compounds (such as oxides and nitrides) can all be included in the wide variety of materials that can be sputtered using different techniques, such as magnetron, RF diode, and DC diode sputtering (Altammar, 2023; Baig et al., 2021; Khan et al., 2022; Mekuye & Abera, 2023).

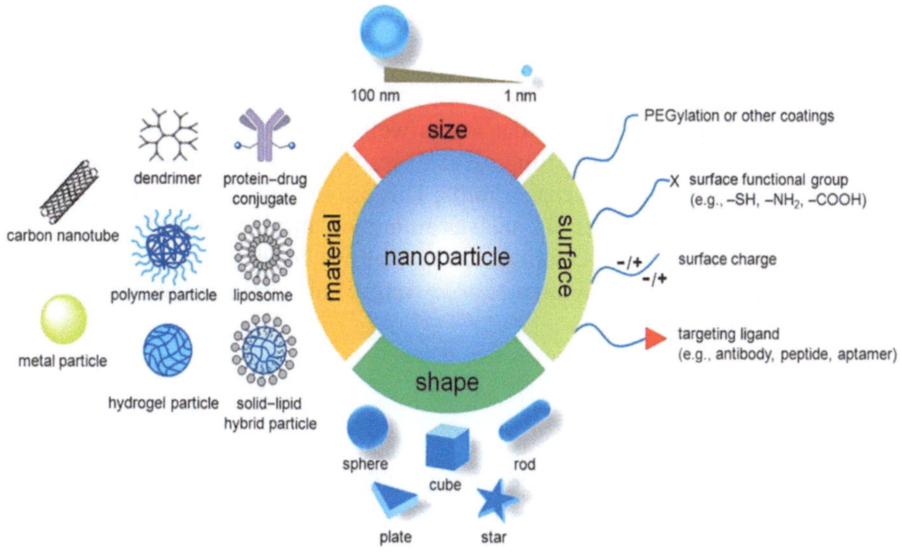

Fig. 9.1 Properties and applications of NPs (Reprint with permission from Tiple et al. (2020) Copyright 2020 Elsevier)

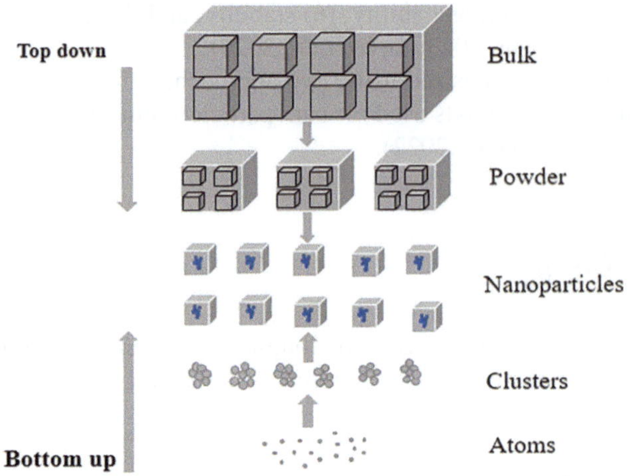

Fig. 9.2 The top-down approach and the bottom-up approach (Reprint with permission from Abid et al. (2022) Copyright 2022 Elsevier)

9.2.1.2 Electron Explosion

This method involves exposing a thin metal wire to a high-current pulse, which causes the metal to ionize, evaporate, and explode. The metal is vaporized and ionized during this process, and when it interacts with the surrounding gas or liquid media, it expands and cools. Ultimately, the condensed vapor causes nanoparticles to develop (Altammar, 2023).

9.2.1.3 Mechanical Milling

A high-energy impact technique that works well for creating nanomaterials from bulk materials is mechanical milling, which uses balls inside containers in various mills, including planetary and shaker mills. This process makes it possible to create multiple nanocomposite materials, such as aluminum alloys enhanced by oxide and carbide, wear-resistant spray coatings, and nanoalloys made of copper, nickel, magnesium, and aluminum (Altammar, 2023; Baig et al., 2021; Khan et al., 2022; Mekuye & Abera, 2023).

9.2.1.4 Electrospinning

Electrospinning is one simple top-down method for producing nanofibers from various materials, especially polymers. This process creates fibers with widths in the nanoscale range by extracting charged threads from polymer melts or solutions. Coaxial electrospinning, which uses a spinneret composed of two coaxial capillaries, is a noteworthy development in this area. By using two viscous liquids—one as the shell and a non-viscous liquid as the core—this arrangement enables the creation of core-shell nanoarchitecture (Altammar, 2023; Baig et al., 2021).

9.2.1.5 Laser Ablation

By using a strong laser beam to evaporate a target material, laser ablation synthesis is an effective way to produce noble metal nanoparticles without the need for stabilizing agents and in an environmentally friendly manner. Numerous nanomaterials, such as metals, carbon nanomaterials, oxides, and ceramics, can be produced using this method. This procedure is improved by pulsed laser ablation in liquids (LASiS), which yields monodisperse colloidal nanoparticle solutions devoid of surfactants and permits control over the characteristics of the nanoparticles (Altammar, 2023; Baig et al., 2021; Khan et al., 2022; Mekuye & Abera, 2023).

9.2.1.6 Sonication

Sonication is the most essential stage in creating nanofluids. Sonication is done using a mechanical homogenizer, ultrasonic path, or ultrasonic vibrator after the mixture has been stirred magnetically. This technique is quite effective for processing a variety of nanomaterials, such as metal oxides, graphene, carbon nanotubes, and inks (Altammar, 2023).

9.2.1.7 Lithography

Lithography is essential for producing nanoparticles with concentrated light or electron beams. It comes in two primary varieties: maskless lithography, which is more economical because it permits arbitrary patterning without masks, and masked lithography, which transfers nanopatterns using templates (e.g., photolithography, nanoimprint lithography). Using various methods, nanolithography makes it easier to fabricate 2D metal arrays with regulated dimensions precisely (Altammar, 2023; Baig et al., 2021; Khan et al., 2022; Mekuye & Abera, 2023).

9.2.1.8 Thermal Decomposition Method

Thermal breakdown is an endothermic technique that uses heat to disrupt chemical bonds in metals to produce stable, monodispersed nanoparticles. This approach tackles the difficulty of accurately regulating the size and form of metal nanoparticles and is feasible for their production. Thermal breakdown provides a straightforward method without requiring raw materials, but its adaptability is limited (Mekuye & Abera, 2023).

9.2.1.9 The Arc Discharge Method

Many nanostructured materials, especially carbon-based ones including fullerenes, carbon nanohorns (CNHs), carbon nanotubes, few-layer graphene (FLG), and amorphous spherical carbon nanoparticles, can be produced using the arc discharge process. This process is significant for producing nanostructures of fullerene. Two graphite rods are submerged in a chamber of pure helium during the procedure to avoid oxygen or moisture, which could prevent the synthesis of fullerene. These nanoparticles are made more accessible by the carbon being vaporized by the arc discharge between the rods. The arc discharge technique is essential for developing carbon nanotechnology (Baig et al., 2021; Khan et al., 2022; Mekuye & Abera, 2023).

9.2.2 Bottom-Up

Bottom-up procedures assemble small atoms and molecules to create nanostructured particles.

9.2.2.1 Chemical Vapor Deposition (CVD)

By reacting vapor-phase precursors, the technique known as chemical vapor deposition (CVD) produces thin coatings on substrate surfaces. High volatility, stability, purity, affordability, non-hazardous nature, and the absence of pollutants upon breakdown are all characteristics of suitable precursors. High-quality, uniform nanoparticles may be produced via CVD, which is especially useful for carbon-based materials like graphene and carbon nanotubes (Altammar, 2023; Baig et al., 2021).

9.2.2.2 Sol-Gel Process

The sol-gel approach is a flexible wet chemical process frequently employed to create superior metal-oxide nanostructures. It entails hydrolyzing liquid precursors—usually metal alkoxides—to create a sol, then condensing it to make a gel. Increasing viscosity creates a porous structure that ages and changes porosity and other characteristics. The gel is aged, then dried and calcinated to create nanoparticles. Low processing temperatures, environmental friendliness, and the capacity to produce intricate nanostructures with consistent quality are just a few benefits of this technique. The final product is greatly influenced by variables such as the type of precursor, rate of hydrolysis, aging period, pH, and water-to-precursor ratio (Altammar, 2023; Baig et al., 2021; Khan et al., 2022).

9.2.2.3 Co-precipitation

Nanoparticles are produced via a wet chemical process called the solvent displacement technique. It uses non-solvent polymers, which can be either natural or manufactured, and solvents, including ethanol, acetone, and hexane. Nanoparticles are created when a polymer solution is combined because of the quick diffusion between the non-solvent phase and the polymer-solvent (Altammar, 2023).

9.2.2.4 Inert Gas Condensation/Molecular Condensation

This method can produce tiny nanoparticles by vaporizing a metallic source in an inert gas atmosphere. The metal gradually evaporates at a regulated temperature, and when the evaporated atoms come out, they quickly lose energy through interaction with the inert gas. The gases are cooled using liquid nitrogen, which promotes the development of nanoparticles with sizes ranging from 2 to 100 nm (Altammar, 2023; Baig et al., 2021).

9.2.2.5 Hydrothermal and Solvothermal

A popular technique for creating nanoparticles is hydrothermal synthesis, which can be carried out at temperatures ranging from room to very. Solvothermal techniques use a non-aqueous environment, the procedure entails a heterogeneous reaction in an aqueous media. These methods, which are usually used in closed systems, have drawn interest because they may produce a variety of nanostructures, such as nanowires, nanorods, nanosheets, and nanospheres (Altammar, 2023; Baig et al., 2021; Mekuye & Abera, 2023).

9.2.2.6 Soft and Hard Templating Methods

Both hard and soft template techniques are frequently used to create nanoporous materials. Using various templates, such as block copolymers, flexible organic molecules, and surfactants, which interact via hydrogen bonds and van der Waals forces, the soft template method is simple. Modifying surfactant characteristics enables various morphologies and customizable pore diameters. In contrast, the hard template method—also referred to as nano-casting—creates nanostructures by filling solid templates with precursor molecules. It is important to choose a template, fill it with a precursor, and then remove the template to expose the finished mesoporous structure. Both techniques allow creating a wide range of special nanomaterials, such as metal oxides, nanowires, and nanorods (Baig et al., 2021; Mekuye & Abera, 2023).

9.2.2.7 Reverse Micelle Method

The reverse micelle method is one efficient way to create nanomaterials with particular sizes and shapes. In this process, water-in-oil emulsions create reverse micelles, with the hydrophilic heads pointing toward a water-containing core. This core facilitates the creation of nanoparticles by functioning as a nanoreactor. The size of these nanoreactors—and hence the nanoparticles—can be regulated by varying the water-to-surfactant ratio; smaller droplets and nanoparticles result from a lower water concentration (Baig et al., 2021; Mekuye & Abera, 2023).

9.2.2.8 Spinning Method

Nanoparticle synthesis using the spinning method is conducted in a spinning disc reactor (SDR), which features a rotating disc within a controlled chamber. Key parameters, such as temperature, disc surface, liquid-to-precursor ratio, rotation speed, and flow rate, influence the process. Liquid reagents are directed at the center of the rapidly rotating disc (300 to 3000 rpm), creating a thin film that enhances heat and mass transfer (Khan et al., 2022; Mekuye & Abera, 2023).

9.2.3 Biological Synthesis

Biological nanoparticle synthesis uses bioactive substances like microorganisms, plant materials, and biowastes (such as fruit and vegetable peels, eggshells, and algae). Simpleness, affordability, and scalability for large-scale production are some advantages of this approach. Additionally, it produces low waste and high-stability nanoparticles with non-toxic byproducts. To prevent dangerous byproducts, it is essential to develop dependable, sustainable green synthesis techniques (Altammar, 2023).

9.3 Application of Nanomaterials

9.3.1 Nanomaterials in Water, Soil, and Air

Nanomaterials play a crucial role in detecting and remediating pollutants in water, soil, and air because of their unique properties, such as high surface area, reactivity, and the ability to interact with a wide range of substances at the molecular level. Nanomaterials can improve water quality by enhancing the removal of pollutants, including heavy metals and organic compounds, via adsorption or catalysis. For soil applications, they aid in the cleanup of contaminated sites by enabling the targeted delivery of remediation agents or by enhancing soil fertility through improved nutrient retention and microbial activity. In air quality management, nanomaterials are used in sensors that can detect toxic gases at low concentrations, allowing for real-time monitoring and swift responses to pollution events. Their small size and tunable properties allow for tailored designs that optimize their effectiveness in various environmental contexts, making nanomaterials invaluable for advancing environmental protection and public health initiatives. In Table 9.1, we will explore various current applications of different nanomaterial's in water, soil, and air.

9.3.2 Nanomaterials in Medicine

Nanotechnology plays a transformative role in medicine and pharmaceuticals by allowing the development of nanoparticles ranging in size from 1 to 100 nm. These nanoscale materials, including lipid-based nanoparticles, carbon nanotubes, dendrimers, and quantum dots, have distinct features that enhance their application in drug delivery, diagnostics, and therapeutics (Khan et al., 2018). NPs can improve targeted drug delivery, reduce side effects, and enable early disease detection. Nanomedicine's interdisciplinary approach integrates chemistry, biology, and engineering, resulting in innovative solutions that address difficult medical challenges and pave the way for personalized medicine and improved patient care (Sharma et al., 2018). NMs enhance cancer detection, improve cell targeting, and reduce off-target effects in biomedical applications. This accuracy enables more tailored medicines, such as multidrug approaches, that successfully target cancer cells while reducing adverse effects. These improvements help to make oncology treatments more effective and individualized (Begines et al., 2020; Vodyashkin et al., 2021, 2024). Fabricating NMs to enhance aggregation at specific sites makes achieving targeted responsiveness for drug release while minimizing undesirable effects possible. Fine-tuning the synthesis of these specialized NMs can greatly improve the efficacy of precise pharmacological therapies, broadening access for more patients and expediting the path to desired therapeutic outcomes. This method promises to advance the precision and efficacy of cancer treatments (Mitchell et al., 2021). Nanotechnology effectively overcomes biological obstacles that inhibit diagnosis and treatment, allowing for focused interventions in damaged parts of the body. This accuracy improves treatment effectiveness while limiting the impacts on healthy tissues, resulting in better patient outcomes (Pope-Harman et al., 2007). In Table 9.2, we will explore various current applications of different nanomaterials in medicine.

9.3.3 Nanomaterial in Electronics

Nanomaterials in electronics are driving significant technological innovation, enabling the design of components with enhanced electrical, mechanical, and thermal properties. By manipulating materials at the nanoscale, engineers and scientists

Table 9.1 Application of different nanomaterial's in water, soil, and air

Nanoparticle	Size	Method	Application	References
Kraft lignin@ Ni@g-C_3N_4	–	The nanocomposite was synthesized and then drop-cast onto a graphite electrode	Detection of cypermethrin pesticide residues in drinking water	Razzaque et al. (2024)
MOF @ starch@ cotton nanocomposite	~30 nm	Conventional solvothermal MOF synthesis method	Removal of fluorinated herbicide from aqueous solutions	Ghosh et al. (2024)
Cu-SG NCs	~0.7 nm	One-pot chemical reduction technique	Detection of p-nitrophenol	Oyebanji et al. (2024)
GDH/Gox-CQD-loaded chitosan-based thin-film biosensors	~2–5 nm	Synthesis of CQDs via hydrothermal process, followed by enzyme loading onto CQDs for biosensor fabrication	Detection of aluminum and copper ions in water samples using fiber-optic biosensors	Vyas et al. (2023)
CuO nanoparticles	~10–100 nm	Co-precipitation, hydrothermal synthesis	Suppression of fusarium in soil, modulation of soil microbiome, or improving soil health	Li, Guo, et al. (2024)
BIS@ZIF-8	~6.40 nm	In situ fabrication process of ZIF-8 by one-pot method	Structural stability and adhered to weeds through electrostatic interactions during rainy conditions	Zhang et al. (2025)
Iron nanoparticles	~20–30 nm	Synthesis via chemical reduction method	Foliar application on maize plants to enhance cadmium stabilization in wastewater-irrigated soil and promote growth	Razzaq et al. (2024)
Ferric oxide nanomaterials	~10–50 nm	Hydrothermal synthesis	Applied in soil, role in cogenerating iron plaque with plant-rhizobacteria symbionts to remove chlorinated contaminants in dryland soils	Zheng et al. (2024)
Nano-MoS_2	~20–100 nm	Exfoliation from bulk MoS_2 or hydrothermal synthesis	Applied in a soil-plant system to enhance soybean growth and improve soil properties through interaction with plant roots	Li et al. (2024)
Lanthanum silicate nanomaterials	~10–100 nm	Sol-gel synthesis	Sheath blight resistance in rice and soil health improvement via interaction with microbes and enhancement of plant growth	Cao et al. (2023)

(continued)

Table 9.1 (continued)

Nanoparticle	Size	Method	Application	References
nHAP, platelet-shaped NMs	~10–100 nm	Wet chemical synthesis	Fertigation to japonica rice (Taipei 309)	Priyam et al. (2022)
Iron and zinc nanomaterials	~10–50 nm	Chemical precipitation method	Applied to soil as a source of bioavailable iron and zinc, enhancing their availability as micronutrient fertilizers to improve plant uptake and growth	Chandrika et al. (2022)
Carbonaceous nanomaterials	~1–100 nm	Solvothermal method	Applied to soybean crops to improve nitrogen availability in the soil and to maintain nitrogen stoichiometry, offset impaired dinitrogen fixation in nitrogen-deficient soils, helping plants assimilate soil nitrogen	Wang et al. (2019)
g-C3N4, and r-GO-modified filters	~1–50 nm	Sol-gel process	Applied in air purification filters to remove and detect particulate matter (PM) from ambient air	Attia et al. (2024)

create faster, smaller, and more energy-efficient devices, ranging from transistors and sensors to screens and batteries. As our world becomes increasingly interconnected, nanomaterials play a crucial role in enhancing the performance and functionality of electronic devices. They also pave the way for advancements in flexible electronics, quantum computing, and sustainable energy solutions, setting the stage for the next generation of innovative products.

9.3.3.1 Application of Gold Nanoparticles

Gold nanoparticles (AuNPs) are becoming increasingly popular in electronics due to their unique electrical, optical, and chemical capabilities. They are efficient conductors in various applications, including as sensors, conductive inks, and electrical circuit components. Their nanoscale dimension enables high surface area-to-volume ratios, which improves performance in devices like transistors and capacitors. AuNPs can also increase the efficiency of solar cells and data storage capabilities, making them useful in the growth of modern electronic technology. In Table 9.3, we will explore various current applications of gold nanoparticles in electronics.

Table 9.2 Application of different nanomaterial's in medicine

Nanoparticle	Size	Method	Application	References
Gold, magnetic and silica-based nanoparticles	~10–200 nm	Solvent evaporation, sol-gel process, co-precipitation, microfluidic methods, emulsion polymerization, laser ablation, microemulsions	Drug delivery, gene delivery, diagnostics, imaging etc.	Singh (2024)
Mesoporous polydopamine nanoparticles	–	–	Synergistic treatment of the tumors, photodynamic therapy, gene therapy.	Yin et al. (2024)
Silver nanoparticles	~10 nm	Plant-mediated synthesis of AgNPs followed by incorporation into alginate hydrogels	Sustained-release antibacterial wound healing hydrogel	Dai et al. (2024)
Gold-based conjugated microporous polymers	–	Synthesis of gold-coordinated porphyrin conjugated microporous polymers (au-CMP), followed by PEGylation to form nanoparticles	Combined photodynamic therapy (PDT) and radiotherapy (RT) for cancer treatment	Pan et al. (2024)
CeO_2 nanoparticles	Typically 5–20 nm	Synthesis through wet chemical methods, functionalization with azacrown ligands, radiolabeling with radionuclides	Radiopharmaceuticals for imaging and targeted drug delivery.	Khabirova et al. (2024)
Lanthanide-based scintillator@ MOF nanocomposites	~10–50 nm	Core-shell synthesis: Lanthanide-based scintillator as core, MOF as shell, followed by controlled growth in a solvothermal method	X-ray-induced photodynamic therapy (PDT) for cancer treatment.	Zhang, Gao, et al. (2024)
Iron-doped mesoporous silica nanoparticles loaded with doxorubicin	~50–200 nm	Co-synthesis of mesoporous silica with iron doping, followed by DOX loading through pore filling, and controlled release under acidic conditions	Chemodynamic therapy (CDT) and chemotherapy for breast cancer	Tang et al. (2024)

(continued)

Table 9.2 (continued)

Nanoparticle	Size	Method	Application	References
Carbon dots	~1.74–9.78 nm	Carbonization of Rubia cordifolia L. at 350 °C for 1 h, followed by extraction with distilled water	Hemostatic bioactivity for promoting blood clotting and wound healing	Zhang et al. (2024)
Ag/ZnO nanoparticles	~281.7 nm	Eco-friendly synthesis using Hylocereus costaricensis stem extract, followed by microwave irradiation to form bimetallic Ag/ZnO nanoparticles	Antibacterial treatment, effective against *Staphylococcus aureus* and *Escherichia coli*	Xaviour et al. (2024)
Macrocyclic receptors	–	Synthesis of macrocyclic receptors through chemical or enzymatic methods for specific biomolecule recognition	Precision medicine (biosensing, targeted drug delivery, therapeutics)	Pan et al. (2023)
Self-adaptive nanomaterials	~10–100 nm	Co-precipitation method, sol-gel method	Targeted drug delivery in cancer therapy, enhancing treatment efficacy	Yang et al. (2022)
Selenium nanoparticles	~20–50 nm	Biogenic synthesis using plant extracts (e.g., from *Ocimum sanctum* leaves) via green synthesis techniques, often involving reduction by plant-derived compounds	Treatment of oxidative stress and inflammation in rheumatoid arthritis (RA) and associated complications	Rehman et al. (2021)
Berberine-based nanostructures	–	Self-assembly of berberine (BBR) and flavonoid glycosides into NPs and NFs via electrostatic and hydrophobic interactions	Antibacterial therapy	Li et al. (2019)
Ruthenium-based nanomaterials conjugated with l-cysteine	~5 nm	Simple wet chemical route under physiological conditions	Anticancer therapy	Ganguly et al. (2018)

Table 9.3 Application of gold nano particle in electronics

Size (nm)	Method	Application	References
~10–50	In situ growth with organosulfur compounds	Flexible electronics, including flexible screens and wearable electronics, improved conductivity, flexibility, high-performance materials	Zhou et al. (2024)
~5–50	Click sintering (tailoring via sintering process)	Sensors, electrical devices, advanced nanostructured materials with improved conductivity and stability	Urban et al. (2024)
~5–40	Inkjet printing of polymer nanocomposites with hybrid gold nanoparticles	Mechanically robust, inkjet-printable electronics with metal-like conductivity	Klos et al. (2024)
~5–20	Synthesis via chemical reduction with capping agents (e.g., citrate) to ensure stability and control size.	Humidity sensors for breath monitoring and smart irrigation systems	Ramesh et al. (2024)

Table 9.4 Application of silver nano particle in electronics

Size (nm)	Method	Application	References
~5–20	Electrochemical exfoliation of graphene, silver nanoparticle-assisted ink	Soft electronics, self-healing polymer composites, PVA-based coatings	Simsek et al. (2024)
~10–30	Inkjet printing of silver nanoparticle ink, followed by photonic sintering	Large-area inkjet-printed flexible hybrid electrodes with high-conductive polymer for flexible electronics	Kant et al. (2024)
~20–50	Preparation of ag nanoparticle pastes, followed by sintering	High-power electronic packaging using sintered silver nanoparticle pastes	Zhao et al. (2024)

9.3.3.2 Application of Silver Nanoparticles

AgNPs are commonly utilized in electronics due to their high conductivity and optical characteristics. They enable the development of conductive inks for printed electronics, hence improving flexible circuits and wearable gadgets. AgNPs enhance sensor sensitivity and are integrated into solar cells, capacitors, and transistors, making them indispensable in sophisticated electronic applications. In Table 9.4, we will explore various current applications of silver nanoparticles in electronics.

9.3.3.3 Application of Copper Nanoparticles

Copper nanoparticles (CuNPs) are gaining popularity in electronics due to their excellent electrical conductivity and cost-effectiveness. They are used as conductive inks for printed circuit boards and flexible electronics to substitute expensive metals

Table 9.5 Application of copper nanoparticles in electronics

Size	Method	Application	References
~10–50 nm	Dry printing of pure copper with high conductivity and adhesion	Flexible electronics with improved conductivity and adhesion properties	Ahmadi et al. (2024)
Nano: ~20–50 nm, micro: ~1–10 μm	Screen printing of copper nano- and microparticles onto paper	Electrodes with good electrical conductivity and adhesion for flexible and disposable electronics, sensors, and other electronic devices	Nyabadza et al. (2024)
~20–50 nm	Preparation of anti-oxidation copper-based conductive inks	Low porosity copper films for power electronics with enhanced oxidation resistance	Yi et al. (2024)
~10–50 nm	Green synthesis using flower extracts	Enhanced microelectronics packaging	Perumal et al. (2024)
~20–100 nm	Preparation of highly loaded CuO nanoparticle suspensions for LIFT-printing	Electronics printing by LIFT, enhancing printed device quality and efficiency	Dimitriou and Michailidis (2024)
~10–50 nm	InkJet printing of copper ink on heated paper substrate, followed by flashlight sintering	Electronics printed on paper substrates with enhanced conductivity	Choi et al. (2024)

such as silver. CuNPs improve the performance of sensors and solar cells by increasing efficiency and reaction time. Furthermore, their antibacterial qualities enable them to safeguard electronic components from microbial contamination, extending their use in the electronics industry. In Table 9.5, we will explore various current applications of copper nanoparticles in electronics.

9.3.3.4 Application of ZnO Nanoparticles

ZnO NPs are employed in electronics due to their semiconducting characteristics and large bandgap. They serve an important role in transparent conductive films used in touch screens, displays, and gas sensors. Furthermore, ZnO NPs improve the efficiency of solar cells and light-emitting devices, making them useful in advancing electrical technology. In the following sections, we will explore various current applications of zinc oxide nanoparticles in electronics, highlighting their impact on performance and functionality.

Esa et al. (2024) studied the compatibility and performance of electrohydrodynamic printing using zinc oxide inkjet ink. They evaluated ink formulation, printing conditions, and the printed structures' resulting electrical and morphological properties. Results showed that this technique effectively produced well-defined zinc oxide patterns with enhanced performance characteristics. Overall, the findings highlighted its potential for applications in electronics and optoelectronics. Yeh and Feng (2024) created a micromachined capacitance-sensitive device that detects glucose and uric acid using immobilized functional ZnO NPs. The study described the device's construction method and assessed its sensitivity and response time performance. The results showed that the device successfully identified the target analytes, demonstrating ZnO nanoparticles' potential for biosensing applications. The findings demonstrate the benefits of employing micromachined devices to accurately and efficiently monitor biochemical indicators.

9.3.3.5 Application of TiO_2 Nanoparticles

TiO_2 NPs are commonly employed in electronics due to their semiconducting characteristics and photocatalytic activity. They are critical in solar cells because they boost light absorption and energy conversion efficiency. TiO_2 NPs are used in sensors to detect gases and contaminants, providing great sensitivity and stability. They also improve the performance of dielectric materials and are utilized in antireflective coatings for displays, which contributes to the advancement of various electronic applications. In Table 9.6, we will explore various current applications of titanium dioxide nanoparticles in electronics.

9.3.4 Nanomaterial in Food

The application of nanomaterials in food technology is revolutionizing our understanding of nutrition, food safety, and preservation. Through the utilization of the distinct characteristics of materials at the nanoscale, scientists are creating inventive approaches to improve food quality and prolong its shelf life. Nanotechnology makes better packaging with spoiling detection capabilities, more efficient nutrient delivery through nanoparticles, and antimicrobial compounds that help prevent foodborne illnesses possible. An increasingly efficient and robust food system can be achieved through the application of nanomaterials, which show great potential

Table 9.6 Application of TiO_2 nanoparticles in electronics

Size (nm)	Method	Application	References
10–50	Laser ablation synthesis in solution (LASiS)	TiO_2 nanoparticle ink production for printed electronics	Duryea et al. (2024)
10–50	Carbon doping method for enhancing electronic properties	Enhancing electronic properties of TiO_2 for various electronic applications	Kurban et al. (2024)
~10–50	Sol-gel method	Influence of phase structure of TiO_2 nanoparticles in resistive switching devices	Tunhoo et al. (2024)

for improving food security and safety as the world's population continues to expand and the demand for sustainable food production rises.

9.3.4.1 Application of Silver Nanoparticles

Ag NPs have received a lot of interest in the food sector because of their remarkable antibacterial capabilities, which make them useful for food preservation and safety. Their capacity to limit bacterial growth increases food goods' shelf life while reducing foodborne diseases. As a result, Ag NPs are increasingly being used in packaging materials and food coatings, representing a promising alternative for improving food quality and safety. In Table 9.7, we will discuss the application of silver nanoparticle in food sector.

Table 9.7 Application of silver nanoparticles in food

Size (nm)	Method	Application	References
~10–50	Green synthesis using plant extracts (e.g., banana pseudo-stem)	Polylactic acid-cellulose silver nanoparticle-based nanocomposite films for food storage	Rejiniemon et al. (2024)
~10–50	Green synthesis using pomegranate waste extract	Antioxidant, anticancer, antiviral, antimicrobial activities, and food preservation	Alharbi et al. (2024)
~10–50 nm	Gamma-irradiation-based synthesis	Cross-linked gelatin–riboflavin-based film incorporated with essential oils and AgNPs for extending shelf life of meat	Sarmast et al. (2024)
~20–50 nm	Chemical reduction method (using sodium citrate as a reducing agent)	Visual quantification of salt in foods on paper analytical devices	Thipwimonmas et al. (2024)
~10–30 nm	Green synthesis using *Salvia tiliifolia* leaf extract	Colorimetric detection of Hg(II) in food and environmental samples	Mume et al. (2024)
~10–40 nm	Chemical reduction method with zeolite support	Chemical reduction method with zeolite support	Benhacine et al. (2024)
~15–50 nm	Chemical reduction method (using sodium citrate)	Electrochemical sensor for detection of carbamate pesticides in food	Yang and Cao (2024)
~10–50 nm	Green synthesis using carbamate starch and cellulose nanofibrils	Eco-friendly food packaging with antibacterial properties	Xia et al. (2024)
~15–40 nm	Green synthesis using chitosan biopolymers	Eco-friendly food packaging with antimicrobial properties	Chandran et al. (2024)

Table 9.8 Application of titanium dioxide nanoparticles in food

Size (nm)	Method	Application	References
~30–60	Deposition on nanofiber films along with cyclodextrin	Food packaging for ripening and preserving	Qin et al. (2024)
~10–40	Green synthesis using chitosan-based matrices	White bread preservation in smart bio-nanocomposite films	Devi et al. (2024)
~25–50	Sol-gel method	Improve the nutritional quality of strawberry under drought stress	Javan et al. (2024)
~20–70	Commercially available (no specific preparation method detailed)	Migration assessment from food packaging materials to food simulants	Bastardo-Fernández et al. (2024)
~30–50	Sol-gel method	Active biodegradable packaging films for food packaging applications	Medina-Jaramillo et al. (2024)
~15–30	Sol-gel method	Active food packaging in gelatin films enhanced with TiO_2-loaded reduced graphene oxide aerogel microsphere	Wang et al. (2024)
~10–30	Sol-gel method	Photocatalytic reusability and SERS signal amplification for food safety detection on cotton fabric	Ma et al. (2024)

9.3.4.2 Application of TiO_2 Nanoparticles

In order to improve humidity sensing, Madbouly et al. (2024) have investigated the use of artificial TiO_2 nanoparticles (~20–50 nm) and torrefied date stones in PVA/PVP nanocomposites. Results demonstrated that the addition of date stones increases the mechanical and thermal properties, whereas TiO_2 nanoparticles dramatically boost humidity sensitivity and response time. The potential of agricultural waste in producing innovative materials is demonstrated by these improved nanocomposites, which are positioned as useful materials for biomedical applications and smart food packaging. In the following sections, we will explore various current applications of titanium dioxide nanoparticles in food, highlighting their impact on safety and preservation. The other food applications of application of TiO_2 nanoparticles are listed in Table 9.8.

9.3.4.3 Application of ZnO Nanoparticles

ZnO NPs are gaining popularity in the food business due to their antibacterial qualities, which serve to limit the growth of bacteria and fungi in food items. They are used in food packaging materials to increase shelf life and prevent spoiling. In addition, ZnO NPs can be used to coat fruits and vegetables, protecting them from diseases and extending their freshness. Their potential as a food preservative emphasizes their importance in enhancing food quality and safety. In Table 9.9, we will explore various applications of zinc oxide nanoparticles in food.

Table 9.9 Application of zinc oxide nanoparticles in food

Size (nm)	Method	Application	References
~30–50	Sol-gel method	Antimicrobial food packaging films with chitosan/ZnO nanoparticles/lecithin-montmorillonite	Sun et al. (2025)
~20–40	Sol-gel method	Sustainable active food packaging with poly(lactic acid) hybrid nanocomposites	Songtipya et al. (2024)
~50–100	Precipitation method	Active food packaging films with Ɛ-Polylysine/ZnO nanoparticles for antimicrobial properties	Kalajahi et al. (2024)
~20–50	Sol-gel method	Active food packaging film with starch-chitosan-Taro mucilage for antibacterial properties and eco-toxicity assessment	Priyanka et al. (2024)
~50	Green preparation method (green synthesis)	Multifunctional chitosan-based food packaging films	Gao et al. (2024)
~25–50	Sol-gel method	Active food packaging film with antimicrobial properties	Zhang et al. (2024)
ZnO coating: ~10–20 nm (on MWCNTs)	Hydrothermal method	Sustainable food packaging to extend shelf life	Ge et al. (2024)
~25	Precipitation method	Functionalized biodegradable pectin/starch films for sustainable packaging	do Nascimento et al. (2024)

9.3.5 Nanomaterial in Textile

Nanomaterials in textiles are a significant advancement in fabric technology, combining science and innovation to improve performance and functionality. Researchers and producers can impart unique features to textiles by modifying materials at the nanoscale, such as enhanced durability, moisture-wicking capacities, and antibacterial effects. These improvements extend the life and comfort of common fabrics and create new opportunities in specialist applications ranging from sportswear to medical textiles. As the need for high-performance and multifunctional fabrics rises, the incorporation of nanotechnology into the textile industry promises to transform how we perceive and interact with the materials around us.

9.3.5.1 Application of Silver Nanoparticles

AgNPs have attracted substantial attention in the textile industry due to their remarkable antibacterial qualities, making them suitable for improving fabric performance and longevity. AgNPs unusual size and high surface area-to-volume ratio allow them to successfully inhibit the growth of bacteria, fungi, and other pathogens, making treated textiles appropriate for various applications such as medical textiles, sportswear, and home furnishings. Recent advances in nanotechnology have resulted

Table 9.10 Application of silver nanoparticles in textile

Size (nm)	Method	Application	References
~10–50	In situ synthesis (reduction method)	Antimicrobial cotton fabric via k-carrageenan coatings	Dumas et al. (2024)
~10–40	Chemical reduction method with polyol process	Antimicrobial polypropylene yarns for textile and fabric applications	Demirhan et al. (2024)
~20–50	Chemical reduction method (using silver nitrate) with polycaprolactone coating	Hygienic textiles for healthcare industries, antimicrobial fabrics	Shilpa et al. (2024)
~10–30	Green synthesis using safflower natural dye as both reducing agent and stabilizer	Colorimetric and fastness studies on wool yarns for textile applications	Sadeghi-Kiakhani and Hashemi (2024)
~20–50	In situ green synthesis using plant extracts as reducing agents	UV protection, antioxidant activity, antimicrobial properties, and sensing response in viscose fibers	Rehan et al. (2024)
~10–50	Functionalization via chemical reduction method using silver nitrate ($AgNO_3$) and a reducing agent	Antimicrobial activity on printed cotton fabric	Rabbi et al. (2024)
~10–20	Green synthesis using Verbascum thapsus L. (mullein) extract	Antibacterial properties on polyamide fabrics	Sadeghi-Kiakhani et al. (2024)
~10–30	Chemical reduction method using silver nitrate ($AgNO_3$) in the presence of a reducing agent	Dyeing properties on cellulosic fabrics, improving antibacterial, color intensity, and fastness	Siddiqua et al. (2024)
~10–50	Hydrothermal synthesis method, using silver nitrate ($AgNO_3$) and a reducing agent at high temperature and pressure	Antimicrobial coating for cotton fabrics to enhance antibacterial properties for use in healthcare textiles and functional clothing	Tooklang et al. (2024)

in novel methods for integrating silver nanoparticles into textiles, enhancing their functional capabilities, durability, and washability. In Table 9.10, we will look at many current applications of silver nanoparticles in textiles.

9.3.5.2 Application of Carbon Nanotubes

Carbon nanotubes (CNTs) have emerged as a revolutionary material in the textile industry due to their exceptional mechanical, electrical, and thermal capabilities. These nanostructures, recognized for their excellent tensile strength and flexibility, enable the fabrication of lightweight and long-lasting fabrics. Smart textiles, wearable technology, enhanced protective gear, and even self-cleaning fabrics can benefit from adding CNTs to improve their conductivity, UV protection, and other capabilities. Recent advances in the synthesis and integration of carbon nanotubes into textiles have paved the way for creative products that meet the changing needs of consumers. In the following sections, we will explore several recent applications of carbon nanotubes in textiles, showcasing their transformative potential in the industry.

Faggio et al. (2024) examined bio-based epoxy resin and CNT coatings on cotton fabrics for smart wearable devices. The research focused on creating eco-friendly coatings that improve the functional features of fabrics, such as electrical conductivity and mechanical strength. The findings showed that the bio-based epoxy resin effectively combines with cotton fibers, whereas the addition of CNTs significantly increases electrical conductivity. The coated fabrics have promising mechanical characteristics and flexibility, making them ideal for wearable applications. Overall, the findings demonstrate the potential of bio-based epoxy resin with CNT coatings to build sophisticated textile materials that will aid in developing smart wearable technology. Islam et al. (2024) analyzed textile filaments manufactured from kraft pulp-based carboxymethylated cellulose (CMC) that have been covalently cross-linked with CNTs to improve antibacterial and UV protection capabilities. The results showed that adding CNTs considerably increases the mechanical strength and flexibility of CMC filaments. The treated filaments also have significant antibacterial activity against numerous pathogens and provide excellent UV protection, making them ideal for use in protective textiles. Overall, the results show that CMC/CNT composite filaments have the potential to be used to build sophisticated textiles with improved functionality for health and safety applications.

9.3.5.3 Application of Graphene-Based Nanomaterials

Graphene-based nanomaterials have quickly acquired popularity in the textile sector due to their remarkable features, such as high strength, electrical conductivity, and thermal stability. Graphene and its derivatives, which are made up of a single layer of carbon atoms arranged in a two-dimensional lattice, allow for the creation of new textiles with advanced characteristics such as increased durability, lightweight construction, and excellent breathability. Incorporating graphene into fabrics brings new opportunities for smart textiles, such as wearable electronics, health monitoring systems, and self-heating materials. Recent research has focused on optimizing the incorporation of graphene into diverse textile substrates, yielding products that match esthetic requirements and give considerable performance gains. In Table 9.11, we will discuss several recent applications of graphene-based nanomaterials in textiles.

9.3.5.3.1 Application of TiO_2 Nanoparticles

TiO_2 nanomaterials have emerged as a powerful tool in the textile industry, primarily due to their photocatalytic properties and UV protection capabilities. Known for their ability to degrade organic pollutants and inhibit bacterial growth when exposed to light, TiO_2 nanoparticles are increasingly used in textile treatments to create

Table 9.11 Application of graphene-based nanomaterials in textile

Size (nm)	Method	Application	References
<10	Water-based dispersion preparation	Conductive coatings for textiles, improving electrical properties	Çaylak et al. (2024)
5–10	Different coating methods (e.g., spray coating, dip-coating, electrophoretic deposition)	Conductive textiles for wearable electronics, smart textiles, energy storage systems	Abdi et al. (2024)
1–10	Liquid-phase exfoliation and dispersion in polymer matrix	Sports clothing to enhance moisture absorption, breathability, wear resistance, and antimicrobial properties	Zhang (2024)
1–10	CVD or liquid-phase exfoliation	Conductive fibers for touch-sensitive gloves	Gao (2024)
1–10	CVD or liquid-phase exfoliation	Graphene-based knitted textile sensing sleeve for joint angle monitoring and gesture differentiation	Zhou, et al. (2024)

Table 9.12 Application of TiO_2 nanoparticles in textile

Size	Method	Application	References
10–50 nm	Coating method (dip-coating or sol-gel)	Fabric coating to enhance thermal insulation and air permeability for workwear fabrics	Akhlaghi Pirposhteh et al. (2024)
10–50 nm	Treatment method (impregnation method)	Treatment of acrylic/cotton fabrics for enhanced properties like UV protection and antimicrobial activity	El Messiry et al. (2024)
TiO_2: ~10–50; ZnO: ~30–60	Surface modification (impregnation method)	Surface modification of jute-cotton union fabric for multifunctional properties such as UV protection, antibacterial, and water repellent	Hassan and Mou (2024)
TiO_2: ~20–50; AgNPs: ~10–30	Cross-linking method using succinic acid as a cross-linking agent	Medical textiles with enhanced antibacterial, UV protection, and biocompatibility properties	Ali et al. (2024)
~20–30	Sol-gel method followed by coating onto textiles	Enhancement of antimicrobial properties for textiles, especially for hygiene-sensitive applications	Salama et al. (2024)

self-cleaning fabrics and enhance durability. Additionally, TiO_2 provides excellent UV shielding, making textiles safer for outdoor use. Recent advancements in incorporating TiO_2 into various fabric types have led to innovative applications in areas such as sportswear, home textiles, and medical textiles, where hygiene and performance are paramount. In Table 9.12, we will discuss several recent applications of TiO_2 nanomaterials in textiles industry.

9.3.5.3.2 Application of ZnO Nanoparticles

Due to their multifunctional capabilities, ZnO NPs have gained popularity in the textile sector. ZnO NPs, known for their strong UV-blocking properties, protect materials from harmful ultraviolet radiation, making them perfect for outdoor and sun-protective clothing. They also have antibacterial qualities that prevent bacterial growth, which improves textile hygiene and durability. Recent advances in integration techniques have assured that ZnO NPs remain effective even after numerous washes, making them suitable for usage in sportswear, medical textiles, and daily clothing. In Table 9.13, we will look at recent applications of zinc oxide nanoparticles in textiles industries.

9.4 Ecological Impact

9.4.1 Effect on Human and Other Territorial Organism

NPs because of their small size, high reactivity, and enormous surface area have many benefits but can present serious health hazards to humans and the environment (Fig. 9.3) (El-Kalliny et al., 2023). Toxicological effects can result from these particles entering an organism through eating or inhalation and moving to different organs (Xuan et al., 2023). Because inhaled NPs can enter the alveoli and travel throughout the body through the bloodstream, they pose a unique risk to the respiratory system (Kumah et al., 2023; Oberdörster et al., 2005). They may result in oxidative stress, inflammation, fibrosis, and even lung cancer (Xuan et al., 2023). Because of their association with lung inflammation, NPs like TiO_2 and SiO_2 may make underlying illnesses like asthma or COPD worse (Ma et al., 2019).

Furthermore, NPs can penetrate the blood–brain barrier, build up in the brain, and cause neurotoxicity through inflammation and oxidative stress (Ray et al., 2009). This can result in cognitive impairment and neurodevelopmental problems, particularly in children (Hersh et al., 2022). By imitating hormones or attaching to hormone receptors, NPs can interfere with the endocrine system's ability to regulate hormones, impacting thyroid and reproductive processes (Bole-Feysot et al., 1998; Zhu et al., 2021). Ag, ZnO, and TiO_2 are examples of metal-based NPs that can produce reactive oxygen species, which can lead to cellular damage and endocrine dysfunction (Lihui et al., 2021; Ma et al., 2018; Pinho et al., 2020). NPs also affect the immune system because they can cause cytokine production and inflammatory reactions. Some NPs, such as carbon black and cadmium (Cd), can also impair immunological function, making people more vulnerable to autoimmune diseases and infections (Bruneau et al., 2015; Yamawaki & Iwai, 2006).

Table 9.13 Application of zinc oxide nanoparticles in textile

Size (nm)	Method	Application	References
~50–100	Sol-gel method followed by dip-coating onto cotton fabrics	Creation of superhydrophobic surfaces on cotton fabrics, enhancing water repellency	Ebrahim and Saffour (2024)
~30–50 (ZnO), ~20–40 (silica)	Sol-gel method followed by dip-coating onto cotton fabric	Enhancing flammability resistance in cotton protective clothing	Saleemi et al. (2024)
~30–50	Sol-gel method or hydrothermal synthesis, followed by coating on cotton textiles	UV protection for cotton textiles to reduce occupational and environmental exposure to UV radiation	Nasirzadeh, Monazam Esmaeelpour, et al. (2024)
~20–40	Synthesized by using pectin as an organic framework to stabilize ZnO nanoparticles, followed by the application onto printed textiles	Enhances the functionality of textile fabrics, including antimicrobial properties and UV protection	Hassabo and Saad (2024)
~20–30	Green synthesis method using chitosan as a stabilizing agent for the ZnO nanoparticles	Enhances multifunctional properties of cellulosic fabrics, such as antimicrobial activity, UV protection, and moisture management	Ibrahim et al. (2024)
~15–20	Biogenic synthesis mediated by Burkholderia sp. EIKU21, using biological resources to produce ZnO nanoparticles	Coating cotton fabric with biogenic ZnO nanoparticles to impart antibacterial properties for textile applications	(Roy et al., 2024)
~10–30	Bio-reduction method	Finishing of cotton fabric to impart multifunctional properties such as antimicrobial activity, UV protection, and water resistance	Nivedha et al. (2024)
Submicron (~200–500)	Sol-gel method	Enhancing the UV protection properties of textiles, improving their use in protective clothing and occupational safety applications	Nasirzadeh, Monazam Esmaielpour, and Golbabaei (2024)

Regarding reproductive health, NPs can penetrate the placental and blood–testis barriers, resulting in inflammation, oxidative stress, and cellular damage. In males, this can lead to decreased sperm motility and count, while in females, it can cause ovarian damage or problems with embryonic development (Pask, 2016). Furthermore, it has been demonstrated that certain NPs, especially metal-based ones like ZnO and nickel (Ni), are carcinogenic (Kwon et al., 2014; Magaye & Zhao, 2012). They cause genetic damage by generating reactive oxygen species

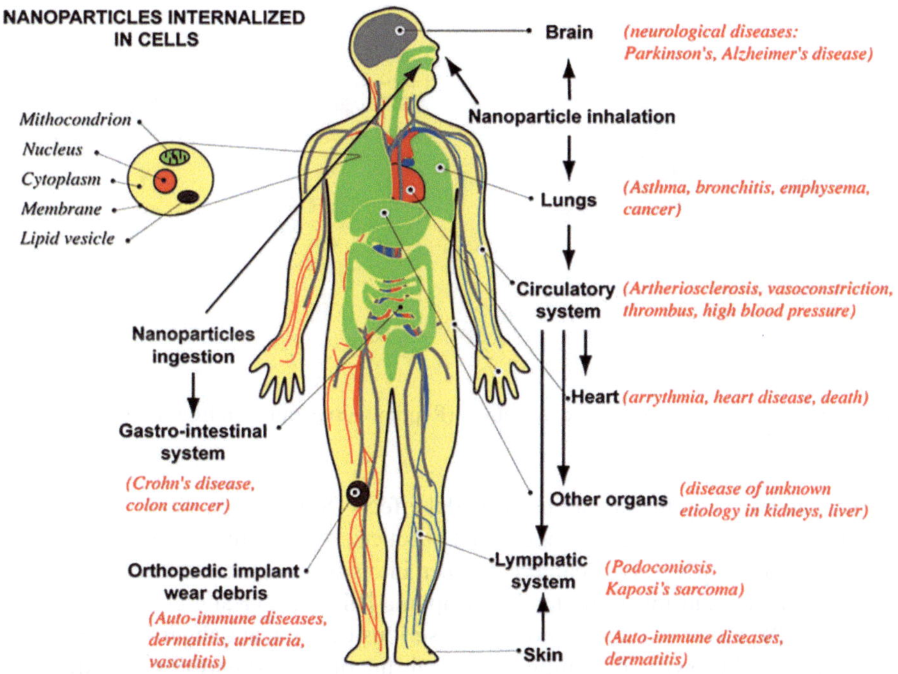

Fig. 9.3 Schematics of the human body with pathways of exposure to NPs, affected organs, and associated diseases from epidemiological, in vivo and in vitro studies (Reprint with permission from El-Kalliny et al. (2023) Copyright 2023 Elsevier)

(ROS), which results in mutations, cell death, and increased cytokine secretion—all of which contribute to cancer development (Kumar & Dhawan, 2013; Sharifi et al., 2012). The size and surface characteristics of NPs affect their carcinogenic potential, and more study is required to completely comprehend their long-term dangers and create mitigation plans for their detrimental impacts.

9.4.2 Effect on Aquatic Life

Water chemistry, pH, salinity, and the presence of organic matter are some variables that affect the behavior and toxicity of NPs in aquatic environments. These variables can also affect the NPs aggregation, bioavailability, and overall impact (Xuan et al., 2023). Studies are still being conducted to determine nanomaterials quantities and toxicological effects. However, much is still unclear about the entire scope of their effects despite growing worry over their environmental fate. Both direct and indirect entry of nanomaterials into water bodies can occur through surface runoff, wastewater treatment effluents, and industrial discharge (Dhall et al., 2024). According to toxicological studies, NPs, particularly those based on metals, can damage fish and

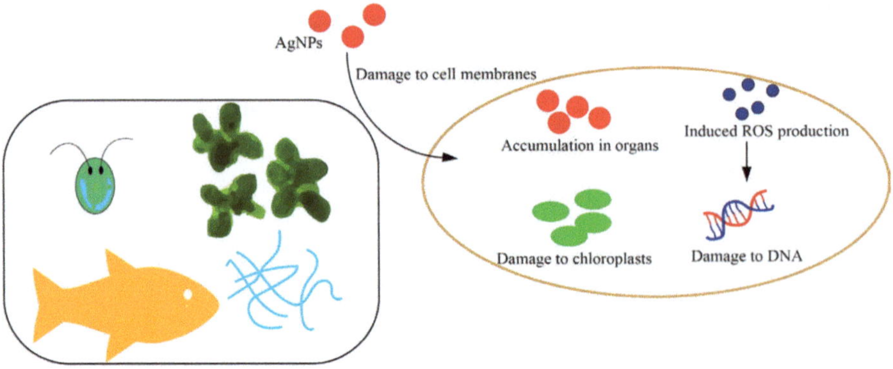

Fig. 9.4 Toxicity to typical aquatic organisms (Reprint with permission from Kang et al. (2023) Copyright 2023 Elsevier)

plants and have mutagenic and bactericidal effects (Singh et al., 2024). Due to their extensive use in consumer and industrial products, Ag NPs have been released into aquatic habitats, where they can break down into silver ions that are harmful to fish, bacteria, algae, and invertebrates like Daphnia (Fig. 9.4) (Kang et al., 2023; Kusi et al., 2020). Silver ion dissolution poses serious ecological dangers by interfering with microbial growth, disrupting biological functions, and harming aquatic plants and animals (Khursheed et al., 2023). Further research is desperately needed to understand the environmental concentrations, biological effects, and pathways by which nanomaterials accumulate and affect aquatic life, including their potential to migrate through food webs, even though risk assessments of their releases into aquatic systems are still in their early stages. Given the growing amount of nanomaterials being released into the environment, carefully considering their long-term ecological effects is necessary.

9.4.3 Toxicological Effects

NPs size, shape, charge, surface properties, and functionalized groups all affect how hazardous they are. Their capacity to interact with proteins in biological systems is one of the essential mechanisms of NP toxicity. Protein unfolding, fibrillation, thiol cross-linking, and decreased enzyme activity can result from NPs assembling around protein molecules (Khan et al., 2019). These impacts can harm cells, interfere with metabolic processes, and interfere with cellular functioning (Fig. 9.5) (Arora et al., 2012). The release of harmful ions is a significant contributing factor to NP toxicity. The release of ions that can seriously damage cells and tissues occurs when the thermodynamic characteristics of nanoparticles promote their breakdown in a biological media or suspending medium (Nie et al., 2023). For example, when silver nanoparticles dissolve, silver ions are released, which can cause oxidative

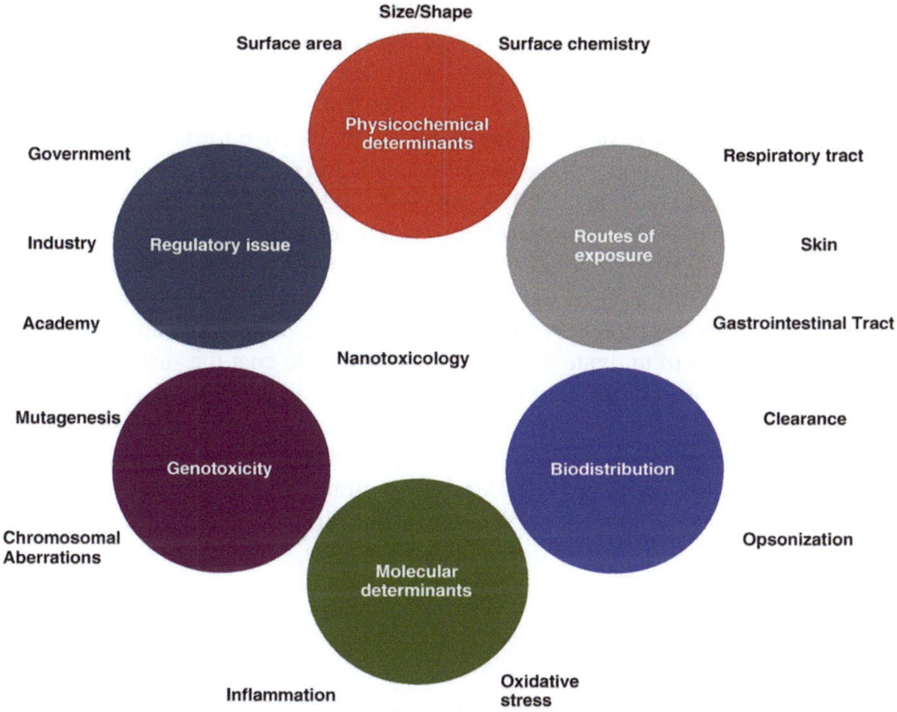

Fig. 9.5 Complex array of issues surrounding toxicity of nanoparticles. (Reprint with permission from Arora et al. (2012) Copyright 2012 Elsevier)

stress, damage cellular membranes, and impair enzyme activity (Patlolla et al., 2015). Understanding how NPs affect living things depends on these toxicological mechanisms, emphasizing the importance of carefully weighing the hazards to human health and the environment.

9.5 Mitigation of Nanomaterials

NMs have been used for many positive ways like in cleaning air, water, soil, in medicine, food, and agriculture sectors. However, the environmental impact of using NMs doesn't come with a cost to the ecosystem or living organisms. Therefore, the mitigation of NMs is a necessity. The used NMs which are been released into the environment as waste have been found to cause a significant negative impact on water organisms (Kumar et al., 2022). The excess NMs have found to increase the cytotoxicity and oxidative stress in the water organisms such as microalgae. The increased accumulation of NMs, such as copper oxide, lanthanum oxide, cerium oxide, and nickel oxide, in the soil has been known to reduce the rate of plant growth and reduce rate of photosynthesis, transpiration, and productivity (Ahmed et al., 2021; Li et al., 2024; Rastogi et al., 2017). It's not only the waste of NMs which hampers the environment but also the workers involved in manufacturing the NMs since they are more exposed to NMs than those not (Bergamaschi et al., 2015). Even though NMs are known for their ultrafine particle size, they are a boon to workers since they are easily penetrable and highly reactive toward biological tissues, cells, organs, and sub-cellular structures, which could lead to nanotoxicity and detrimental events (Thu et al., 2023; Xuan et al., 2023). Factors which can be used for the mitigation of NMs during designing is their degree of aggregation or agglomeration. It has been put in front that NMs toxicity increases if the particles are aggregated or agglomerated after its use (Thu et al., 2023). The workers should be given proper regulations and protocols for the safe handling, production, and disposal of NMs. Another way to mitigate the toxicity of NMs is to coat the surface with biological materials which are compatible to the NMs in use (Xuan et al., 2023). Furthermore, developing predictive methods which can accurately measure or study the toxic behavior of NMs on the bases of their surface or physiochemical properties will help in the mitigation of NMs. These strategies can help to mitigate the risks associated with NMs by harnessing their benefits in various fields.

9.6 Conclusion

In this era of development, life without the use of nanoparticles is strenuous as they have various applications in different sectors like food, agriculture, textile, medicine, electronics, and many more, which are discussed in this chapter. Despite its multiple applications, the various methods of synthesizing NPs, like physical and chemical methods, possess harm to the environment and the workers. These disadvantages, such as the use of hash chemicals, massive use of energy, and resources, and so on, can be overcome by using biological methods as this method is greener and eco-friendly than physical and chemical methods. However, the bigger problem lies in the remediation and its waste being released into the environment. When these nanoparticles are released into the environment without proper treatment, it disturb the ecological balance and can affect the human and aquatic environment. One of the most toxic traits of NPs is their agglomeration behavior, that is, due to their small size, they easily agglomerate. Therefore, the mitigation of NPs is a necessary step to reduce its effect pre and post treatment. One of the ways to mitigate the effect of NPs waste is to relay on the use of magnetic NPs in place of non-magnetic NPs as the changes in reusability increase with magnetic NPs. They can also be fabricated with an organic metal framework so that they do not cause adverse reactions to the living organism before and after use. As biological methods employ biological materials for synthesizing the NPs, the waste of such NPs degrades biologically when released into the environment. Additionally, creating predictive methods to accurately evaluate the toxic behavior of NPs based on their surface and physicochemical properties will contribute to the effective mitigation of NPs. Ultimately, a holistic approach combining safer synthesis methods, effective waste management, and predictive toxicity assessments is essential to ensure that the benefits of NPs can be realized while minimizing their potential harm.

References

Abdi, B., Tarhini, A., Baniasadi, H., & Tehrani-Bagha, A. R. (2024). Developing graphene-based conductive textiles using different coating methods. *Advanced Materials Technologies, 9*, 2301492.

Abid, N., Khan, A. M., Shujait, S., Chaudhary, K., Ikram, M., Imran, M., Haider, J., Khan, M., Khan, Q., & Maqbool, M. (2022). Synthesis of nanomaterials using various top-down and bottom-up approaches, influencing factors, advantages, and disadvantages: A review. *Advances in Colloid and Interface Science, 300*, 102597.

Ahmadi, Z., Patel, A., Hill, C., Peeples, S. R., Jones, J. M., Boebinger, M. G., & Mahjouri-Samani, M. (2024). Dry printing pure copper with high conductivity and adhesion for flexible electronics. *ACS Applied Electronic Materials, 6*(5), 3933–3940.

Ahmed, B., Rizvi, A., Ali, K., Lee, J., Zaidi, A., Khan, M. S., & Musarrat, J. (2021). Nanoparticles in the soil–plant system: A review. *Environmental Chemistry Letters, 19*, 1545–1609.

Akhlaghi Pirposhteh, E., Mortazavi, S. B., Farhang Dehghan, S., Khaloo, S. S., & Montazer, M. (2024). Optimization and development of workwear fabric coated with TiO2 nanoparticles in order to improve thermal insulation properties and air permeability. *Journal of Industrial Textiles, 54*, 15280837241258168.

Alharbi, A. A., Alghamdi, A. M., Al-Goul, S. T., Allohibi, A., Baty, R. S., Qahl, S. H., & Beyari, E. A. (2024). Valorizing pomegranate wastes by producing functional silver nanoparticles with antioxidant, anticancer, antiviral, and antimicrobial activities and its potential in food preservation. *Saudi Journal of Biological Sciences, 31*, 103880.

Ali, M. A.-S., Abdel-Rahim, E. A.-M., Mahmoud, A. A.-A., & Mohamed, S. E. (2024). Innovative textiles treated with TiO2-AgNPs with succinic acid as a cross-linking agent for medical uses. *Scientific Reports, 14*, 8045.

Altammar, K. A. (2023). A review on nanoparticles: Characteristics, synthesis, applications, and challenges. *Frontiers in Microbiology, 14*, 1155622.

Ameta, S. K., Rai, A. K., Hiran, D., Ameta, R., & Ameta, S. C. (2020). Use of nanomaterials in food science. In *Biogenic nano-particles and their use in agro-ecosystems* (pp. 457–488). Springer.

Apoorva, M. S., Kumar, S., & Bhatia, A. (2023). Nanotechnological interventions for climate change mitigation. *Journal of Agricultural Physics, 23*, 1–14.

Arora, S., Rajwade, J. M., & Paknikar, K. M. (2012). Nanotoxicology and in vitro studies: The need of the hour. *Toxicology and Applied Pharmacology, 258*, 151–165.

Ashfaq, A., Khursheed, N., Fatima, S., Anjum, Z., & Younis, K. (2022). Application of nanotechnology in food packaging: Pros and cons. *Journal of Agriculture and Food Research, 7*, 100270.

Attia, Y. A., Ezet, A. E., Saeed, S., & Galmed, A. H. (2024). Nano carbon-modified air purification filters for removal and detection of particulate matters from ambient air. *Scientific Reports, 14*, 621.

Ayan, S., Aranci-Ciftci, K., Ciftci, F., & Ustundag, C. B. (2023). Nanotechnology and COVID-19: Prevention, diagnosis, vaccine, and treatment strategies. *Frontiers in Materials, 9*, 1059184.

Baig, N., Kammakakam, I., & Falath, W. (2021). Nanomaterials: A review of synthesis methods, properties, recent progress, and challenges. *Materials Advances, 2*, 1821–1871.

Bastardo-Fernández, I., Chekri, R., Oster, C., Thoury, V., Fisicaro, P., Jitaru, P., & Noireaux, J. (2024). Assessment of TiO2 (nano) particles migration from food packaging materials to food simulants by single particle ICP-MS/MS using a high efficiency sample introduction system. *NanoImpact, 34*, 100503.

Begines, B., Ortiz, T., Pérez-Aranda, M., Martinez, G., Merinero, M., Argüelles-Arias, F., & Alcudia, A. (2020). Polymeric nanoparticles for drug delivery: Recent developments and future prospects. *Nanomaterials, 10*, 1403.

Benhacine, F., Lounis, F. M., & Hadj-Hamou, A. S. (2024). Development of a nanocomposite material based on PCL/zeolite-supported silver nanoparticles for active food packaging. *Polymer Science, Series A, 66*, 240–252.

Bergamaschi, E., Murphy, F., Poland, C. A., Mullins, M., Costa, A. L., McAlea, E., Tran, L., & Tofail, S. A. M. (2015). Impact and effectiveness of risk mitigation strategies on the insurability of nanomaterial production: Evidences from industrial case studies. *Wiley Interdisciplinary Reviews: Nanomedicine and Nanobiotechnology, 7*, 839–855.

Bole-Feysot, C., Goffin, V., Edery, M., Binart, N., & Kelly, P. A. (1998). Prolactin (PRL) and its receptor: Actions, signal transduction pathways and phenotypes observed in PRL receptor knockout mice. *Endocrine Reviews, 19*, 225–268.

Bruneau, A., Fortier, M., Gagne, F., Gagnon, C., Turcotte, P., Tayabali, A., Davis, T. A., Auffret, M., & Fournier, M. (2015). In vitro immunotoxicology of quantum dots and comparison with dissolved cadmium and tellurium. *Environmental Toxicology, 30*, 9–25.

Cao, X., Chen, X., Liu, Y., Wang, C., Yue, L., Elmer, W. H., White, J. C., Wang, Z., & Xing, B. (2023). Lanthanum silicate nanomaterials enhance sheath blight resistance in rice: Mechanisms of action and soil health evaluation. *ACS Nano, 17*, 15821–15835.

Çaylak, S., Demirel, O., Javadzadehkalkhoran, M., Navidfar, A., Yacsacan, M., & Trabzon, L. (2024). Preparation and characterization of high-performance water-based graphene dispersions for conductive coating on textiles. *The Journal of The Textile Institute, 116*, 446–454.

Chandran, S., Pricillia, A., Natarajan, A., & Venkatesan, G. (2024). Enhancing sustainability: Chitosan biopolymers with ag nanoparticles for eco-friendly applications in food packaging. *Biomass Conversion and Biorefinery*. https://doi.org/10.1007/s13399-024-05325-7

Chandrika, K. S. V. P., Qureshi, A. A., Singh, A., Sarada, C., & Gopalan, B. (2022). Fe and Zn metal nanocitrates as plant nutrients through soil application. *ACS Omega, 7*, 45481–45492.

Chausali, N., Saxena, J., & Prasad, R. (2023). Nanotechnology as a sustainable approach for combating the environmental effects of climate change. *Journal of Agriculture and Food Research, 12*, 100541.

Choi, S., Jang, Y.-R., Kim, H.-S., & Lee, C. S. (2024). Flashlight sintering characteristics of the inkjet-printed nanosized copper ink on an auxiliary heated paper substrate. *International Journal of Precision Engineering and Manufacturing-Green Technology, 11*, 365–379.

Dai, Q., Jia, R., Li, H., Yang, J., & Qin, Z. (2024). Preparation and application of sustained-release antibacterial alginate hydrogels by loading plant-mediated silver nanoparticles. *ACS Sustainable Chemistry & Engineering, 12*, 1388–1404.

Demirhan, A. D., Tuugral, S., Sariicsik, M., Göktacs, D., & Kartal, G. E. (2024). Investigation of polypropylene yarns containing silver nanoparticles: Fiber characteristics, yarn properties, and fabric performance analysis. *Journal of Applied Polymer Science, 141*, e55770.

Devi, N., Bansal, N., Sharma, S., Dubey, S. K., & Kumar, S. (2024). Novel chitosan-based smart bio-nanocomposite films incorporating TiO_2 nanoparticles for white bread preservation. *International Journal of Biological Macromolecules, 267*, 131367.

Dhall, S., Nigam, A., Harshavardhan, M., Mukherjee, A., & Srivastava, P. (2024). A comprehensive overview of methods involved in nanomaterial production and waste disposal from research labs and industries and existing regulatory guidelines for handling engineered nanomaterials. *Environmental Chemistry and Ecotoxicology, 6*, 269–282.

Dimitriou, E., & Michailidis, N. (2024). Preparation of highly loaded copper (II) oxide nanoparticle suspensions for LIFT-printing of electronics. *European Journal of Materials, 4*, 2316918.

do Nascimento, W. J., da Costa, J. C. M., Alves, E. S., de Oliveira, M. C., Monteiro, J. P., Souza, P. R., Martins, A. F., & Bonafe, E. G. (2024). Zinc oxide nanoparticle-reinforced pectin/starch functionalized films: A sustainable solution for biodegradable packaging. *International Journal of Biological Macromolecules, 257*, 128461.

Dumas, L., de Souza, M. C., Bonafe, E. G., Martins, A. F., & Monteiro, J. P. (2024). Optimized incorporation of silver nanoparticles onto cotton fabric using k-carrageenan coatings for enhanced antimicrobial properties. *ACS Applied Bio Materials, 7*, 6908–6918.

Duryea, D., McKibben, N., Manzi, J., Brennan, T., Subbaraman, H., Estrada, D., & Kandadai, N. (2024). Titanium dioxide nanomaterial ink production through laser ablation synthesis in solution for printed electronics applications. *Advanced Engineering Materials, 26*, 2400721.

Ebrahim, M., & Saffour, Z. (2024). Synthesis and characterization of ZnO nanoparticles and their application on cotton fabrics to obtain superhydrophobic surfaces. *Baghdad Science Journal, 11*, 8934.

El Messiry, M., Issa, M., & Abdellatif, S. (2024). Investigation of acrylic/cotton single Jersey knitted fabrics treated with titanium dioxide (TiO2) nanoparticles. *Egyptian Journal of Chemistry, 67*, 363–372.

El-Kalliny, A. S., Abdel-Wahed, M. S., El-Zahhar, A. A., Hamza, I. A., & Gad-Allah, T. A. (2023). Nanomaterials: A review of emerging contaminants with potential health or environmental impact. *Discover Nano, 18*, 68.

Esa, Z., Nauman, M. M., Ullah, M., Khalid, M. U., Mehdi, M., Abid, M., Iqbal, A., Zaini, J. H., & Ali, K. (2024). Compatibility and performance study of electrohydrodynamic printing using zinc oxide inkjet ink. *Scientific Reports, 14*, 16813.

Faggio, N., Olivieri, F., Bonadies, I., Gentile, G., Ambrogi, V., & Cerruti, P. (2024). Bio-based epoxy resin/carbon nanotube coatings applied on cotton fabrics for smart wearable systems. *Journal of Colloid and Interface Science, 670*, 337–347.

Fytianos, G., Rahdar, A., & Kyzas, G. Z. (2020). Nanomaterials in cosmetics: Recent updates. *Nanomaterials, 10*, 979.

Ganguly, B. N., Maity, B., Maity, T. K., Manna, J., Roy, M., Mukherjee, M., Debnath, S., Saha, P., Shilpa, N., & Rana, R. K. (2018). L-cysteine-conjugated ruthenium hydrous oxide nanomaterials with anticancer active application. *Langmuir, 34*, 1447–1456.

Gao, W., Mu, B., Yang, F., Li, Y., & Wang, A. (2024). Green preparation of licorice flavonoids/ZnO/attapulgite nanocomposites for multifunctional chitosan-based food packaging films. *LWT, 201*, 116273.

Gao, X. (2024). Material selection of conductive fibers for touchable gloves and application of graphene. *Science and Technology of Engineering, Chemistry and Environmental Protection, 1*, 51.

Ge, F.-F., Yao, W.-H., Potiyaraj, P., Lin, J.-H., Wu, C.-S., Meng, L., Zhang, X., Preuksarattanawut, C., Ma, Z.-L., & Tsou, C.-H. (2024). Enhanced PBS nanocomposites with ZnO-coated MWCNT for extending shelf life in sustainable food packaging applications. *Journal of Polymer Research, 31*, 278.

Ghosh, S., Laha, P., & Biswas, S. (2024). A superhydrophobic nanocrystalline MOF embedded starch@ cotton composite for fast, selective and nanomolar sensing and adsorptive removal of a fluorinated herbicide from aqueous medium. *Environmental Science: Nano, 11*, 3599–3614.

Gupta, V., Mohapatra, S., Mishra, H., Farooq, U., Kumar, K., Ansari, M. J., Aldawsari, M. F., Alalaiwe, A. S., Mirza, M. A., & Iqbal, Z. (2022). Nanotechnology in cosmetics and cosmeceuticals a review of latest advancements. *Gels, 8*, 173.

Hassabo, A. G., & Saad, M. A. (2024). Pectin as organic framework for ZnONPs to enhance the functionality of textile printed fabrics. *Journal of Textiles, Coloration and Polymer Science, 21*, 509–518.

Hassan, M. N., & Mou, M. B. (2024). Surface modification of jute-cotton union fabric using TiO2 and ZnO nanoparticles for multifunctional properties. *Heliyon, 10*, e29970.

Hersh, A. M., Alomari, S., & Tyler, B. M. (2022). Crossing the blood-brain barrier: Advances in nanoparticle technology for drug delivery in neuro-oncology. *International Journal of Molecular Sciences, 23*, 4153.

Ibrahim, N. A., Ameen, H. A., & Eid, B. M. (2024). Green synthesized chitosan and ZnO nanoparticles for sustainable use in multifunctionalization of cellulosic fabrics. *Polymer Bulletin, 81*, 3621–3640.

Iqbal, R., Khan, S., Ali, H. M., Khan, M., Wahab, S., & Khan, T. (2022). Application of nanomaterials against SARS-CoV-2: An emphasis on their usefulness against emerging variants of concern. *Frontiers in Nanotechnology, 4*, 1060756.

Islam, M. S., Alam, M. N., van de Ven, T. G. M., Barbeau, B., & Rahaman, M. S. (2024). Antibacterial and UV protection properties of textile filaments fabricated from Kraft pulp-based carboxymethylated cellulose covalently cross-linked with carbon nanotubes. *Polymer Bulletin, 81*, 6785–6800.

Javan, M., Selahvarzi, Y., Sayyad-Amin, P., & Rastegar, S. (2024). Potential application of TiO2 nanoparticles to improve the nutritional quality of strawberry cv. Camarosa under drought stress. *Scientia Horticulturae, 330*, 113055.

Kalajahi, N. J., Ghanbarzadeh, B., Dehghannya, J., & Ostadrahimi, A. (2024). The physicomechanical optimization and antimicrobial properties of the polylysine/ZnO nanoparticles loaded active packaging films. *Food and Humanity, 3*, 100401.

Kang, J., Zhou, N., Zhang, Y., Wang, Y., Song, C., Gao, X., Song, G., Guo, J., Huang, L., & Ma, T. (2023). Synthesis, multi-site transformation fate and biological toxicity of silver nanoparticles in aquatic environment: A review. *Environmental Technology Innovation, 32*, 103295.

Kant, C., Mahmood, S., Seetharaman, M., & Katiyar, M. (2024). Large-area inkjet-printed flexible hybrid electrodes with photonic sintered silver grids/high conductive polymer. *Small Methods, 8*, 2300638.

Khabirova, S., Menshikov-Tonyan, M., Aleshin, G., Prikhodko, A., Kozlov, D., Anokhin, E., Babeshkin, K., Titchenko, N., Zubenko, A., & Shchukina, A. (2024). Assessing the biocompatibility and stability of CeO 2 nanoparticle conjugates with azacrowns for use as radiopharmaceuticals. *RSC Medicinal Chemistry, 15*, 4100–4110.

Khan, H. A., Sakharkar, M. K., Nayak, A., Kishore, U., & Khan, A. (2018). Nanoparticles for biomedical applications: An overview. In *Nanobiomaterials* (pp. 357–384). Elsevier.

Khan, I., Saeed, K., & Khan, I. (2019). Nanoparticles: Properties, applications and toxicities. *Arabian Journal of Chemistry, 12*, 908–931.

Khan, Y., Sadia, H., Ali Shah, S. Z., Khan, M. N., Shah, A. A., Ullah, N., Ullah, M. F., Bibi, H., Bafakeeh, O. T., & Ben, K. N. (2022). Classification, synthetic, and characterization approaches to nanoparticles, and their applications in various fields of nanotechnology: A review. *Catalysts, 12*, 1386.

Khursheed, S., Dutta, J., Ahmad, I., Rather, M. A., Badroo, I. A., Bhat, T. A., Ahmad, I., Amin, A., Shah, A., & Qadri, T. (2023). Biogenic silver nanoparticles: Synthesis, applications and challenges in food sector with special emphasis on aquaculture. *Food Chemistry: X, 20*, 101051.

Klos, M. A. H., González-Garcia, L., & Kraus, T. (2024). Mechanically robust, inkjet-printable polymer nanocomposites with hybrid gold nanoparticles and metal-like conductivity. *ACS Applied Materials & Interfaces, 16*(24), 31576–31585.

Kumah, E. A., Fopa, R. D., Harati, S., Boadu, P., Zohoori, F. V., & Pak, T. (2023). Human and environmental impacts of nanoparticles: A scoping review of the current literature. *BMC Public Health, 23*, 1059.

Kumar, A., & Dhawan, A. (2013). Genotoxic and carcinogenic potential of engineered nanoparticles: An update. *Archives of Toxicology, 87*, 1883–1900.

Kumar, C. M. V., Karthick, V., Kumar, V. G., Inbakandan, D., Rene, E. R., Suganya, K. S. U., Embrandiri, A., Dhas, T. S., Ravi, M., & Sowmiya, P. (2022). The impact of engineered nanomaterials on the environment: Release mechanism, toxicity, transformation, and remediation. *Environmental Research, 212*, 113202.

Kumar, J. E., Tsungom, M., Wanshanlang, K., Rajeshwar, S., & Kumar, S. M. (2021). The removal of binary mixture of dyes by heterogeneous Fenton oxilation: Kinetics, product identification and toxicity assessment. *Acta Chimica Slovenica, 68*, 833–848.

Kurban, M., Polat, C., Serpedin, E., & Kurban, H. (2024). Enhancing the electronic properties of TiO2 nanoparticles through carbon doping: An integrated DFTB and computer vision approach. *Computational Materials Science, 244*, 113248.

Kusi, J., Scheuerman, P. R., & Maier, K. J. (2020). Emerging environmental contaminants (silver nanoparticles) altered the catabolic capability and metabolic fingerprinting of microbial communities. *Aquatic Toxicology, 228*, 105633.

Kwon, J. Y., Koedrith, P., & Seo, Y. R. (2014). Current investigations into the genotoxicity of zinc oxide and silica nanoparticles in mammalian models in vitro and in vivo: Carcinogenic/genotoxic potential, relevant mechanisms and biomarkers, artifacts, and limitations. *International Journal of Nanomedicine, 9*, 271–286.

Li, H., Guo, Y., Liang, A., Xu, X., Hao, Y., Shang, H., Li, C., Cai, Z., Han, L., & Zhao, J. (2024). Commonly used engineered nanomaterials improve soil health via suppressing soil-borne fusarium and positively altering soil microbiome. *ACS EST Engineering, 4*, 915–927.

Li, M., Zhang, P., Guo, Z., Zhao, W., Li, Y., Yi, T., Cao, W., Gao, L., Tian, C. F., & Chen, Q. (2024). Dynamic transformation of nano-MoS2 in a soil-plant system empowers its multifunctionality on soybean growth. *Environmental Science & Technology, 58*, 1211–1222.

Li, P., Xia, Y., Song, K., & Liu, D. (2024). The impact of nanomaterials on photosynthesis and antioxidant mechanisms in gramineae plants: Research progress and future prospects. *Plants, 13*, 984.

Li, T., Wang, P., Guo, W., Huang, X., Tian, X., Wu, G., Xu, B., Li, F., Yan, C., & Liang, X.-J. (2019). Natural berberine-based Chinese herb medicine assembled nanostructures with modified antibacterial application. *ACS Nano, 13*, 6770–6781.

Lihui, X., Jinming, G., Yalin, G., Hemeng, W., Hao, W., & Ying, C. (2021). Albicanol inhibits the toxicity of profenofos to grass carp hepatocytes cells through the ROS/PTEN/PI3K/AKT axis. *Fish & Shellfish Immunology, 120*, 325–336.

Ma, X., Wang, H., Chen, Y., Fu, L., Zhou, J., Zhang, L., Xing, Z., Zhang, Q., & Xia, L. (2024). Application of ag@g-C3N4/TiO2 cotton fabric flexible substrate with dual functionality: Photocatalytic reusability and SERS signal amplification for food safety detection. *Applied Surface Science, 661*, 160068.

Ma, Y., Guo, Y., Ye, H., Huang, K., Lv, Z., & Ke, Y. (2019). Different effects of titanium dioxide nanoparticles instillation in young and adult mice on DNA methylation related with lung inflammation and fibrosis. *Ecotoxicology and Environmental Safety, 176*, 1–10.

Ma, Y.-B., Lu, C.-J., Junaid, M., Jia, P.-P., Yang, L., Zhang, J.-H., & Pei, D.-S. (2018). Potential adverse outcome pathway (AOP) of silver nanoparticles mediated reproductive toxicity in zebrafish. *Chemosphere, 207*, 320–328.

Madbouly, A., Morsy, M., & Moustafa, H. (2024). Utilization of torrefied date stones with synthesized TiO2 nanoparticles for promoting humidity sensing of PVA/PVP nanocomposites for smart food packaging and biomedical applications. *Ceramics International, 50*, 38522–38531.

Magaye, R., & Zhao, J. (2012). Recent progress in studies of metallic nickel and nickel-based nanoparticles genotoxicity and carcinogenicity. *Environmental Toxicology and Pharmacology, 34*, 644–650.

Martinez, G., Merinero, M., Pérez-Aranda, M., Pérez-Soriano, E. M., Ortiz, T., Villamor, E., Begines, B., & Alcudia, A. (2020). Environmental impact of nanoparticles application as an emerging technology: A review. *Materials, 14*, 166.

Medina-Jaramillo, C., Quintero-Borregales, L. M., Goyanes, S., Bernal, C., & Famá, L. (2024). A biodegradable and active bilayer nanocomposite obtained from starch-yerba mate extract and starch-TiO2 nanoparticles films for packaging applications. *Starch, 76*, 2300048.

Mekuye, B., & Abera, B. (2023). Nanomaterials: An overview of synthesis, classification, characterization, and applications. *Nano Select, 4*, 486–501.

Mhetre, H., Kanse, D., & Patil, D. (2021). Nanomaterials: Applications in electronics. *International Journal of Advanced Engineering and Nano Technology, 4*, 7–19.

Mitchell, M. J., Billingsley, M. M., Haley, R. M., Wechsler, M. E., Peppas, N. A., & Langer, R. (2021). Engineering precision nanoparticles for drug delivery. *Nature Reviews Drug Discovery, 20*, 101–124.

Mume, L., Kebede, M., Bekana, D., Tan, Z., & Amde, M. (2024). Salvia tiliifolia leaf extract-based silver nanoparticles for colorimetric detection of Hg (II) in food and environmental samples. *Journal of Food Composition and Analysis, 133*, 106426.

Naseer, B., Srivastava, G., Qadri, O. S., Faridi, S. A., Islam, R. U., & Younis, K. (2018). Importance and health hazards of nanoparticles used in the food industry. *Nanotechnology Reviews, 7*, 623–641.

Nasirzadeh, N., Monazam Esmaeelpour, M., Naseri, N., & Omari Shekaftik, S. (2024). Improving ultraviolet protection properties of cotton textiles using zinc oxide (ZnO) nanomaterials: An approach for controlling occupational and environmental exposures. *International Journal of Environmental Health Research, 34*, 2067–2087.

Nasirzadeh, N., Monazam Esmaielpour, M. R., & Golbabaei, F. (2024). The role of submicron zinc oxide particle size in improving UV protection by textiles. *International Journal of Environmental Health Research, 34*, 3068–3071.

Nie, P., Zhao, Y., & Xu, H. (2023). Synthesis, applications, toxicity and toxicity mechanisms of silver nanoparticles: A review. *Ecotoxicology and Environmental Safety, 253*, 114636.

Nivedha, K., Kalaiarasi, K., Poonkothai, M., & Swathilakshmi, A. V. (2024). Multifunctional finishing of cotton fabric using zinc oxide nanoparticles synthesized by bio-reduction method. *Biomass Conversion and Biorefinery*. https://doi.org/10.1007/s13399-024-05761-5

Nyabadza, A., Plouze, A., Heidarinassab, S., Vazquez, M., & Brabazon, D. (2024). Screen-printed electrodes on paper using copper nano-and micro-particles. *Journal of Materials Research and Technology, 29*, 5189–5197.

Oberdörster, G., Oberdörster, E., & Oberdörster, J. (2005). Nanotoxicology: An emerging discipline evolving from studies of ultrafine particles. *Environmental Health Perspectives, 113*, 823–839.

Onyeaka, H., Passaretti, P., Miri, T., & Al-Sharify, Z. T. (2022). The safety of nanomaterials in food production and packaging. *Current Research in Food Science, 5*, 763–774.

Oyebanji, M., Chen, L., Qian, R., Tu, M., Zhang, Q., Yang, X., Yu, H., & Zhu, M. (2024). Ultra-small water-soluble fluorescent copper nanoclusters for p-nitrophenol detection. *Analytical Methods, 16*, 6688–6695.

Ozcakir, G. (2023). Applications of nanomaterials in food industry: A review. *Materials Proceedings, 14*, 22.

Pan, X., Cui, S., Fan, S., Cao, C., Jiao, Y., Fu, Y., Niu, J., Lin, S., Zhu, Y., & Liu, Y. (2024). Au-based conjugated microporous polymers for combined photodynamic and radiation therapy in cancer treatment. *Materials Chemistry Frontiers, 8*, 3925–3934.

Pan, Y.-C., Tian, J.-H., & Guo, D.-S. (2023). Molecular recognition with macrocyclic receptors for application in precision medicine. *Accounts of Chemical Research, 56*, 3626–3639.

Papi, M., De Spirito, M., & Palmieri, V. (2023). Nanotechnology in the COVID-19 era: Carbon-based nanomaterials as a promising solution. *Carbon, 210*, 118058.

Pask, A. (2016). The reproductive system. In *Non-coding RNA and the reproductive system*. Springer.

Patel, U., Rathnayake, K., Hunt, E. C., & Singh, N. (2022). Role of nanomaterials in COVID-19 prevention, diagnostics, therapeutics, and vaccine development. *Journal of Nanotheranostics, 3*, 151–176.

Patlolla, A. K., Hackett, D., & Tchounwou, P. B. (2015). Silver nanoparticle-induced oxidative stress-dependent toxicity in Sprague-Dawley rats. *Molecular and Cellular Biochemistry, 399*, 257–268.

Perumal, M. V., Nagarajan, S., Srinivasan, R., Jayamoorthy, K., & Usharani, T. (2024). Green synthesis of copper nanoparticles using flower extracts: A promising route for enhanced microelectronics packaging. *Journal of Molecular & Engineering Materials, 12*, 2450010.

Pinho, A. R., Martins, F., Costa, M. E. V., Senos, A. M. R., da Cruz Silva, O. A. B., Pereira, M. L., & Rebelo, S. (2020). In vitro cytotoxicity effects of zinc oxide nanoparticles on spermatogonia cells. *Cells, 9*, 1081.

Pope-Harman, A., Cheng, M. M.-C., Robertson, F., Sakamoto, J., & Ferrari, M. (2007). Biomedical nanotechnology for cancer. *Medical Clinics of North America, 91*, 899–927.
Priyam, A., Yadav, N., Reddy, P. M., Afonso, L. O. B., Schultz, A. G., & Singh, P. P. (2022). Uptake and benefits of biogenic phosphorus nanomaterials applied via fertigation to japonica rice (Taipei 309) in low-and high-calcareous soil conditions. *ACS Agricultural Science & Technology, 2*, 462–476.
Priyanka, S., Namasivayam, S. K. R., Kennedy, J. F., & Moovendhan, M. (2024). Starch-chitosan-Taro mucilage nanocomposite active food packaging film doped with zinc oxide nanoparticles–Fabrication, mechanical properties, anti-bacterial activity and eco toxicity assessment. *International Journal of Biological Macromolecules, 277*, 134319.
Qin, Z., Jiang, Q., Chen, M., Li, J., & Zhang, H. (2024). Food packaging for ripening and preserving banana based on ethylene-loaded nanofiber films deposited with nanosized cyclodextrin metal-organic frameworks and TiO2 nanoparticles. *Food Packaging and Shelf Life, 45*, 101332.
Rabbi, M. B., Tania, I. S., Sani, A. A., & Uddin, M. Z. (2024). Facile approach for preparing printed cotton fabric with antimicrobial activity by utilizing the functional characteristics of nano-silver. *Journal of Inorganic and Organometallic Polymers and Materials, 34*, 3675–3688.
Raj, S., Jose, S., Sumod, U. S., & Sabitha, M. (2012). Nanotechnology in cosmetics: Opportunities and challenges. *Journal of Pharmacy and Bioallied Sciences, 4*, 186–193.
Ramesh, A. M., Rajesh, M., Chandran, A., & Surendran, K. P. (2024). Gold-nanoparticle-based flexible humidity sensor for breath monitoring and smart irrigation systems. *ACS Applied Nano Materials, 7*, 15593–15605.
Rastogi, A., Zivcak, M., Sytar, O., Kalaji, H. M., He, X., Mbarki, S., & Brestic, M. (2017). Impact of metal and metal oxide nanoparticles on plant: A critical review. *Frontiers in Chemistry, 5*, 78.
Ray, P. C., Yu, H., & Fu, P. P. (2009). Toxicity and environmental risks of nanomaterials: Challenges and future needs. *Journal of Environmental Science and Health Part C, 27*, 1–35.
Razzaq, S., Zhou, B., Ullah, Z., Zia-ur-Rehman, M., Guo, H., Adil, M., Xiaopeng, C., & Qian, L. W. (2024). Exploring the impact of organic and inorganic amendments, with foliar application of iron nanoparticles, on cadmium stabilization and growth of maize in wastewater irrigated-soil. *Journal of Hazardous Materials Letters, 5*, 100111.
Razzaque, S., Abubakar, M., Farid, M. A., Zia, R., Nazir, S., Razzaque, H., Ali, A., Ali, Z., Mahmood, A., & Al-Masry, W. (2024). Detection of toxic cypermethrin pesticides in drinking water by simple graphitic electrode modified with Kraft lignin@ Ni@gC3N4 nano-composite. *Journal of Materials Chemistry B, 12*, 9364–9374.
Rehan, M., Mashaly, H. M., Abdel-Aziz, M. S., Abdelhameed, R. M., & Montaser, A. S. (2024). Viscose fibers decorated with silver nanoparticles via an in-situ green route: UV protection, antioxidant activities, antimicrobial properties, and sensing response. *Cellulose, 31*, 5899–5930.
Rehman, A., John, P., & Bhatti, A. (2021). Biogenic selenium nanoparticles: Potential solution to oxidative stress mediated inflammation in rheumatoid arthritis and associated complications. *Nanomaterials, 11*, 2005.
Rejiniemon, T. S., Raishy, H. R., Bhamadevi, R., Binobead, M. A., Aljowaie, R. M., & Rathi, M. A. (2024). Banana pseudo-stem and cattle manure for lactic acid production and the application of polylactic acid-cellulose silver nanoparticle-based nanocomposite films in food storage. *BioResources, 19*, 5654.
Roy, T., Basak, N., Mainak, S., Das, S., Ali, S. I., & Islam, E. (2024). Burkholderia sp. EIKU21 mediated synthesis of biogenic ZnO nanoparticle–based pigment for development of antibacterial cotton fabric through nanocoating. *Biomass Conversion and Biorefinery, 15*, 6927–6942.
Sadeghi-Kiakhani, M., & Hashemi, E. (2024). Silver nanoparticles synthesized on the wool yarns with safflower natural dye: Colorimetric and fastness studies. *Progress in Color, Colorants and Coatings, 17*, 325–332.
Sadeghi-Kiakhani, M., Hashemi, E., & Norouzi, M.-M. (2024). Clean synthesis of silver nanoparticles (AgNPs) on polyamide fabrics by Verbascum thapsus L.(mullein) extract: Characterization, colorimetric, antibacterial, and colorfastness studies. *Environmental Science and Pollution Research, 31*, 32637–32648.

Salama, K. F., AlJindan, R., Alfadhel, A., Akhtar, S., & Al-Suhaimi, E. A. (2024). Enhanced antimicrobial performance of textiles coated with TiO2 nanoparticles. *Journal of Industrial Textiles, 54*, 15280837241233744.

Saleemi, S., Mannan, H. A., Riaz, T., Hai, A. M., Zeb, H., & Khan, A. K. (2024). Optimizing synergistic silica-zinc oxide coating for enhanced flammability resistance in cotton protective clothing. *Fibers, 12*, 44.

Sarmast, E., Shankar, S., Salmieri, S., Rahmouni, S. A., Mahmud, J., & Lacroix, M. (2024). Cross-linked gelatin–riboflavin-based film incorporated with essential oils and silver nanoparticle by gamma-irradiation: A novel approach for extending the shelf life of meat. *Food Hydrocolloids, 147*, 109330.

Sharifi, S., Behzadi, S., Laurent, S., Forrest, M. L., Stroeve, P., & Mahmoudi, M. (2012). Toxicity of nanomaterials. *Chemical Society Reviews, 41*, 2323–2343.

Sharma, D., Sharma, N., Pathak, M., Agrawala, P. K., Basu, M., & Ojha, H. (2018). Nanotechnology-based drug delivery systems: Challenges and opportunities. In *Drug targeting and stimuli sensitive drug delivery systems* (pp. 39–79). Elsevier.

Shilpa, S. A., Arthi, C., Hikku, G. S., Jeyasubramanian, K., Veluswamy, P., & Ikeda, H. (2024). Silver nanosphere/polycaprolactone coated cotton fabrics as hygienic textiles for health care industries. *Biointerface Research in Applied Chemistry, 14*, 11.

Siddiqua, U. H., Riaz, A., Faheem, M. S., Batool, R., Ullah, I., & Sabir, Q. U. A. (2024). Effect of silver nanoparticles finishing on dyeing properties of newly synthesized reactive dye applied on cellulosic fabric. *Fibers and Polymers, 25*, 987–995.

Simsek, B., Ruhkopf, J., Plachetka, U., Rademacher, N., Belete, M., & Lemme, M. C. (2024). Silver nanoparticle-assisted electrochemically exfoliated graphene inks coated on PVA-based self-healing polymer composites for soft electronics. *ACS Applied Materials & Interfaces, 16*, 7838–7849.

Singh, G., Thakur, N., & Kumar, R. (2024). Nanoparticles in drinking water: Assessing health risks and regulatory challenges. *Science of the Total Environment, 949*, 174940.

Singh, S., Arya, V., Mishra, R. K., Rajput, S. K., & Dhanasekaran, M. (2024). Nanotechnology in the diagnosis and management of polycystic ovary syndrome: A systematic scoping analysis to improve patient care. *Next Research, 1*, 100004.

Skwarek, A., Gkasecka, A., Jaguszewski, M. J., Szarpak, Ł., Dziecikatkowski, T., & Filipiak, K. J. (2021). Nanoparticles: A breakthrough in COVID-19 prevention, diagnosis and treatment. *Archives of Medical Science: AMS, 19*, 1410.

Soliman, S. (2023). Nanomedicine: Advantages and disadvantages of nanomedicine. *Journal of Nanomedicine and Nanotechnology, 14*, 666.

Songtipya, P., Prodpran, T., Songtipya, L., & Sengsuk, T. (2024). Synergism of zinc oxide/organoclay-loaded poly (lactic acid) hybrid nanocomposite plasticized by triacetin for sustainable active food packaging. *Journal of Renewable Materials, 12*, 951–967.

Subramanian, K. S., Karthika, V., Praghadeesh, M., & Lakshmanan, A. (2020). Nanotechnology for mitigation of global warming impacts. In *Global climate change: Resilient and smart agriculture* (pp. 315–336). Springer.

Sun, X., Wang, H., Liang, H., Meng, N., & Zhou, N. (2025). Fabrication of antimicrobial chitosan/ZnO nanoparticles/lecithin-montmorillonite films for food packaging application. *Food Hydrocolloids, 159*, 110686.

Tang, L., Chen, M., Wang, D., He, Y., Ge, G., Zeng, Z., Shu, J., Guo, W., Wu, S. X., & Xiong, W. (2024). Doxorubicin and iron-doped mesoporous silica nanoparticles for chemodynamic therapy and chemotherapy of breast cancer. *New Journal of Chemistry, 48*, 17294–17309.

Thipwimonmas, Y., Kongkaew, S., Phua, C. H., & Limbut, W. (2024). A distance-based paper analytical device: Visual quantification of salt in foods with silver nanoparticles on paper. *Microchemical Journal, 199*, 109853.

Thu, H. E., Haider, M., Khan, S., Sohail, M., & Hussain, Z. (2023). Nanotoxicity induced by nanomaterials: A review of factors affecting nanotoxicity and possible adaptations. *OpenNano, 14*, 100190.

Tiple, A. D., Badwaik, V. J., Padwad, S. V., Chaudhary, R. G., & Singh, N. B. (2020). A review on Nanotoxicology: Aquatic environment and biological system. *Materials Today: Proceedings, 29*, 1246–1250.

Tooklang, P., Audtarat, S., Chaisen, K., Thepsiri, J., Chingsungnoen, A., Jittabut, P., & Dasri, T. (2024). Functionalization of silver nanoparticles coating cotton fabrics through hydrothermal synthesis for improved antimicrobial properties. *Nano Express, 5*, 25009.

Tunhoo, B., Chantarawong, D., Thiwawong, T., & Onlaor, K. (2024). Influence of phase structure of TiO2 nanoparticles on resistive switching devices. *Current Applied Science and Technology, 77*, e0258184.

Urban, M., Rosati, G., Maroli, G., Della, P. F., Bonini, A., Sajti, L., Fedel, M., & Merkoçi, A. (2024). Nanostructure tuning of gold nanoparticles films via click sintering. *Small, 20*, 2306167.

Vodyashkin, A., Stoinova, A., & Kezimana, P. (2024). Promising biomedical systems based on copper nanoparticles: Synthesis, characterization, and applications. *Colloids and Surfaces B: Biointerfaces, 237*, 113861.

Vodyashkin, A. A., Rizk, M. G. H., Kezimana, P., Kirichuk, A. A., & Stanishevskiy, Y. M. (2021). Application of gold nanoparticle-based materials in cancer therapy and diagnostics. *ChemEngineering, 5*, 69.

Vyas, T., Choudhary, S., Sharan Rathnam, S., & Joshi, A. (2023). Fiber-optic detection of aluminum and copper in real water samples using enzyme–carbon quantum dot (CQD)-based thin-film biosensors. *ACS ES&T Engineering, 4*, 694–705.

Wang, J., Cheng, Y., Li, S., Liu, B., Yang, L., Geng, F., Xie, S., Qi, R., Zhang, Y., & Liu, D. (2024). Enhanced properties of gelatin films incorporated with TiO2-loaded reduced graphene oxide aerogel microspheres for active food packaging applications. *International Journal of Biological Macromolecules, 261*, 129772.

Wang, Y., Schimel, J. P., Nisbet, R. M., Gardea-Torresdey, J. L., & Holden, P. A. (2019). Soybeans grown with carbonaceous nanomaterials maintain nitrogen stoichiometry by assimilating soil nitrogen to offset impaired dinitrogen fixation. *ACS Nano, 14*, 585–594.

Wolfram, J., Zhu, M., Yang, Y., Shen, J., Gentile, E., Paolino, D., Fresta, M., Nie, G., Chen, C., & Shen, H. (2015). Safety of nanoparticles in medicine. *Current Drug Targets, 16*, 1671–1681.

Xaviour, J., Sreelekshmi, S., Joseph, J., Fathima, S. A., & Sajini, T. (2024). Eco-friendly synthesis and enhanced antibacterial action of bimetallic ag/ZnO nanoparticles using Hylocereus costaricensis stem extract. *RSC Sustainability, 2*, 3077–3089.

Xia, Y., Wang, S., Meng, F., Xu, Z., Fang, Q., Gu, Z., Zhang, C., Li, P., & Kong, F. (2024). Eco-friendly food packaging based on paper coated with a bio-based antibacterial coating composed of carbamate starch, calcium lignosulfonate, cellulose nanofibrils, and silver nanoparticles. *International Journal of Biological Macromolecules, 254*, 127659.

Xuan, L., Ju, Z., Skonieczna, M., Zhou, P.-K., & Huang, R. (2023). Nanoparticles-induced potential toxicity on human health: Applications, toxicity mechanisms, and evaluation models. *MedComm, 4*, e327.

Yamawaki, H., & Iwai, N. (2006). Mechanisms underlying nano-sized air-pollution-mediated progression of atherosclerosis carbon black causes cytotoxic injury/inflammation and inhibits cell growth in vascular endothelial cells. *Circulation Journal, 70*, 129–140.

Yang, G., Liu, Y., Chen, J., Ding, J., & Chen, X. (2022). Self-adaptive nanomaterials for rational drug delivery in cancer therapy. *Accounts of Materials Research, 3*, 1232–1247.

Yang, Z., & Cao, L. (2024). Electrochemical sensor based on silver nanoparticles/carboxylated multi-walled carbon nanotubes modified glassy carbon electrode for detection of carbamate pesticide in food. *Journal of Food Measurement and Characterization, 18*, 6225–6236.

Yeh, C.-T., & Feng, G.-H. (2024). Micromachined capacitance-sensitive device with immobilized functional ZnO nanoparticles detecting glucose and uric acid. *Journal of Applied Electrochemistry, 54*, 2135–2150.

Yi, M., Zhu, W., Wu, P., Qiu, Y., Sun, G., & Wang, L. (2024). Enhanced anti-oxidation copper-based conductive inks for low porosity copper films in power electronics. *IEEE Transactions on Components, Packaging and Manufacturing Technology, 14*, 1149–1155.

Yin, Q., Xu, L., Chen, Z., Cui, Y., Zhao, W., Geng, F., & Tao, C. (2024). Study on the application of hollow mesoporous polydopamine as a nanoparticle drug carrier in tumor combination therapy. *New Journal of Chemistry, 48*, 17525–17537.

Zhang, D., Guo, X., Rao, W., Pan, D., Cao, F., Zhai, T., Zheng, W., Abubakar, Y. S., Guan, X., & Chen, Z. (2025). A multi-stimuli–response metal–organic framework nanopesticide for smart weed control in agriculture. *Environmental Science: Nano, 73*, 18480.

Zhang, H. (2024). Application of graphene nanoparticles in sports clothing for enhancing moisture absorption, breathability, wear resistance, and antimicrobial properties. *Materials Express, 14*, 721–726.

Zhang, L., Gao, F., Liu, S., Ju, M., Sun, C., Sun, G., Ju, Q., Yang, K., & Fang, Z. (2024). Synthesis of lanthanide-based scintillator@ MOF nanocomposites for X-ray-induced photodynamic therapy. *Inorganic Chemistry Frontiers, 11*, 1607–1615.

Zhang, X., Zhang, B., Mao, R., Huang, Z., Jing, K., Jin, C., Yang, B., Qi, J., Yu, M., & Xiong, G. (2024). A novel multilayer film based on sodium alginate/k-carrageenan-gelatin incorporated with ZnO nanoparticles and oregano essential oil for active food packing. *Progress in Organic Coatings, 187*, 108170.

Zhang, Z., Hu, W., Yu, A., Kuang, H., & Wang, M. (2024). Hemostatic bioactivity and mechanism of novel Rubia cordifolia L.-derived carbon dots. *Nanoscale Advances, 6*, 6229–6238.

Zhao, K., Guan, X., Zhao, J., Cao, W., Dai, B., & Zhu, J. (2024). Rapid preparation of sintered ag nanoparticles pastes for high-power electronic packaging. *Materials Letters, 355*, 135562.

Zheng, T., Hou, J., Wu, T., Jin, H., Dai, Y., Xu, J., Yang, K., & Lin, D. (2024). Ferric oxide nanomaterials and plant-rhizobacteria symbionts cogenerate iron plaque for removing highly chlorinated contaminants in dryland soils. *Environmental Science & Technology, 58*(25), 11063–11073.

Zhou, J., Xia, M., Hu, L., & Li, Y. (2024). In situ growth of gold nanofilms with branched structures in the presence of organosulfur for high-performance flexible electronics. *Langmuir, 40*, 18225–18232.

Zhou, Y., Sun, Y., Li, Y., Shen, C., Lou, Z., Min, X., & Stewart, R. (2024). A highly durable and UV-resistant graphene-based knitted textile sensing sleeve for human joint angle monitoring and gesture differentiation. *Advanced Intelligent Systems, 6*, 2400124.

Zhu, B., Han, J., Lei, L., Hua, J., Zuo, Y., & Zhou, B. (2021). Effects of SiO2 nanoparticles on the uptake of tetrabromobisphenol a and its impact on the thyroid endocrine system in zebrafish larvae. *Ecotoxicology and Environmental Safety, 209*, 111845.

Chapter 10
Life Cycle Assessment and Environmental Impact of Graphene and Carbon Nanotube Production

B. Lokeshwari, P. Saranraj, M. Manigandan, K. Gayathri, and K. Kesavardhini

10.1 Introduction

Nanotechnology is one of the most extensive innovative fabrication materials with mechanical, electrical, and thermal properties. Graphene and its compounds, like the carbon Nanotube, are the most used nanomaterials due to their unique structure and properties that promote applying graphene-based materials in electronics, energy storage, medicine, aerospace, and composite materials. Graphene is a single layer of sp^2-bonded carbon atoms packed into a two-dimensional honeycomb lattice and displaying extraordinary electrical conductivity, mechanical and chemical durability, flexibility, and a large area for use in flexible electronics, supercapacitors, and detectors. Si-based nanomaterials are employed extensively for bio-imaging and drug delivery for improved mechanical properties due to their toughness, strength, ductility, and thermal resistance engineering, which can also include tensile strength, thermal and electrical stability, and drug delivery (Finnveden et al., 2009). The resource and hazardous material-dependent large-scale production of graphene and CNTs causes acute environmental problems. There are many methods for producing graphene, including chemical vapour deposition (CVD), mechanical and chemical exfoliation, electrochemical synthesis, and chemical reduction with unique energy requirements, carbon footprints, and chemical waste generation of chemical waste.

Chemical exfoliating and reducing techniques are associated with environmental and aquatic pollution derived from the toxic agents and reducing agents used (García García et al., 2021). Various processes involved in carbon nanotube synthesis include arc discharge, laser ablation, and catalytic CVD processes, which affect

B. Lokeshwari · P. Saranraj (✉) · M. Manigandan · K. Gayathri · K. Kesavardhini
PG and Research Department of Microbiology, Sacred Heart College (Autonomous), Tirupattur, Tamil Nadu, India

S. Kulkarni (ed.), *Next-Generation Nanomaterials for Sustainable Engineering*, Lecture Notes in Nanoscale Science and Technology 43,
https://doi.org/10.1007/978-3-032-05719-8_10

environmental processes. Life cycle assessment (LCA), an ISO standard framework for the environmental footprints associated with the life cycle of a product, can be used for sustainability assessment of the technical and economic aspects of graphene and CNTs production. When used in the analysis of the life cycle of a material, LCA examines the extraction of raw materials, the synthesis, processing, end-use, and disposal of the material. The life cycle assessment (LCA) showed the environmental impact of large-scale production of graphene and CNTs is very energy demanding and that CVD synthesis approaches are the most carbon-emission-intensive methods in the production of electricity and precursor gas (Aben et al., 2014).

Non-green chemical synthesis methods with toxic reagents are responsible for waste generation and the need for waste management and safer alternatives. The production of graphene and CNT can be done using sustainable methods by utilising low-energy fabrication strategies. The production techniques increase the very high temperatures and create toxic chemicals, which are responsible for the emission of greenhouse gases and environmental toxicity and thus play a pivotal role in making these materials sustainable. The environmental sustainability of these graphene and CNT-based products is often addressed regarding their disposal and recycling, often deferring to potential toxicity and stability in the environment. New processes with these nanomaterials are adding to greenhouse gas emissions and resource consumption. The sustainability of graphene and CNT production can be enhanced by using renewable energy sources, developing green synthesis methods, and using efficient recycling techniques. The innovation, regulations, and waste management techniques are key points that should be addressed to minimise the environmental hazards of graphene and CNT manufacturing and their technological impacts for a sustainable future.

10.2 Life Cycle Stages of Graphene

Life cycle assessment (LCA) examines the various stages in the life of graphenes, including the environmental impact of raw material extraction and the final disposal of graphene. Graphene, a two-dimensional (2D) structure in which monolayer sp^2-hybridised carbon atom hexagonal lattices, has demonstrated electrical conductivity, mechanical strength, and heat resistance, and its diverse applications range from electronics and energy storage to biomedical devices and composite materials. The sustainability of graphene is mainly based on the specific properties of graphene and the environmental impact of graphene (Klinglmair et al., 2013). Graphene can be derived from various sources like graphite, graphene oxide, other feedstock, or raw materials with varying thermodynamic energy and their respective effects.

Graphene production processes can be divided into top-down or bottom-up processes based on the energy inputs. Functionalisation of graphene to obtain suitable properties is necessary for the application requirements. The incorporation of graphene in applications like batteries, photovoltaic cells, nano-composite materials, and many more, and the raw material extraction involves the sourcing of materials rich in carbon, or carbon feedstocks, including graphite, hydrocarbons, or biomass-derived carbon precursors, with widely varying impacts on the environment. The conventional pathway to graphene production involves mining natural graphite, which raises the issues of resource depletion, energy-intensive extraction, and land degradation due to mining. However, carbon-rich precursors from renewable sources have been explored as more sustainable alternatives. The synthesis of graphene is the central environmentally intensive stage because the various production methods have diverse energy consumption, GHG emissions, and chemical waste (Jacquemin et al., 2012).

Standard synthesis methods can be divided into chemical vapour deposition (CVD), mechanical exfoliation, liquid-phase exfoliation, and chemical reduction of graphene oxide (GO). The most used method for high-quality graphene is CVD, which adopts the thermal decomposition of hydrocarbon precursors (methane or acetylene) on metal substrates at high temperatures, resulting in high energy consumption and carbon emissions. The liquid-phase and mechanical exfoliation methods depend on organic solvents or surfactants, resulting in chemical waste and potential aquatic toxicity. Material processing and functionalisation refers to modifying graphene to make it more compatible, often involving chemical treatments, doping, or integration of polymers, secondary environmental impacts in the form of solvents and releases of hazardous waste (Baumann & Tillman, 2004). The product integration and use of graphene in commercial products, such as batteries, sensors, coatings and composites, may affect their overall life cycle sustainability.

Downstream applications of graphene-enabled systems can also have significant environmental benefits, such as enhanced energy efficiency for electronic devices or longer battery life. The overall sustainability of graphene materials is challenged by two key factors: the potential release of nanoparticles into ecosystems, with associated health and environmental risks, and the critical need for effective end-of-life management. End-of-life management is a key factor for the overall sustainability of all graphene materials (Mannan & Al-Ghamdi, 2020). Graphene has conventional materials with their recyclability, biodegradability, and waste treatment. Disposal methods include incineration, landfilling, or chemical degradation of graphene-containing materials in bioaccumulation and environmental biotoxicity. The synthesis methods for graphene materials are improving sustainability, synthesis techniques, using renewable carbon footprints, and effective waste management strategies, which can all increase the sustainability of graphene-synthesised materials. The life cycle phases of graphene production and its environmental impacts have sustainable alternatives (Hossain & Poon, 2018).

10.2.1 Raw Material Acquisition

Graphene is a two-dimensional nanomaterial constructed of sp^2-hybridised carbon atoms. It is predominantly fabricated from carbon-rich precursors, such as natural graphite, hydrocarbon gases, and biomass-derived carbon sources (Peters & Weil,

2016). The environmental impacts associated with resource extraction, energy depletion, GHG emissions and waste. Top graphite-producing countries, China, Brazil, and Canada, use open-pit or underground mining methods that can contribute to soil degradation, biodiversity loss, and groundwater contamination. The conversion of mined graphite to graphene precursors is a series of chemical modifications, creating solid waste and effluents with hazardous compounds like concentrated acids and oxidising agents that may endanger the environment and human health (Curran, 1996). Synthetic graphite is a cheaper alternative to natural graphite produced through the high-temperature graphitisation of petroleum coke or coal tar. Synthetic Graphite (SG) provides granular, higher purity precursor materials for graphene in a small amount of carbon emissions and energy consumption during its production, as its molecular precursors are derived from fossil fuels.

Chemical vapour deposition (CVD) synthesis is the most common technique for high-quality graphene production (Ando et al., 2000). It uses hydrocarbon gases, including methane, acetylene, or benzene as carbon feedstocks. CVD allows the manufacture of large-area, defect-free graphene, and it occurs at high temperatures (>1000 °C), with energy consumption and CO_2 emissions. The CVD growth of graphene utilises transition metal catalysts like copper or nickel, which need to be extracted and processed, potentially causing both resource depletion and environmental impacts. Biomass-derived carbon feedstocks have been considered an eco-friendly alternative to graphene synthesis. Biomass is renewable and carbon-neutral and can be applied for graphene synthesis through pyrolysis or hydrothermal carbonisation. The biomass sources include lignocellulosic materials, agricultural waste, and organic residues. These processes minimise the non-renewable feedstock and waste valorisation and make graphene production more sustainable (Cassell et al., 1999).

The precursors for graphene synthesis have an effect on water usage, chemical toxicity, waste production, and the environmental footprints. Some processes, such as chemical exfoliation and reduction, involve toxic reagents such as sulphuric acid, nitric acid, and potassium permanganate that generate toxic effluents, inducing aquatic pollution and ecological contamination. Graphite mining and processing consume a significant amount of water, which provides challenges in water-scarce regions. Adopting sustainable raw material strategies is essential to reducing environmental impacts (Dunens et al., 2010). Life cycle assessment shows the low-impact graphene production by the raw materials. The feedstocks, processes, and optimisations for a sustainable graphene nanotechnology sector (Goedkoop et al., 2008).

10.2.1.1 Extraction Processes

The pretreatment and extraction process are used to obtain the precursor materials for life cycle assessment of graphene production. Sources such as natural graphite, hydrocarbons, and biomass have specific extraction methods that have distinct properties. Many of these processes require graphite mining and can contribute significantly to land degradation, deforestation and hydro-geological disruptions (Hornbostel et al., 2006). Natural graphite must be treated with chemicals to be refined and toxic waste by-products of acidic effluent are harmful to soil and water systems. The chemical vapour deposition (CVD) synthesis utilises hydrocarbon gases, like methane or acetylene, that are obtained through natural gas processing or petroleum refining that generates GHGs and fossil fuels. The eco-friendly method includes carbon extraction from biomass created by pyrolysis or hydrothermal carbonisation of agricultural waste or other organic materials to produce graphene precursors. This has the potential for sustainable alternatives to minimise ecological footprints by the various forms of the extraction methods in graphene production (Chaing et al., 2001).

10.2.1.2 Transportation Impacts

In the transportation phase, raw materials like natural graphite, hydrocarbon gases, or biomass-derived precursors are transported from mining sites, petroleum refineries, or agricultural processing facilities to graphene fabrication plants. It also requires long-distance transportation by trucks, ships, or trains that produces emissions of carbon dioxide (CO_2), nitrogen oxides (NO_x), and particulate matter (PM) that pollute the air and cause global warming (Maxwell et al., 2006). The high bulk density and weight of graphite ores increase fuel consumption during transportation. The graphene is synthesised and transported to downstream industries for electronics, energy storage, biomedical devices, composite materials, and packaging to environmental waste (Vazquez et al., 2002). Low-carbon transportation methods, including electric or hydrogen-powered trucks, can reduce graphene transport carbon footprints. The sustainability challenges in the life cycle of graphene to minimise environmental impacts (Martinez et al., 2002).

10.2.2 Production

The production stage of graphene in its life cycle assessment (LCA) is one of the most environmentally costly stages, as well as the processes involved in energy usage, the treatment of raw materials, the usage of chemicals, and the generation of waste. Standard methods include mechanical exfoliation, chemical vapour deposition (CVD), liquid-phase exfoliation, and chemical reduction (Beguin & Frackowiak, 2009). Mechanical exfoliation is more from graphite through the mechanical process for industrial production. Chemical vapour deposition (CVD) is the most commonly used method for high-quality graphene synthesis. It relies on the deposition of hydrocarbon gases like methane or acetylene on metal substrates at temperatures of over 1000 °C, making it energy demanding and contributing to CO_2 emissions (Madadrang et al., 2012). Liquid-phase exfoliation uses solvents or surfactants to split graphite into graphene sheets, but it also involves chemical waste and potential aquatic toxicity. The chemical reduction of graphene oxide, the mass production method, has oxidising and reducing agents (Gutowski et al., 2009).

10.2.3 Utilisation

The diverse properties of graphene, including high electrical conductivity, mechanical strength, thermal stability, and large surface area, leading to energy storage, biomedical devices, and advanced composites, the utilisation stage in the life cycle assessment (LCA) of graphene becomes vital in its environmental benefits (Shaffer et al., 2006). Graphene is a lightweight, strong, and environmental-friendly material; its impact on the environment during application is influenced by energy use in electronic applications, compatibility with other materials and the lifespan of products. Graphene reduces resistance and improves charge storage in energy storage devices, such as lithium-ion batteries and supercapacitors, resulting in increased energy efficiency and indirectly contributing to lower carbon emissions by the lifespan of batteries and decreasing dependence on conventional, resource-intensive materials. The corrosion resistance and mechanical properties of graphene increase material lifetime and decrease the need for and resource consumption from the replacement of materials (Curran et al., 2007). The toxicity and environmental applications released into ecosystems may affect aquatic and soil organisms. The graphene interactions with various polymers, metals, and other nanomaterials also impact the degree of recycling and disposal efficiency (Kohler et al., 2008). It also analyses the benefits, sustainability potential, and long-term environmental impact of graphene usage in various industries.

10.2.4 Disposal

The disposal stage of graphene in the LCA is vital to its environmental sustainability. The processing of graphene-based materials can cause significant ecological and human health threats. Graphene differs from traditional bulk materials as it exists at a nanoscale dimension, which causes its degradation, accumulation, and interactions with biological and environmental systems (Tillman, 2000). Graphene-containing products, such as landfilling, incineration, recycling, and biodegradation, have different environmental factors. Landfilling is still the predominant disposal method for leaching graphene nanoparticles into the soil and water systems (Baumann & Tillman, 2004). Incineration can be efficient for reducing the volume of graphene waste; it has the potential to generate airborne particulate emissions and toxic by-products; graphene is in polymeric or composite materials.

Recycling and material recovery can be a more sustainable solution since graphene can be reprocessed or chemically treated to restore functionality and reduce the production of fresh raw materials (Olivier et al., 2003). The separation of graphene from complex waste streams interacts with multiple materials, making recovery. Biodegradable graphene composites and enzyme degradation are being developed to mitigate changes in the industrial adoption of graphene, disposal, and waste management approaches with environmental impact. It is more critical to graphene, which minimises its potential for the environment. The disposal issues and risks of graphene by-products and their end-of-life have the recycling technologies to safeguard the long-term environmental compatibility of graphene.

10.3 Life Cycle of Carbon Nanotubes

The life cycle of carbon nanotubes (CNTs) is similar to that of graphene, and both belong to the class of carbon-based nanomaterials, leading to different production processes. The acquisition of raw materials, production, use, and disposal has significant implications for their environmental impacts. The raw materials of CNT synthesis are graphite, hydrocarbons, and biomass-derived feedstocks, and the most common synthesis method is chemical vapour deposition (CVD) for high quality (Gregory et al., 2016). Some hydrocarbon gases like methane, acetylene, or ethylene emit significant carbon and consume energy. Other techniques, like arc discharge and laser ablation, still depend on electricity and gas-based precursors that are potentially less energy-intensive than CVD.

The source of raw materials, such as graphite, uses metals to prepare catalysts for resource depletion and toxic waste (Arvidsson et al., 2013). With high mechanical strength, electrical conductivity, and thermal properties, CNTs are used in various industries. They may also be toxic towards humans and the environment throughout their life cycle, including during use and end-of-life (EOL) stages. The disposal and recycling of CNTs are not sufficiently well-developed to deal with nanomaterials. Innovative and sustainable end-of-life solutions such as recycling, safe disposal or biodegradation to minimise carbon nanotubes for the environmental impacts (Benoit et al., 2009).

10.3.1 Raw Material Acquisition

The production stages of graphene have approximately four distinct life cycle phases. The acquisition of raw materials as a feedstock selection can impact the environment throughout the production process. Graphene is a nanomaterial consisting of a monolayer of carbon atoms and can be obtained from various carbonaceous precursors (Swarr et al., 2011). The most widely used raw materials for manufacturing graphene are natural graphite, synthetic graphite, and gaseous carbon compounds, such as methane, acetylene, or carbon obtained from biomass. The extraction of graphite, necessary for mechanical exfoliation and chemical oxidation methods, is also required on a large scale through open-pit methods.

Graphite's energy-consuming refinement and purification significantly contribute to carbon emissions with their chemical treatments in the acids or solvents (Arvidsson et al., 2011). Synthetic graphite is derived from petroleum coke or coal tar that emits fossil fuels, carbon footprints, and depleting resources. Carbon vapour deposition (CVD) hydrocarbon gas synthesis is a precursor for extracting from natural gas or petroleum sources, depleting fossil fuels and emitting greenhouse gases. Biomass-derived feedstocks are the sustainable alternative because their productivity involves food production resources and exhibits variability in feedstock quality (Valdivia et al., 2013). The sustainability potential of raw material sourcing in graphene production on optimising renewable resources and mining methods to mitigate the footprints of graphene production.

10.3.2 Manufacturing of Carbon Nanotube Products

Carbon nanotubes (CNTs) are tubular structures made of carbon atoms and are usually produced through chemical vapour deposition (CVD), arc discharge, and laser ablation methods (Arvidsson et al., 2014). CVD is the most widely used method; the hydrocarbon gases, like methane or acetylene, will decompose at high temperatures and deposit CNTs on, a metal catalyst substrate. This method is highly energy-intensive and contributes to carbon emissions as the reaction temperatures. Arc discharge and laser ablation were also high-energy processes; in the discharge, high voltages were employed for plasma generation in laser ablation, and carbon-containing materials were vapourised using laser objects. It is most widely used for high-purity CNT production. They are energy-intensive and result in significant emissions. The production of CNTs in carbon precursors has been known to deplete resources and carbon emissions derived from petroleum-based feedstocks. The chemical waste produced during the manufacturing processes may create toxic by-products. The environmental impacts need to develop more energy-efficient production techniques for renewable carbon feedstocks in CNT manufacturing to enhance sustainability (Suh & Huppes, 2005).

10.3.3 Carbon Nanotube Products and Their Applications

The different life cycle stages of graphene and carbon nanotube (CNT) production and the phase of CNTs are crucial as they affect their environmental performance and sustainability. The mechanical strength, electrical conductivity, and thermal stability of CNTs are widely used in electronics, energy storage, aerospace, and biomedical applications (Hauschild & Huijbregts, 2015). CNTs are used in electronic devices like transistors, sensors, and displays. It also enhances the charge storage capacity and life of batteries and supercapacitors, which can significantly reduce carbon emissions through more efficient renewable energy systems. They are used in composite materials, for example, in the aerospace and automotive industries, to reinforce materials without adding considerable weight, which leads to increased fuel economy and reduced overall environmental impact (Colvin, 2003). CNTs can potentially infiltrate biological systems, having adverse effects if released into the atmosphere as electronic recycling or industrial waste (Goedkoop et al., 2008). The risk management strategies involve safe handling, encapsulation methods and responsible product design to minimise the release of CNT during use. The integration of eco-design strategies performs optimisation and decreases the risk to human health and ecosystems in the sustainability of CNTs.

10.3.4 End-of-Life (EOL) Management of Carbon Nanotubes

The life cycle of production for graphene and CNTs that substantially influences the environmental problems in the material is the end-of-life (EOL) stage. The properties of CNT, such as high tensile strength, chemical stability, and hardness for disposal and recycling techniques, overcome environmental and health hazards. As a result of their superior properties, CNT products are being discovered in several applications such as electronics, power storage, automotive, and aerospace, and their durability to mechanical strength and corrosive in the end-of-life (Auffan et al., 2011). Landfilling (the most common disposal method) can facilitate CNT ingestion within ecosystems. The environmental toxicity of CNTs is due to their nanoscale size and large surface area, which allows for their bioaccumulation in soil, water and marine living organisms. Burning can emit toxic gases or particulate matter, coupled with remaining CNTs, and thus have possible risks to air quality and health. The composite materials can further reprocess the CNT into new products while avoiding environmental impacts associated with producing raw materials (Chae et al., 2010).

10.4 Fundamentals of Life Cycle Assessment

Life cycle assessment (LCA) provides the environmental impact of materials, such as graphene and CNTs, through each lifecycle step, from raw material extraction to disposal after end-use. Life cycle assessment (LCA) is oriented towards analysing the environmental impacts or products where resource depletion, carbon emissions and energy consumption are developed (Menegaki & Damigos, 2018). LCA can also be utilised to evaluate the environmental impacts of some methods of synthesis that have been demonstrated to be variable in energy requirements, material feedstock input and waste output. LCA phase defines ecological or health impacts, including greenhouse gas emissions, water use, and toxicity. The underlying evaluation of raw materials, manufacture, transport, and energy is needed to form a base. This phase will assess environmental impacts, including global warming potential, eutrophication, and human toxicity levels. The production of graphene and CNTs follows environmental-friendly processes to reduce environmental impact across the life cycle (Cremiato et al., 2018).

10.5 Environmental Impact of Graphene and Carbon Nanotube Production

The lifecycle impacts of graphene and CNTs are divergent, from raw material extraction to synthesis, application, and disposal stages. Resource depletion and land degradation are caused by mining during the raw material acquisition stages, and carbon emissions and pollution are caused by petroleum reagents and metal catalysts used in CNT synthesis processes, such as the arc discharge and laser ablation phases (Ju et al., 2021). Dense processes (CVD, arc discharge in the production phase) lead to a more significant carbon footprint for classical materials. This process can create toxic by-products and hazardous waste that must be disposed of or managed to limit environmental harm. Graphene, in addition to CNTs, is critical for electronics, energy storage and composites (>20% global market, estimated to be >$5 billion), but the benefit of these nanomaterials may come with an unfortunate toxicological burden as nanomaterials are released into the environment during their use and/or disposal. In terms of the end-of-life stage, waste management is an emerging issue because recycling processes of graphene and CNTs potentially lead to environmental persistence in landfills (De Buzin et al., 2017). Chemical processes involved in the manufacture of graphene and CNT have significant implications for the sustainability of these nanomaterials as they will contribute to the environmental impact of the production process. The environmental impacts of the process, including ecotoxicity potential, energy use, human toxicity potential, and blue water footprint, are summarised in Fig. 10.1.

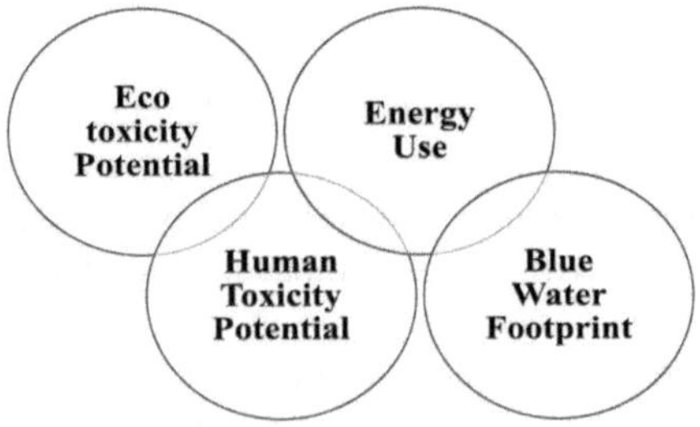

Fig. 10.1 Environmental impact of graphene and carbon nanotube synthesis processes

10.5.1 Energy Use

Graphene and carbon nanotubes (CNTs) describe essential properties and applications in various fields. These materials must be evaluated on their ecological impact to make them sustainable (Chen et al., 2019). Life cycle assessment (LCA) is one of the frequently employed approaches to evaluate the environmental burden of the production of graphene and CNT throughout their entire life cycle, that is, from raw material extraction to disposal. Energy consumption disproportionately affects the ecological footprint of graphene and CNT production. All synthesis processes, like chemical vapour deposition (CVDs) and laser ablation, are also very energy-intensive processes requiring high temperatures and sophisticated equipment, making them energy demanding. Energy sources are also critical in considering the overall environmental impact, as emissions from renewable sources are significantly lower than those from fossil fuel sources (Kiddee et al., 2013). Greenness: Another critical factor in the life cycle energy consumption of graphene and CNTs is the recycling and reusability potential. Hence, the low-carbon production of graphene and CNT and its future sustainable development greatly benefit from improving energy efficiency at every life cycle stage to alleviate environmental pollution.

10.5.2 Blue Water Footprint

The water use in production was in the life cycle assessment (LCA) of graphene and CNT production, from raw material extraction to final products (Jiang et al., 2021). Graphene and CNTs are synthesised in CVD or arc discharge, which requires much water for cooking, cleaning, and carbon source extraction, thus increasing the blue water footprints (Owen & Handy, 2007). The life cycle stages in water consumption can reduce the ecological footprints of nanomaterials and environmental-friendly applications in multiple sectors (Cui & Forssberg, 2003).

10.5.3 Human Toxicity Potential

Human toxicity potential determines the potential adverse effects of graphene and CNT production on human health. Life cycle assessment (LCA) analyses the hazardous emissions from raw material extraction to end-of-life (Deng et al., 2021). The production of graphene and CNTs involves the manufacturing of graphene, and CNTs involve using potentially hazardous chemicals such as solvents, reactants, and toxic gases with potential risk factors. For example, the graphene oxide sample to carbon monoxide, benzene, or polycyclic aromatic hydrocarbons in the CVD or synthesis method, chronic exposure to these components can cause health problems, including respiratory illnesses, skin irritation, or carcinogenic effects. The inhalation risk of human exposure resulting in adverse health effects such as lung inflammation, fibrosis, or another respiratory disease from releasing these materials into the air during processing or applications is also associated with the small particles and nanoscale components of graphene and CNTs. Potential human toxicity related to the contaminations that occur with the disposal and recycling of these materials in the environment and human population via food or water supply contamination (Fa et al., 2020). Thus, to ensure the green production of graphene and CNTs, it is necessary to continuously estimate their human toxicity potential.

10.5.4 Ecotoxicity Potential

Ecotoxicity potential (ECOP) in graphene and CNT production measures environmental harm throughout their life cycle (Burkhardt et al., 2011). Sustainability concerns associated with graphene and CNT production are evaluated through life cycle assessment (LCA) in releasing toxic chemicals in their life cycle, like the production, usage, and disposal phases that could threaten aquatic, terrestrial, and microbial habitats. Nanoparticles during production and disposal can disrupt aquatic organisms' growth, reproduction, and survival. The graphene and CNTs make them highly reactive molecules. This increases the probability of these materials interacting with biological systems, resulting in bioaccumulation and ecological disruption towards the bulk environmental toxicity of the synthetic materials (Urbansky et al., 2001). Graphene and CNTs are non-biodegradable; they cannot degrade in the environment, causing chronic contamination and adversely affecting the growth of plants and animals in the environment. As potential ecotoxicity can be derived from the life cycle stages of graphene and CNTs, these concerns sustainable production methods and appropriate waste management strategies. Using toxic chemicals improves recyclability and reduces the use of poisonous substances (Leonelli & Mason, 2010).

10.6 Potential of Life Cycle Assessment to Improve the Sustainability of Carbon Nanotubes and Graphene Production

Carbon nanotubes (CNTs) and graphene have unique and valuable macroscopic and microscopic properties, and life cycle assessment (LCA) has been an efficient way to enhance their production environmental performance. It has been done to quantify their environmental costs and benefits throughout their life cycle stages. Life cycle assessment (LCA) impacts energy, water, human toxicity, ecotoxicity, and carbon footprint, quantifying the environmental impacts (Hoekstra et al., 2011). Synthesis of CNTs and graphene involves energy-cost stages, including raw material extraction, high-temperature synthesis, and post-processing treatments. It can also identify the stages of energy consumption and analyse the processes involved that can be replaced or whether the energy can be sourced from renewable energy sources to reduce carbon emissions. Life cycle assessment (LCA) is an analytical tool that optimises resource utilisation and recycling and evaluates toxicity hazards in graphene and CNT production (Hauschild et al., 2008).

Harmful elements like heavy metals or solvents, during synthesis, transport or disposal could help manufacturing and disposal practices. The application minimises the threats posed to human health and the environment by synthesising graphene and CNTs. LCA may also contribute to green production processes, such as solvent-free synthesis or the more eco-efficient purification process of the environment and material. LCA also addresses sustainable material design and development based on the environmental impacts of each life cycle stage (Hagman et al., 2013). With the increasing demand for CNTs and graphene in numerous applications, adopting LCA will be essential to the responsible production of these materials. LCAs have significant potential to improve CNT and graphene sustainability and offer an environmental evaluation to minimise resource consumption, decrease the power of toxicity and minimise the production of nanomaterials (Rosenbaum et al., 2008).

10.7 Conclusion

The life cycle stages of graphene, from raw material acquisition to disposal, underscore both its transformative potential and environmental challenges. Life cycle assessment (LCA) reveals critical impacts, including resource depletion from graphite mining, high energy consumption in synthesis methods like chemical vapour deposition (CVD), and chemical waste from exfoliation processes. While graphene's application in energy storage, electronics, and composites offers sustainability benefits through enhanced efficiency and durability, risks such as nanoparticle release and end-of-life management complexities—landfilling, incineration, and limited recyclability pose significant ecological and health concerns. Comparatively, carbon nanotubes share similar life cycle challenges but differ in synthesis pathways and toxicity profiles. Sustainable alternatives, such as biomass-derived feedstocks, renewable energy integration, and biodegradable composites, highlight pathways to mitigate environmental footprints. LCA emerges as a pivotal tool for identifying hotspots, optimising resource use, and guiding eco-design innovations. To realise graphene's full potential, advancing green synthesis techniques, circular economy strategies, and stringent disposal protocols is imperative, ensuring alignment with global sustainability goals.

References

Aben, N. V. D., Voll, P., Peters, M., & Bardow, A. (2014). Life cycle assessment of CO_2 capture and utilization: A tutorial review. *Chemical Society Reviews, 43*(23), 7982–7994.

Ando, Y., Zhao, X., Hirahara, K., Suenaga, K., Bandow, S., & Lijima, S. (2000). Mass production of single-wall carbon nanotubes by the arc plasma jet method. *Chemical Physics Letters, 323*, 580–585.

Arvidsson, R., Kushnir, D., Sandén, B. A., & Molander, S. (2014). Life cycle assessment (LCA) of emerging technologies. *Environmental Science and Technology, 48*, 8–15.

Arvidsson, R., Molander, S., & Sandén, B. A. (2013). Review of potential environmental and health risks of the nanomaterial graphene. *Human and Ecological Risk Assessment: An International Journal, 19*(4), 873–887.

Arvidsson, R., Molander, S., Sandén, B. A., & Hassellöv, M. (2011). Challenges in exposure modelling of nanoparticles in aquatic environments. *Human and Ecological Risk Assessment: An International Journal, 17*(1), 245–262.

Auffan, M., Rose, J., Chaneac, C., Jolivet, J. P., Wiesner, M. R., & Bottero, J. Y. (2011). *Nanotoxicology and nanoethics*. Springer.

Baumann, H., & Tillman, A. M. (2004). The hitchhiker's guide to LCA: An orientation in life cycle assessment methodology and application. *The International Journal of Life Cycle Assessment, 11*(2), 142–142.

Beguin, F., & Frackowiak, E. (2009). *Carbons for electrochemical energy storage and conversion systems*. CRC Press.

Benoit, C., Mazijn, B., Andrews, E. S., Barthel, L. P., Beck, T., Ciroth, A., Cucuzzella, C., Gensch, C. O., Hébert, J., Lesage, P., Manhart, A., Mazeau, P., Methot, A. L., Moberg, A., Norris, G., Parent, J., Prakash, S., Reveret, J. P., Spillemaeckers, S., Ugaya, C. M. L., Valdivia, S., & Weidema, B. P. (2009). *Guidelines for social life cycle assessment of products*. United Nations Environment Programme and Society of Environmental Toxicology and Chemistry.

Burkhardt, J. J., Heath, G. A., & Turchi, C. S. (2011). Life cycle assessment of a parabolic trough concentrating solar power plant and the impacts of key design alternatives. *Environmental Science & Technology, 45*(6), 2457–2464.

Cassell, A. M., Raymakers, J. A., Kong, J., & Dai, H. (1999). Large-scale CVD synthesis of single-walled carbon nanotubes. *The Journal of Physical Chemistry B, 103*, 6484–6492.

Chae, S. R., Badireddy, A. R., Budarz, J. F., Lin, S. H., Xiao, Y., Therezien, M., & Wiesner, M. R. (2010). Heterogeneities in fullerene nanoparticle aggregates affecting reactivity, bioactivity and transport. *ACS Nano, 4*(9), 5011–5018.

Chaing, I. W., Brinson, B. E., Smalley, R. E., Margrave, J. L., & Hauge, R. H. (2001). Purification and characterisation of single wall carbon nanotubes. *The Journal of Physical Chemistry B, 105*, 1157–1161.

Chen, Z., Yang, M., Shi, Q., Kuang, X., Qi, H. J., & Wang, T. (2019). Recycling waste circuit board efficiently and environmentally friendly through small-molecule assisted dissolution. *Scientific Reports, 9*(1), 1–9.

Colvin, V. L. (2003). The potential environmental impact of engineered nanomaterials. *Nature Biotechnology, 21*(10), 1166–1170.

Cremiato, R., Mastellone, M. L., Tagliaferri, C., Zaccariello, L., & Lettieri, P. (2018). Environmental impact of municipal solid waste management using life cycle assessment: The effect of anaerobic digestion, materials recovery and secondary fuels production. *Ren Energy, 124*, 180–188.

Cui, J., & Forssberg, E. (2003). Mechanical recycling of waste electric and electronic equipment: A review. *Journal of Hazardous Materials, 99*(3), 243–263.

Curran, M. A. (1996). *Environmental life cycle assessment*. McGraw-Hill.

Curran, M. A., Frankl, P., Heijungs, R., Kohler, A., & Olsen, S. I. (2007). *Nanotechnology and life cycle assessment*. Elsevier.

De Buzin, P. J., Heck, N. C., & Vilela, A. C. (2017). EAF dust: An overview on the influences of physical, chemical and mineral features in its recycling and waste incorporation routes. *Journal of Materials Research and Technology, 6*(2), 194–202.

Deng, Y., Dewil, R., Appels, L., Ansart, R., Baeyens, J., & Kang, Q. (2021). Reviewing the thermochemical recycling of waste polyurethane foam. *Journal of Environmental Management, 278*, 111527.

Dunens, O. M., Mackenzie, K. J., & Harris, A. T. (2010). Large scale synthesis of double-walled carbon nanotubes in fluidised beds. *Industrial and Engineering Chemistry Research, 49*, 4031–4035.

Fa, S., Yamamoto, M., Nishihara, H., Sakamoto, R., Kamiya, K., & Nishina, Y. (2020). Carbon-rich materials with three-dimensional ordering at the angstrom level. *Chemical Science, 11*(23), 5866–5873.

Finnveden, G., Hauschild, M. Z., Ekvall, T., Guinée, J., Heijungs, R., Hellweg, S., Koehler, A., Pennington, D., & Suh, S. (2009). Life cycle assessment (LCA) and its applications in environmental management. *Journal of Environmental Management, 91*, 1–10.

García García, G., Fernandez, M. C., Armstrong, K., Woolass, S., & Styring, P. (2021). Analytical review of life-cycle environmental impacts of carbon capture and utilization technologies. *ChemSusChem, 14*(4), 995–1015.

Goedkoop, M., Heijungs, R., Huijbregts, M., De Schryver, A., Struijs, J., & van Zelm, R. (2008). *A life cycle impact assessment method which comprises harmonised category indicators at the midpoint and endpoint level.* Dutch Ministry of Housing.

Gregory, J. R., Noshadravan, A., Olivetti, E. A., & Kirchain, R. E. (2016). Life cycle assessment (LCA) and sustainability in materials and environmental impact assessments. *Environmental Science & Technology, 50*, 12–22.

Gutowski, T. G., Branham, M. S., Dahmus, J. B., Jones, A. J., & Thiriez, A. (2009). Thermodynamic analysis of resources used in manufacturing processes. *Environmental Science & Technology, 43*, 1584–1590.

Hagman, J., Nerentorp, M., Arvidsson, R., & Molander, S. (2013). Dobiofuels require more water than fossil fuels? Life cycle-based assessment of jatropha oil production in rural Mozambique. *Journal of Cleaner Production, 53*, 176–185.

Hauschild, M. Z., Huijbregts, M., Jolliet, O., Macleod, M., Margni, M., van de Meent, D., Rosenbaum, R. K., & McKone, T. E. (2008). Building a model based on scientific consensus for life cycle impact assessment of chemicals: The search for harmony and parsimony. *Environmental Science & Technology, 42*(19), 7032–7037.

Hauschild, M. Z., & Huijbregts, M. A. (2015). *LCA compendium - the complete world of life cycle assessment. Life cycle impact assessment.* Springer.

Hoekstra, A. Y., Chapagain, A. K., Aldaya, M. M., & Mekonnen, M. M. (2011). *The water footprint assessment manual: Setting the global standard.* Earthscan.

Hornbostel, B., Haluska, M., Cech, J., Dettlaff, U., & Roth, S. (2006). Arc discharge and laser ablation synthesis of single walled carbon nanotubes. *Carbon Nanotubes NATO Science Series, 222*(1), 1–18.

Hossain, M. D. U., & Poon, C. S. (2018). Comparative LCA of wood waste management strategies generated from building construction activities. *Journal of Cleaner Production, 177*, 387–397.

Jacquemin, L., Pontalier, P.-Y., & Sablayrolles, C. (2012). Life cycle assessment (LCA) applied to the process industry: A review. *International Journal of Life Cycle Assessment, 17*(8), 1028–1041.

Jiang, Q., Leeabai, N., Dilixiati, D., & Takahashi, F. (2021). Perceptive preference toward recycling bin designs: Influential design item depending on waste type, the impact of past perception experiences on design preference, and the effect of colour design on waste separation. *Waste Management, 127*, 130–140.

Ju, S., Shin, G., Lee, M., Koo, J. M., Jeon, H., & Ok, Y. S. (2021). Biodegradable chito-beads replacing non-biodegradable microplastics for cosmetics. *Green Chemistry, 23*, 6953–6965.

Kiddee, P., Naidu, R., & Wong, M. H. (2013). Electronic waste management approaches: An overview. *Waste Management, 33*(5), 1237–1250.

Klinglmair, M., Sala, S., & Brandão, M. (2013). Assessing resource depletion in LCA: A review of methods and methodological issues. *International Journal of Life Cycle Assessment, 19*(3), 580–592.

Kohler, A. R., Som, C., Helland, A., & Gottschalk, F. (2008). Studying the potential release of carbon nanotubes throughout the application life cycle. *Journal of Cleaner Production, 16*, 927–937.

Leonelli, C., & Mason, T. J. (2010). Microwave and ultrasonic processing: Now a realistic option for industry. *Chemical Engineering and Processing: Process Intensification, 49*(9), 885–900.

Madadrang, C. J., Kim, H. Y., Gao, G., Wang, N., Zhu, J., Feng, H., Gorring, M., Kasner, M., & Hou, S. (2012). Adsorption behavior of EDTA-graphene oxide for Pb (II) removal. *ACS Applied Materials & Interfaces, 4*(3), 1186–1193.

Mannan, M., & Al-Ghamdi, S. G. (2020). Environmental impact of water-use in buildings: Latest developments from a life-cycle assessment perspective. *Journal of Environmental Management, 261*, 110198.

Martinez, M. T., Callegas, M. A., Benito, A. M., Maser, W. K., Cochet, M., Andres, J. M., Schreiber, J., Chauvet, O., & Fierro, J. L. G. (2002). Microwave single walled carbon nanotubes purification. *Chemical Communications, 12*, 1000–1001.

Maxwell, D., Sheate, W. R., & Vorst, R. V. D. (2006). Functional and systems aspects of the sustainable product and service development approach for industry. *Journal of Cleaner Production, 14*(17), 1466–1479.

Menegaki, M., & Damigos, D. (2018). A review on current situation and construction and demolition waste management challenges. *Current Research in Green and Sustainable Chemistry, 13*, 8–15.

Olivier, J., Manuele, M., Raphael, C., Sébastien, H., Jerome, P., Gerald, R., & Ralph, R. (2003). *International Journal of Life Cycle Assessment, 8*, 6–10.

Owen, R., & Handy, R. (2007). Viewpoint: Formulating the problems for environmental risk assessment of nanomaterials. *Environmental Science & Technology, 41*(16), 5582–5588.

Peters, J., & Weil, M. (2016). A critical assessment of the resource depletion potential of current and future lithium-ion batteries. *Resources, 5*(4), 46–54.

Rosenbaum, R. K., Bachmann, T. M., Gold, L. S., Huijbregts, M. A. J., Jolliet, O., Juraske, R., Koehler, A., Larsen, H. F., MacLeod, M., Margni, M., McKone, T. E., Payet, J., Schuhmacher, M., van de Meent, D., & Hauschild, M. Z. (2008). USEtox-the UNEP-SETAC toxicity model: Recommended characterisation factors for human toxicity and freshwater ecotoxicity in life cycle impact assessment. *International Journal of Life Cycle Assessment, 13*(7), 532–546.

Shaffer, M. S. P., Sandler, J. K. W., & Advani, S. G. (2006). *Processing and properties of nanocomposites* (pp. 1–419). World Scientific Publishing Company.

Suh, S., & Huppes, G. J. (2005). Life cycle assessment (LCA) and sustainability in production systems. *Journal of Cleaner Production, 13*, 7–15.

Swarr, T. E., Hunkeler, D., Klopffer, W., Pesonen, H.-L., Ciroth, A., Brent, A. C., & Pagan, R. (2011). *Environmental life cycle costing: A code of practice*. Society for Environmental Toxicology and Chemistry.

Tillman, A. M. (2000). Life cycle assessment: An overview. *Environmental Impact Assessment Review, 20*, 1–12.

Urbansky, E. T., Brown, S. K., Magnuson, M. L., & Kelty, C. A. (2001). Perchlorate levels in samples of sodium nitrate fertiliser derived from Chilean caliche. *Environmental Pollution, 112*(3), 299–302.

Valdivia, S., Ugaya, C. L., Hildenbrand, J., Traverso, M., Mazijn, B., & Sonnemann, G. (2013). Social life cycle assessment (S-LCA). *International Journal of Life Cycle Assessment, 18*, 9–15.

Vazquez, E., Georgakilas, V., & Prato, M. (2002). Microwave assisted purification of HIPCO carbon nanotubes. *Chemical Communications, 14*, 2308–2309.

Chapter 11
Circular Economy Strategies for Graphene and Carbon Nanotubes: Recycling, Reuse, and End of Life Management

P. Saranraj, Shrikaant Kulkarni, B. Lokeshwari, K. Gayathri, and L. Charlie Jelura

11.1 Introduction

Graphene and carbon nanotubes (CNTs) have become significant two-dimensional materials due to their high conductivity and elasticity, thus leading to substantial advances in electronics, energy storage, sensors, and many more. Electronic applications are used in extraction, production, usage, and disposal. The benefits of graphene and CNTs should be the adverse effects of recycling, reuse, and end-of-life management. For the efficient recovery and recycling of the materials to reach the end of their life cycle, graphene and carbon nanotubes will be essential and play a significant role in these processes as well the extended use of these raw materials to reduce their ecological impacts (Graham et al., 2013). Recent strategies have also allowed for the recycling of graphene and CNTs and the separation and purification of nanomaterials of waste products for further use. The environmental impact of the nanotubes is an essential factor that concerns recyclability approaches, chemical vapour deposition (CVD) and mechanical exfoliation, and cost-effectiveness and retention of the basic materials after reuse.

Reuse strategies typically rely on repurposing materials or applications with minimal or no processing. CNTs and graphene are examples of these materials that would no longer be applied to help improve the characteristics of lower-value products like construction materials, coatings, or composites. This decreases the need for virgin materials and lowers the energy use and greenhouse gas emissions associated with new nanomaterial manufacture (Kizilkanat, 2016). Applying graphene

P. Saranraj (✉) · B. Lokeshwari · K. Gayathri · L. C. Jelura
PG and Research Department of Microbiology, Sacred Heart College (Autonomous), Tirupattur, Tamil Nadu, India

S. Kulkarni
Faculty of Engineering and Technology, Sanjivani University, Kopargaon, Maharashtra, India

S. Kulkarni (ed.), *Next-Generation Nanomaterials for Sustainable Engineering*, Lecture Notes in Nanoscale Science and Technology 43,
https://doi.org/10.1007/978-3-032-05719-8_11

and CNTs in electronics, energy storage devices, and environmental remediation significantly increases the potential for recycling into high-value applications at the end of life.

The Circular Economy approach to graphene and CNTs is their end-of-life (EOL) management. The increasing demand for these materials has efficient EOL management practices that minimise potential environmental impacts. Carbon nanotube (CNT) and graphene-based product waste from being released into the environment as undesired pollutants (Xie et al., 2021). Integrating life cycle assessment (LCA) tools into the process can provide a better environmental cost of graphene and CNTs over their complete life cycle, from raw material extraction to the product's end-of-life. Appropriate end-of-life management strategies can make such materials much more sustainable.

The great potential of a circular economy regarding the management of graphene and CNTs in its use of nanomaterial-based products, including nanocomposites, bright nanopolymer material, a drug delivery system and high-energy batteries that recover technologies, need to be designed in dimension, form and material by the recycling process (Deng, 2005). Safely handling and disposing of nanomaterials has become a challenge in the circular economy principles. The recycling methods and special machinery may also restrict the incorporation of these processes. The limited waste and eco-footprint through recycling technology improvements, increasing end-of-life (EOL) reuse applications and sustainable EOL management practices (Wang et al., 2006). The circular economy approach to the life cycle of graphene and carbon nanotubes enables sustainability. It prevents these novel materials from contaminating the environment as a by-product of their innovative potential.

11.2 Sources of Solid Waste for Graphene Production

The production process of graphene moves through multiple waste streams in solid form and needs effective management. Graphene is a one-atom-thick allotrope of carbon that exhibits a key material's mechanical, electrical, and thermal properties in several industrial sectors, including electronics, energy storage, aerospace, and biomedicine. The utilisation of graphene advances in the waste produced from graphene synthesis and utilisation (Chen et al., 2004). The characterisation and handling of the solid residues generated during the manufacture of graphene in various processes are crucial for minimising the environmental impact of the substance and enhancing the sustainability of graphene-based industries. The stages of raw material processing, production processes, and product action in solid wastes are produced from graphene production. Upcycling, recycling, and using materials for as long as possible through closed-loop systems with sustainable end-of-life solutions eliminates unnecessary waste and creates value for the by-products.

Solid waste is produced in many stages of graphene manufacturing, particularly in the raw material extraction and preparation stage. Graphite is often found in the earth and harvested for its high-purposed carbon sources to make Graphene. The extraction and purification of graphite releases vast amounts of by-products like rock off-cuts, impurities, and residues from chemical treatments. The chemical deposition of fibres (CVD) and alternatives to other carbon-based raw materials like the coal tar pitch (Dehghani & Aslani, 2020). The materials also produce significant quantities of waste residue, including carbonised by-products and solvents. A small fraction of this waste may be converted into other products, such as tableware, to solid fuel, plus solid waste that requires disposal or recycling.

Graphene is produced through various sources, including chemical vapour deposition (CVD), liquid-phase exfoliation, and thermal reduction, and these methods also generate a wide variety of solid waste. CVD is a widely used method to produce high-quality graphene by depositing carbon-based gases on top of metals above decomposition temperature. This method can also yield by-products such as the remnants of metal catalysts, soot, and different types of particulate matter that are typically trapped in filters or accumulated on solid surface substrates (Liu, 2018). CVD-process-synthesised graphene would be laden with impurities or catalysts of materials that require purification steps through more waste-generation processes. Like ultrasonic liquid-phase exfoliation, surfactant, and exfoliated graphite waste are produced in the solvent residues. The purification steps to remove impurities contribute to additional solid wastes.

The production of graphene is significant in the handling and processing graphene-containing products for solid waste. The raw graphene can also be used for several steps like milling, grinding, cutting, and many more. These processes often generate large quantities of ultra-fine particles of graphene and other composite materials that are waste. Such waste typically involves graphene and other materials, such as polymer matrices or resin systems, which may be used in recycling (Liu et al., 2017). Waste may also play a role in graphene products' degradation or end-of-life (EOL) management (Chiarello & Zinno, 2005). The recovered materials could contain graphene or CNTs that must be recovered and treated to ensure these materials do not become environmental pollutants. Solid waste types produced during the end-of-life (EOL) phase of consumer electronics with graphene or other industrial applications are generated during the treatment and recycling of the contaminated electronics and electronics containing graphene. These different forms of solid waste need to be managed and disposed.

Graphene and carbon nanotubes are usually made or utilised in limited quantities due to their high surface area and chemical reactivity; these materials can present unique challenges for appropriate waste handling and disposal, with inappropriate disposal resulting in possible environmental contamination (Baeza et al., 2013). Industries must be involved in graphene production in the waste management strategies that rely on recycling and reusing. The waste disposal of graphene, including solvent-based purification techniques and mechanical separation methods, is used to collect the valuable graphene from waste sources. The recycling waste generated during graphene production for applications such as additives to strengthen construction materials or the production of low-cost batteries will have the waste out of landfills and reduce the demand for raw materials for new products (Wang et al., 2008). The efficient EOL management methodologies, including waste collection, sorting, and repurposing of these materials, are crucial to their recycling/reuse without environmental pollution since the distribution of graphene and CNTs contributes to pollution. The solid waste generated in graphene production is from the extraction of raw materials, the production, the storage, and the recycling or disposal of final products (Chung, 2017). The graphene industry has the potential to advance sustainability even further through new technologies and practices for reducing waste, recovering valuable materials and reducing the environmental footprint of nanotubes.

11.3 Recycling Waste Materials Into Graphene Nanotubes

The production of graphene nanotubes (GNTs) is crucial to realise circular economy approaches to mitigate resource depletion and environmental pollution caused by the disposal of carbonised nanoparticles. Graphene and carbon nanotubes (CNTs) are usually manufactured by traditional methods based on fossil fuel-derived feedstocks and energy-extensive processes, leading to sustainability problems and waste accumulation. Reusable graphite and nanotube-based composites offer a wide range of nature such as biomolecules, biological species or biomolecules, chemical substances and cooking materials; several new types of reagents and other key reagents have been used in various industries, such as ceramics, nonferrous element holder, catalysts, and so on. There is a direct method of energy by the biofuel from composting municipal waste, hydrothermal carbonisation and controlled thermal treatment of biomass waste like agricultural residues, lignin, and food waste that provides the carbon from an alternative source for the direct synthesis of nanomaterials. The waste materials pollute the environment when producing graphene nanotubes, offering a greener and more cost-effective route to the graphene nanotubes than conventional methods. Using recycled GNTs in industrial sectors, including electronics, energy storage, composites, and environmental remediation, also reduces the feedstock. Recycling waste into graphene nanotubes is environmentally friendly and reduces carbon emissions, landfill accumulation, and energy consumption compared to conventional synthesis methods. It is also economically friendly (Gao et al., 2018).

The significant benefits of waste-derived GNTs are the potential for recycled materials to reduce production costs and nanotechnology development. The recycling technologies of graphene and CNTs are generally in their optimisation and technology development for large-scale industrial applications. AI-based material recovery separated, new catalytic systems with GNT good chemical properties should be in suitable production plants (Stynoski et al., 2015). Graphene nanotube recycling is on the edge of technological innovations. Sustainable and cleaner production technologies can help attain global sustainability goals, such as moving industries towards a resource-efficient, low-carbon, and circular economy. Graphene bearers like graphene nanotubes can retain a certain amounts of mechanical properties in the recycling systems to the circular economy, which can ensure these technologies are made by graphene.

11.3.1 Processing Factors of Graphene Nanotubes

Recycling waste into GNTs is a processing point that has implications for the efficiency, yield and sustainability of processes in a circular economy. The use of the carbon-rich waste raw materials like the plastic waste, biomass, electronic waste, and by-products of industrial processes and the type of raw material is crucial in the final properties of the graphene nanotubes (Khushnood et al., 2018). The waste source's composition, purity, and structure are crucial factors affecting the synthesis process, and pre-treatment methods are needed to remove impurities. Pyrolysis is the most commercially used method of waste-to-graphene nanotube transformation. The static environment, carbon waste feedstock, is placed in the formation of organics that are subsequently converted into carbonaceous structures. Critical approaches include cardiovascular disease (CVD) of chemical vapours from gaseous carbon precursors derived from waste through decomposition, reacting with metal catalysts to have high-quality graphene nanotubes with precise dimensions and properties. This type of a catalyst, such as iron, nickel, or cobalt nanoparticles, affects the nanotubes' yield, quality, and structure properties.

The alternative processes of hydrothermal and solvothermal have also been used to recycle biomass and organic waste into C to N nanotubes by carbonising under hydrothermal and solvothermal conditions at high temperatures and under high pressure. The waste conversion process generates graphene nanotubes, as the specific temperature conditions influence the graphitisation ratio and the nanotube's structural stability (Cheng et al., 2021). Some treatment conditions may produce defects or heterogeneity of the structure, and insufficient treatment may lead to the incomplete transformation of precursors and low-quality materials. Waste feedstocks frequently contain impurities like metals, polymers, and inorganic residuals, enhancing product purity and performance. The recycling of waste into graphene nanotubes is used to reduce the energy consumption for their recovery; these laboratory-scale methods will require adaptations on an industrial scale. These scalable and economical alternatives for biowaste-derived graphene nanotube production may be enabled by certain technologies followed by artificial intelligence-derived process optimisation or by the development through catalyst regeneration. The combination of renewable energy in thermal processing and chemical synthesis would significantly reduce the environmental footprint of the novel method of graphene nanotube production. LCA is crucial for the environmental impact of recycling graphene nanotubes along their life cycle in markets, which mitigates waste and promotes renewable energy usage with the circular economy principles. Nanotechnology, material science, and advancements in process engineering will make this transformation from waste to graphene nanotubes more efficient, practical, and a greener alternative to traditional production (Zheng, 2021). It is used to recycle waste materials such as graphene nanotubes for a sustainable, low-carbon, and higher resource efficiency in the future of advanced nanomaterials.

11.3.2 Applications of Waste-Generated Graphene Nanotubes

The application of waste-treated GNTs is highly beneficial for developing circular economy strategies and encouraging the integration of sustainable nanomaterials across different industrial sectors. Waste-based graphene nanotubes are used in lithium-ion batteries, supercapacitors, and fuel cells deployed in the energy industry, enhancing conductivity, energy storage capacity, and efficiency in charge-discharge processes. These advanced materials offer high surface area and better electron mobility, improving battery performance and leading to longer-lasting and more effective energy storage technologies. By developing next-generation nanotubes in energy sources such as photovoltaics, optoelectronics requires lightweight and flexible electrode components for photovoltaic efficiency (Donnini et al., 2018). The waste graphene nanotubes are used for manufacturing electronic gadgets, such as flexible circuits, eco-friendly, and efficient manufacturing of sensors and transistors. Due to their enhanced conductivity, they decrease weight and increase fuel efficiency. Graphene nanotubes derived from waste in components of vehicles lead to a decrease in fuel consumption, carbon emissions, and imbalance and an increase in the wisdom of materials.

These nanotubes are integrated into high-performance formulations of cement and concrete to improve structural integrity, reduce the potential for cracking and create better thermal insulation properties. With their high surface adsorption capacity, these materials can effectively remove heavy metals, organic pollutants, and contaminants from wastewater in cleaner and more sustainable water resources. The biocompatibility and nanoscale precision of graphene nanotubes derived from waste materials are widely used in drug delivery systems, biosensors, and medical imaging technologies in the healthcare sector. Site-targeted drug delivery systems aim to enhance therapeutic efficacy but may also concentrate adverse reactions at the target site. Guo et al. (2021) developed high-performance antimicrobial coatings and fabrics with protective characteristics. Graphene nanotubes enhance printed structures' strength, conductivity, and flexibility for advanced functional materials in various industrial uses. The waste-derived graphene nanotubes are vast, with industrial scalability, cost-effectiveness, and quality control for large-scale adoption. These commercial applications increase their advances in recycling technologies and process optimisation and support sustainability (Faneca et al., 2018).

11.3.3 Grapheme Nanotubes for Energetic Applications

The electrical conductors have a very high surface area and excellent mechanical strength. Thus, many of their components can be used in energy storage and conversion systems. Graphene nanotubes obtained from waste can be used as conductive additives and structural reinforcements in electrodes of LIB (Chung, 2005). They provide better battery efficiency, faster charging rates, and longer cycle life due to more effective electron transport than traditional batteries, increasing charge transfer kinetics. The distinctive structure and light weight ensure the manufacture of flexible and portable following generation energy storage devices (Butenegro et al., 2021; Wang et al., 2018). The high-surface-area graphene Nanotubes in the electric double layers as perfect charge data storage press can increase energy density and power outputs. In addition, their inclusion in electrode products also promotes their electrochemical stability. The waste-derived graphene nanotubes used in the study also fine-tune fuel cells, primarily as catalyst supports for hydrogen fuel cell electrodes.

The application is in thermoelectric materials, and the graphene nanotubes can improve thermal conductivity, resulting in enhanced electron transport and energy harvesting from waste heat. Due to their excellent electronic properties, graphene nanotubes are used as transparent conductive films and electron transport layers to maximise light absorption and conversion efficiency (Yang, 2002). Graphene nanotubes have also been used as components in next-generation energy storage devices, including sodium-ion and solid-state batteries, which are considered conventional lithium-based technologies. Functionalised nanotubes are often used in electrode and electrolyte materials to enhance ionic/ionic-conductivity, mechanical stability, and device performance. Graphene nanotubes are high-conductivity materials for power cables, enabling energy distribution, reducing energy losses during transportation to the energy screen, and increasing the stability of the power grid. Because of their high surface area and porous structure, they can also be used as hydrogen storage materials that can help clean energy systems by enabling the adsorption and desorption of hydrogen.

Some challenges, including large-scale production, cost reduction, and consistency of nanotube properties, remain to be addressed before mass adoption can occur within the energy sector. These challenges enhance recycling technologies, optimise processes, and functionalise materials. These recent advances in waste-producing graphene nanotubes will also affect commercial energy technologies. However, its use will require a push on the policy front and the concerted efforts of the industry to bring these drip materials into the energy technology ecosystem (Wang et al., 2018). This innovation can minimise environmental harm associated with the specific technology sectors by the processes that are not only energy-efficient but also used for the faster adaptation to sustainable energy for integrating circular economy models with high-technology nanomaterials.

11.3.4 Challenges in Waste-to-Graphene Transformation

GNTs can help promote waste recycling worldwide and expand the application of waste GNTs into circular strategies. The composition and quality of different types of waste contain different levels of impurities, additives, and noncarbon elements that directly affect the structure and properties of the resulting graphene nanotubes. Several methods such as pyrolysis, catalytic CVD and hydrothermal processing hold great potential. These techniques for industrial manufacturing require significant advances in cost, energy, and efficiency. Additionally, the devolvement processes with specific conversion processes and high-temperature treatment can be sustainable and economic depending on the applications of such processes as a recycling process.

The selection and fine-tuning of suitable catalysts are also significant factors in controlling the nanotubes' growth conditions, morphology, and purity (Mendes et al., 2015). However, the technical challenges of catalyst recovery and reusability have the potential to cause an unavoidable compromise between sustainability and the competitive value of the process. The metal catalysts left over in the products may need additional purification steps, increasing processing complexity and cost. The graphene nanotubes come from either waste and/or biomaterials. Thus, the quality and performance of the final material can be determined by different sources depending on the materials used. The graphene and carbon nanotube production processes may have widespread implementation and waste-to-graphene processes in the circular economy (Akbar & Liew, 2020). The conversion of waste materials into graphene nanotubes is more sustainable and resource-friendly.

11.4 Recycling Process of Carbon Nanotubes

CNTs have been used across diverse domains, including electronics and energy storage materials. Collecting and separating waste materials such as polymer composites, electronic components, paint coatings, and waste from industries in recycling CNTs. They extract CNT from their composite without damaging the composites' structural properties (Zhang et al., 2020). Mechanical methods like shredding, milling, and ultrasonication form small particles of the CNT composites through agglomeration to promote the release of nanotubes. Polymeric binders are typically used for CNTs, and thermal treatment methods such as pyrolysis and controlled combustion are used to harvest pure forms of CNTs (Wang et al., 2019). Pyrolysis is a thermal decomposition process in a low-oxygen environment, breaking waste into valuable by-products such as biochar, syngas, and oil (Zhou et al., 2003). CNTs serve as energy storage devices, conductive films, and strength-embedded composites. Sustainable and economic use to the conventional disposal route, the potential of CNT nanomaterials prevents carbon emissions and resource depletion. Circular economy and sustainable global utilisation of CNTs with the technologies on CNT recycling that ensure sustainable material recovery (Akbar et al., 2021). CNT recycling at a low cost generates a critical alternative to environmental damage and reduces high production costs for future sustainable nanotechnology and advanced materials applications. Figure 11.1 outlines sustainable recycling methods for carbon nanotubes, including thermal recycling (e.g., pyrolysis, combustion), mechanical recycling, and chemical recycling (e.g., hydrolysis, solvolysis). The fluidized bed process further highlights innovative approaches to nanotube recovery, emphasising the importance of circular economy strategies in nanotechnology.

11.4.1 Mechanical Recycling

Mechanical recycling of carbon nanotubes (CNTs) is an important method of circular economy approaches, enabling recovery, reuse, and recycling of CNT-containing materials that minimise environmental impacts. The physical decrease in the element size of CNT composites, coatings, and other materials that use nanomaterials changes the solid state without changing its chemical structure. The initial stage of mechanical recycling is the gathering and sorting of CNT-containing waste like polymer composites, conductive films, and structural reinforcements (Rangelov et al., 2016). CNTs are often added into matrices, including plastics or metals, and large-scale materials into small particles through mechanical methods like shredder, mill, and grinder. Another method for CNT dispersion and separation from other surrounding materials is ultrasonication, which utilises high-frequency sound waves. CNTs are separated from ground waste based on size and density. The undesirable CNTs are later utilised to qualify new composite materials, coatings, and structures (Aquatar et al., 2021). Mechanical recycling preserves the CNT structure better than chemical or thermal approaches, but often with high-purity material without degrading it. There are also several solutions for reducing mechanical waste volumes that can be used to enhance the technology. The mechanical recycling of

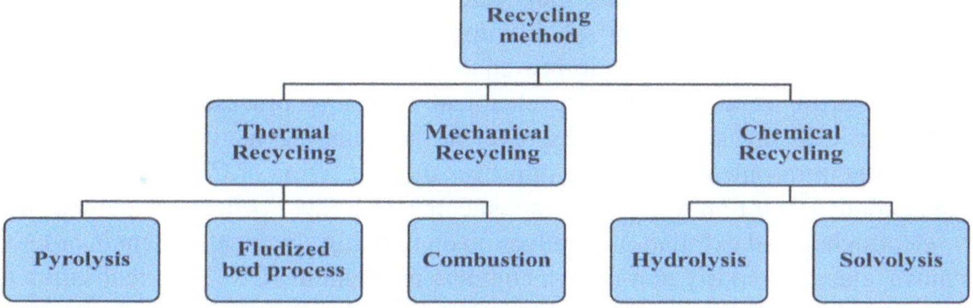

Fig. 11.1 Sustainable recycling of carbon nanotubes

CNT-based products can minimise waste, ensure material sustainability, and encourage the sustainable development of nanotechnology for CNT-based products.

11.4.2 Thermal Recycling

The physical and thermal recycling of carbon nanotubes (CNTs) is among the most important methods in several circular economy pathways. CNTs recover from composite materials and waste products through controlled thermolysis. This method typically utilises pyrolysis, a high-temperature process carried out in an oxygen-deficient atmosphere, whereby CNT-containing products like the polymer composites and electronic wastes undergo thermal decomposition and pyrolysis for fragmentation of the caramel-like polymeric material without damaging the Nanotube Macromolecular Structure (Aquatar et al., 2022). It also eliminates the resins and other non-carbon materials in the CNTs. Pyrolysis must be controlled in temperature and condition to prevent the structural degradation of the CNTs because the fine adjustment during heating may create the defects or morphology changes.

Controlled combustion or thermal oxidation is a thermal method used to purify the CNTs more thoroughly when burning away pollutants or undesired materials. These strategies effectively relieve an incredible amount of waste but cannot simply output high-value CNT with exact temperature administration and an advanced remediation system. Thermal recycling is an efficient approach to recovering CNTs as they have high energy consumption, limited up-scalability of the process, and bulky devices (Pérez-Sequera et al., 2020). These challenges associated with CNT recycling make thermal recycling for the carbon nanotubes and promote sustainability for the environmental issues associated with CNT disposal. Thermal recycling has the potential to act as the life cycle of CNT-based materials and help in the circular economy.

11.4.3 Chemical Recycling

The chemical recycling method also allows the functionalisation of CNTs to enhance surface properties for specific applications. It is possible to build compounds where the dispersion is improved or the reactivity in chemical reactions is greater (Bartoli et al., 2020). The chemical recycling process involves solvent extraction, utilising solvents to dissolve the surrounding matrix material in the CNTs. The ultrasonication process improves the dispersion of CNTs in the solvent and facilitates their removal from composite material. Electrochemical methods like the electrochemical exfoliation or reduction on utilising electrical current in order to remove, eject, or modify non-carbon components from the CNTs the realisation of waste reduction, circular economies and the minimises the raw material consumption (Isahak et al., 2023).

11.5 Synergies Between Circular Economy and Its Technologies

The circular economy approaches combined with the specific technologies in recycling, reuse, and end-of-life treatment technologies, such as graphite and carbon nanotube in the specific areas that are being used, significantly facilitate the sustainable practices in the relevant nanomaterial sectors. Mechanical, thermal, and chemical recycling is essential for recovering and reusing graphene and CNTs to reduce the materials and prevent adverse environmental impacts (Ma et al., 2017). Thermal recycling of carbon polymer composites, primarily through pyrolysis, is used to reduce carbon-rich waste materials, enables the recovery of high-quality CNTs, and decreases carbon footprints from their disposal. CNTs with unique and functionality they are liquids with very expensive and that are acid oxidation or solvent extraction that indicate the presence of chemical recycling progression that purifies the surface and functionalised CNTs so that these substances interface with different high-performance applications such as energy storage devices, composites, electronics, and many more (Zhou & Wang, 2020). The high efficiency, low cost, and low carbon cost, as well as resource use from production and manufacturing and recyclability in the graphene and CNTs, are resource-efficient and sustainable in the environment (Yao et al., 2020).

11.5.1 Waste-to-Resource Model

Waste products in plastic composites, electronics, and industrial by-products can be treated using mechanical, thermal, and chemical recycling technologies to recover CNTs and graphene that can be used as raw materials for new products (Marinova et al., 2020). Mechanical recycling, like shredding or grinding, is generated from the CNTs. The thermal recycling process, like the pyrolysis, comes from the organic raw materials by the purified carbon nanotubes. Acid oxidation or solvent extraction techniques can remove such contaminants and yield CNTs in a form suitable for high-value soluble applications (Harris et al., 2020). The raw-input materials in its ecological footprints and developing a more sustainable life cycle for nanomaterials (Meng et al., 2018).

11.5.2 Dual Impact on Waste and Emissions

Waste management and emissions reduction are due to strategies in the sustainability of graphene and CNT recycling technologies and circular economy strategies based on their environmental benefits. The disposal of CNTs in landfills or incinerators is a problem because CNTs are very valuable nanomaterials, and industries will reduce their waste disposal. Mechanical, thermal, and chemical recycling technologies enable the recovery of high-quality. CNTs can replace raw materials and reduce the environmental impacts of raw material extraction (Karuppannan Gopalraj & Kärki, 2020).

Methods of thermal processing such as pyrolysis involve taking a carbon-based material and breaking it down at very high temperatures under little or no oxygen that can also relieve the necessity to create and operate energy-intensive systems that are not only high energy and materials consumers but also produce emissions. Chemical recycling is based on the less energy-dependent route to recovery by the recycling process's carbon footprint. The lifecycle of graphene and CNTs also has fewer air pollution hazards that can arise from their use and by reducing the potential for disposal to the dangerous gases within our atmosphere (Verma et al., 2018)

11.5.3 Sustainability and Environmental Impact

The end-of-life interaction of graphene and CNTs used for the recycling, reuse, and end-of-life process is sustainable (Danish et al., 2022). Recycling technologies can play different roles in the lifecycle of CNTs and graphene: mechanical, thermal, or chemical processes can ensure that nanomaterials are recovered from waste products into the production cycle. Thermal recycling, especially pyrolysis, permits the conversion of waste materials into valuable CNTs and is less relative to common waste disposal methods such as incineration. Carbon nanotubes (CNTs) can be produced through chemical recycling, a method that circumvents the high energy consumption characteristic of conventional early-stage production. Recycling the CNTs and carbon-based products allows their reuse for high-performance applications like the energy storage devices, functional composites, and electronics that are more durable and high-end products and reduce energy and resource consumption (Pimenta & Pinho, 2011).

11.6 Recent Initiatives in Circular Economy

The circular economy initiatives have developed graphene and CNTs and could increase sustainability and transform the environmental footprint of graphene and CNTs (Li & Englund, 2017). The industrial use of graphene and CNTs increases in application sectors such as electronics, energy storage and manufacture, and effective implementation of a circular economy approach. The composite materials or electro-wastes can be recovered by carbon nanotubes (CNTs) to avoid the resource waste. Mechanical recycling methods include plastic and metal fragmentation. CNTs are obtained with strong and complete separation from polymeric or metallic systems involving controlled combustion and pyrolysis, which may also improve the efficiency of this process, allowing high-purity CNT yields with large-scale recycling and thereby promoting more ecologically responsible practices (Danish & Ozbakkaloglu, 2022). The Graphene and CNTs from production to end-of-life stages can signal relevant areas of waste, energy, and emissions (Morin et al., 2012). The processes of grapheme and CNTs are more sustainable and efficient by recycling and reusing these high-performance materials in a circular low-carbon economy (Borjan et al., 2021).

11.7 Limitations of Circular Economy

The circular economy significantly approaches the recycling, reuse, and end-of-life stages of graphene and CNT materials. CNTs are often used in diverse composites like plastics, metals, electronics, and many more (Aflaki Samani & Jabbari Lak, 2019). The extraction of CNTs from these composites is weeding and energy-intensive, requiring more sophisticated methods, including mechanical separation, thermal treatments, and chemical processing (Li et al., 2022). CNTs recovery for efficient or economical on the industrial scale and the methods like acid oxidation or solvent extraction are used to generate toxic substances in the environmental and safety problems (Belli et al., 2018). The implementation of circular economy principles for CNTs is scalability in the development stage for their capacity to recycle large volumes of CNT waste (Ma et al., 2018). CNT-based products require the collection, sorting and processing infrastructures to handle CNT-based products (Zhandarov & Mäder, 2005). There is a high energy cost associated with recycling CNTs, and the costs involved, specifically the energy costs and labour involved in recycling CNTs, often surpass the cost of virgin material, especially since the market for recycled CNTs is still an emerging one. Thus, industries may not have a compelling incentive for recycled CNTs, leading to the recycling infrastructures is significant demand (Wang et al., 2019). CNTs could be developed through new recycling technologies, various new applications of CNTs in various industries, and the potential for market strategies to develop the recycling of CNTs (Faneca et al., 2018).

11.8 Conclusion

The production of graphene, while revolutionary across industries, generates significant solid waste at various stages from graphite mining and chemical synthesis to end-of-life disposal of graphene-containing products. Key processes like chemical vapour deposition (CVD) and liquid-phase exfoliation yield metal catalyst residues, solvents, and composite by-products, necessitating robust waste management strategies to mitigate environmental harm. Recycling and circular economy approaches, such as converting waste plastics, biomass, and electronic scraps into graphene nanotubes (GNTs), offer promising pathways to reduce resource depletion and landfill accumulation. Techniques like pyrolysis, hydrothermal processing, and catalytic conversion not only valorize waste but also lower the carbon footprint of graphene synthesis compared to conventional methods (Dehghani & Aslani, 2020; Gao et al., 2018). However, challenges persist, including impurities in waste feedstocks, energy-intensive recycling processes, and scalability limitations, which hinder cost-effectiveness and industrial adoption (Akbar & Liew, 2020). Despite these hurdles, advancements in AI-driven process optimisation, catalyst recovery, and policy frameworks for sustainable nanotechnology underscore the potential to align graphene production with circular economy principles. By prioritising waste-to-resource innovations and lifecycle assessments, the graphene industry can transition towards a resource-efficient future, balancing technological progress with ecological stewardship.

References

Aflaki Samani, M., & Jabbari Lak, S. (2019). Experimental investigation on the mechanical properties of recycled aggregate concrete reinforced by waste carbon fibers. *International Journal of Environmental Science and Technology, 16*, 4519–4530.

Akbar, A., Kodur, V. K. R., & Liew, K. M. (2021). Microstructural changes and mechanical performance of cement composites reinforced with recycled carbon fibers. *Cement and Concrete Composites, 121*, 104069.

Akbar, A., & Liew, K. M. (2020). Influence of elevated temperature on the microstructure and mechanical performance of cement composites reinforced with recycled carbon fibers. *Composites. Part B, Engineering, 198*, 108245.

Aquatar, M. O., Bhatia, U., Rayalu, S. S., & Krupadam, R. J. (2022). Reduced grapheme oxide-MnO_2 nanocomposite for CO_2 capture from flue gases at elevated temperatures. *Science of the Total Environment, 816*, 151522.

Aquatar, M. O., Mankar, J. S., Bhatia, U., Rayalu, S. S., & Krupadam, R. J. (2021). Graphene nanosheets from hazardous/solid wastes: An efficient CO_2 capture material. *Journal of Environmental Chemical Engineering, 9*(5), 105839.

Baeza, F. J., Galao, O., Zornoza, E., & Garcés, P. (2013). Effect of aspect ratio on carbon fiber reinforced cement composites strain sensing capacity. *Materials and Design, 51*, 1085–1094.

Bartoli, M., Frediani, M., Kaminsky, W., Rosi, L., İsmail, Y., Kumar, R., Chawla, J., & Kumar, B. (2020). Carbon-based material for environmental protection and remediation. In *Intech open eBooks*. IntechOpen.

Belli, A., Mobili, A., Bellezze, T., Tittarelli, F., & Cachim, P. (2018). Evaluating the self-sensing ability of cement mortars manufactured with graphene Nanoplatelets, virgin or recycled carbon fibers through piezo resistivity tests. *Sustainability, 10*, 4013.

Borjan, D., Knez, Z., & Knez, M. (2021). Recycling of carbon fiber-reinforced composites - Difficulties and future perspectives. *Materials, 14*, 4191.

Butenegro, J. A., Bahrami, M., Abenojar, J., & Martínez, M. A. (2021). Recent progress in carbon fibre reinforced polymers recycling: A review of recycling methods and reuse of carbon fibres. *Materials, 14*, 6401.

Chen, B., Wu, K. R., & Yao, W. (2004). Conductivity of carbon fibre reinforced cement-based composites. *Cement and Concrete Composites, 26*, 291–297.

Cheng, J. Q., Wang, W. G., & Han, J. (2021). Progress on mechanical properties of carbon fiber reinforced cement matrix. *Cement and Concrete Composites, 92*, 135–144.

Chiarello, M., & Zinno, R. (2005). Electrical conductivity of self-monitoring CFRC. *Cement and Concrete Composites, 27*, 463–469.
Chung, D. (2005). Dispersion of short fibers in cement. *Journal of Materials in Civil Engineering, 17*, 379–383.
Chung, D. (2017). Processing-structure-property relationships of continuous carbon fiber polymer-matrix composites. *Materials Science & Engineering R: Reports, 113*, 1–29.
Danish, A., Mosaberpanah, M. A., Salim, M. U., Amran, M., Fediuk, R., Ozbakkaloglu, T., & Rashid, M. F. (2022). Utilization of recycled carbon fiber reinforced polymer in cementitious composites. A critical review. *Journal of Building Engineering, 53*, 104583.
Danish, A., & Ozbakkaloglu, T. (2022). Greener cementitious composites incorporating sewage sludge ash as cement replacement: A review of progress, potentials, and future prospects. *Journal of Cleaner Production, 371*, 133364.
Dehghani, A., & Aslani, F. (2020). The effect of shape memory alloy, steel, and carbon fibres on self-compacting cementitious composites' fresh, mechanical, and electrical properties. *Cement and Concrete Composites, 112*, 103659.
Deng, Z. C. (2005). The fracture and fatigue performance in carbon fiber reinforced concrete flexure. *Cement and Concrete Composites, 27*, 131–140.
Donnini, J., Bellezze, T., & Corinaldesi, V. (2018). Mechanical, electrical and self-sensing properties of cementitious mortars containing short carbon fibers. *Journal of Building Engineering, 20*, 8–14.
Faneca, G., Segura, I., Torrents, J., & Aguado, A. (2018). Development of conductive cementitious materials using recycled carbon fibres. *Cement and Concrete Composites, 92*, 135–144.
Gao, J., Sha, A. M., Wang, Z. J., Hu, L. Q., Yun, D., Liu, Z. Z., & Huang, Y. (2018). Characterization of carbon fiber distribution in cement-based composites by computed tomography. *Construction and Building Materials, 177*, 134–147.
Graham, R. K., Huang, B., Shu, X., & Burdette, E. G. (2013). Laboratory evaluation of tensile strength and energy absorbing properties of cement mortar reinforced with micro-and meso-sized carbon fibers. *Construction and Building Materials, 44*, 751–756.
Guo, Z., Zhuang, C. L., Li, Z. H., & Chen, Y. (2021). Mechanical properties of carbon fiber reinforced concrete (CFRC) after exposure to high temperatures. *Composite Structures, 256*, 113072.
Harris, S., Weinzettel, J., Bigano, A., & Källmén, A. (2020). Low carbon cities in 2050? GHG emissions of European cities using production-based and consumption-based emission accounting methods. *Journal of Cleaner Production, 248*, 119206.
Isahak, W., Shaker, L. M., & Al-Amiery, A. (2023). Oxygenated hydrocarbons from catalytic hydrogenation of carbon dioxide. *Catalysts, 13*(1), 115–122.
Karuppannan Gopalraj, S., & Kärki, T. (2020). A review on the recycling of waste carbon fibre/glass fibre-reinforced composites: Fibre recovery, properties and life-cycle analysis. *SN Applied Sciences, 2*, 433–440.
Khushnood, R. A., Muhammad, S., Ahmad, S., Tulliani, J. M., & Maqsom, A. (2018). Theoretical and experimental analysis of multifunctional high-performance cement mortar matrices reinforced with varying lengths of carbon fibers. *Materiales de Construcción, 68*, 172–180.
Kizilkanat, A. B. (2016). Experimental evaluation of mechanical properties and fracture behaviour of carbon fibre reinforced high strength concrete. *Periodica Polytechnica Civil Engineering, 60*, 289–296.
Li, H., & Englund, K. (2017). Recycling of carbon fiber-reinforced thermoplastic composite wastes from the aerospace industry. *Journal of Composite Materials, 51*, 1265–1273.
Li, H. Y., Liebscher, M., Yang, J., Davoodabadi, M., Li, L. H., Du, Y. H., Yang, B., Hempel, S., & Mechtcherine, V. (2022). Electrochemical oxidation of recycled carbon fibers for an improved interaction toward alkali-activated composites. *Journal of Cleaner Production, 368*, 133093.
Liu, K. (2018). *Preparation and electrothermal properties of carbon fiber/graphene conductive asphalt concrete*. Harbin Institute of Technology.

Liu, K., Lu, L., Wang, F., & Liang, W. (2017). Theoretical and experimental study on multiphase model of thermal conductivity for fibre reinforced concrete. *Construction and Building Materials, 148*, 465–475.

Ma, B. G., Zhang, T., Tan, H. B., Liu, X. H., Mei, J. P., Qi, H. H., Jiang, W. B., & Zou, F. B. (2018). Effect of triisopropanolamine on compressive strength and hydration of cement-fly ash paste. *Construction and Building Materials, 179*, 89–99.

Ma, Q., Yu, Y., Sindoro, M., Fane, A. G., Wang, R., & Zhang, H. (2017). Carbon-based functional materials derived from waste for water remediation and energy storage. *Advanced Materials, 29*(13), 1605361.

Marinova, S., Deetman, S., Van der Voet, E., & Daioglou, V. (2020). Global construction materials database and stock analysis of residential buildings between 1970–2050. *Journal of Cleaner Production, 247*, 119–146.

Mendes, T. M., Hotza, D., & Repette, W. L. (2015). Nanoparticles in cement-based materials: A review. *Reviews on Advanced Materials Science, 40*, 89–96.

Meng, F., Mckechnie, J., & Pickering, S. J. (2018). An assessment of financial viability of recycled carbon fibre in automotive applications. *Composites Part A: Applied Science and Manufacturing, 109*, 207–220.

Morin, C., Loppinet-Serani, A., Cansell, F., & Aymonier, C. (2012). Near- and supercritical solvolysis of carbon fibre reinforced polymers (CFRPs) for recycling carbon fibres as a valuable resource: State of the art. *Journal of Supercritical Fluids, 66*, 232–240.

Pérez-Sequera, A. C., Díaz-Pérez, M. A., & Serrano-Ruiz, J. C. (2020). Recent advances in the electroreduction of CO_2 over heteroatom-doped carbon materials. *Catalysts, 10*(10), 1179–1185.

Pimenta, S., & Pinho, S. T. (2011). Recycling carbon fibre reinforced polymers for structural applications: Technology review and market outlook. *Waste Management, 31*, 378–392.

Rangelov, M., Nassiri, S., Haselbach, L., & Englund, K. (2016). Using carbon fiber composites for reinforcing pervious concrete. *Construction and Building Materials, 126*, 875–885.

Stynoski, P., Mondal, P., & Marsh, C. (2015). Effects of silica additives on carbon nanotube and carbon fiber reinforced Portland cement mortar fracture properties. *Cement and Concrete Composites, 55*, 232–240.

Verma, S., Balasubramaniam, B., & Gupta, R. K. (2018). Recycling, reclamation and re-manufacturing of carbon fibres. *Current Research in Green and Sustainable Chemistry, 13*, 86–90.

Wang, C., Lei, P., Li, B. L., Gao, N., Zhao, L. P., & Li, K. Z. (2018). Influences of molding processes and different dispersants on the dispersion of chopped carbon fibers in cement matrix. *Heliyon, 4*, e00868.

Wang, C., Li, K. Z., Li, H. J., Guo, L. J., & Jiao, G. S. (2008). Influence of CVI treatment of carbon fibers on the electromagnetic interference of CFRC composites. *Cement and Concrete Composites, 30*, 478–485.

Wang, W., Wu, S., & Dai, H. Z. (2006). Fatigue behaviour and life prediction of carbon fiber reinforced concrete under cyclic flexural loading. *Materials Science and Engineering, 434*, 347–351.

Wang, Y., Zhang, S., Li, G., & Shi, X. (2019). Effects of alkali-treated recycled carbon fiber on the strength and free drying shrinkage of cementitious mortar. *Journal of Cleaner Production, 228*, 1187–1195.

Xie, C. P., Cao, M. L., Khan, M., Yin, H., & Guan, J. F. (2021). Review on different testing methods and factors affecting fracture properties of fibre reinforced cementitious composites. *Construction and Building Materials, 273*, 21766.

Yang, Y. (2002). Methods study on dispersion of fibers in CFRC. *Cement and Concrete Research, 32*, 747–750.

Yao, X., Yasmeen, R., Padda, I. U. H., Shah, W. U. H., & Kamal, M. A. (2020). Inequalities by energy sources: An assessment of environmental quality. *PLoS One, 15*(3), e0230503.

Zhandarov, S., & Mäder, E. (2005). Characterization of fiber/matrix interface strength: Applicability of different tests, approaches and parameters. *Composites Science and Technology, 65*, 149–160.

Zhang, L. W., Ji, W. M., Hu, Y., & Liew, K. M. (2020). Atomistic insights into the tunable transition from cavitation to crazing in diamond nanothread-reinforced polymer composites. *Research, 154*, 781–792.

Zheng, S. P. (2021). Study on the effect of carbon fiber on the performance of cement mortar. *Journal of Liaoning University of Petroleum & Chemical Technology, 41*, 34–42.

Zhou, C., & Wang, Y. (2020). Recent progress in the conversion of biomass wastes into functional materials for value-added applications. *Science and Technology of Advanced Materials, 21*(1), 787–804.

Zhou, M. Y., Zhou, J. F., & Liu, C. H. (2003). Mechanical property test analysis of carbon fiber/ cement mortar. *Journal of Hohai University, 4*, 15–18.

Part IV
Beyond Nanotechnology and Case Studies

Chapter 12
Picotechnology for Biological Sciences

Jyoti Chaudhary, Megha Yadav, Giriraj Tailor, and Saurabh Singh

12.1 Introduction to Picotechnology

In the hierarchy of technological scales, human civilization has steadily progressed from manipulating matter at the macroscale down to ever-smaller dimensions: from the millimeter scale of early machine tools, to the micrometer precision of microelectronics, and finally to the nanometer domain of nanotechnology. As our understanding and capabilities in science and engineering have advanced, the potential for exploring and manipulating matter at scales even smaller than the nanometer—specifically at the picometer range (10^{-12} m)—has captured the imaginations of scientists, futurists, and engineers. This emerging concept, often referred to as picotechnology, is situated at a scale three orders of magnitude smaller than nanotechnology. While still largely theoretical, picotechnology holds the promise of enabling precise control over atomic and subatomic particles, potentially unlocking unprecedented advances in materials, computing, energy, and medicine.

The term "picotechnology"is not as widely recognized or rigorously defined as nanotechnology. Nanotechnology generally refers to the manipulation of matter at scales between 1 and 100 nm, dealing with the arrangement of atoms and molecules to create novel materials and devices (Roco & Bainbridge, 2001). In contrast, picotechnology would concern engineering at the picometer level—one thousand

J. Chaudhary · M. Yadav
Department of Chemistry, M.L.S. University, Udaipur, Rajasthan, India

G. Tailor (✉)
Department of Chemistry, M.L.S. University, Udaipur, Rajasthan, India

Department of Chemistry, Mewar University, Chittorgarh, Rajasthan, India

S. Singh
Department of Chemistry, M.L.V. Govt. College, Bhilwara, Rajasthan, India

S. Kulkarni (ed.), *Next-Generation Nanomaterials for Sustainable Engineering*, Lecture Notes in Nanoscale Science and Technology 43,
https://doi.org/10.1007/978-3-032-05719-8_12

times smaller than a nanometer—where one begins to approach the scale of interatomic spacings, electron orbitals, and even nuclear dimensions (Drexler, 2013). At this scale, matter is governed almost entirely by quantum mechanical principles. Atoms themselves have diameters on the order of 100 pm, and their electronic shells and nuclear constituents lie within similar ranges (Atkins & de Paula, 2011).

To put this into perspective: while nanotechnology can organize collections of atoms into complex nanostructures, picotechnology would attempt to alter the fundamental arrangement of particles within or around an atom. This could mean manipulating electron distributions at the picometer scale to create materials with previously unimaginable electronic, optical, or magnetic properties. It might also involve influencing nuclear states or isotopic compositions at will, venturing into territories typically reserved for nuclear physics and high-energy particle accelerators (Freitas Jr., 2005).

Because of this extraordinary level of control, picotechnology is often considered even more speculative than its conceptual sibling, femtotechnology, which deals with manipulation at the scale of 10^{-15} m—approaching the size of atomic nuclei and the realm of quarks and gluons (Merkle, 1997). Picotechnology rests at the cusp between atomic and subatomic realms, where electrons and nuclei define the boundaries of chemical behavior and material properties.

At the heart of picotechnology lies quantum mechanics, the scientific framework that governs behavior at atomic and subatomic scales. Although quantum theory has enabled spectacular technological advances—from the transistor to the laser—it also imposes limitations that challenge our intuition. The Heisenberg Uncertainty Principle, for example, constrains the simultaneous determination of a particle's position and momentum, making precise and deterministic manipulation at the picometer scale inherently difficult (Sakurai & Napolitano, 2017). Additionally, quantum tunneling, zero-point energy fluctuations, and electron-proton coupling effects complicate the idea of "fixing" particles in place at such minuscule dimensions.

Current nanotechnological tools, such as scanning tunneling microscopes (STMs) and atomic force microscopes (AFMs), allow scientists to visualize and manipulate single atoms on surfaces (Binnig & Rohrer, 1985). However, controlling the electron orbitals and nuclear states within an atom remains well beyond our present experimental capabilities. Achieving picometer-scale control would likely require advanced forms of quantum control, coherent states of matter, and perhaps entirely new classes of instrumentation, potentially involving ultrafast laser pulses, intense electromagnetic fields, or novel quantum computing architectures designed to manipulate subatomic configurations (Dowling & Milburn, 2003).

Experimental techniques currently used in precision metrology and spectroscopy, such as X-ray diffraction (Attwood & Sakdinawat, 2017) and scanning transmission electron microscopy (STEM) with sub-angstrom resolution, hint at the possibility of picometer-level structural characterization. Indeed, modern electron microscopes can resolve atomic positions with picometer precision (Erni et al., 2009). However, resolving is not the same as engineering. While we may measure atomic structures to picometer accuracy, intentionally rearranging electrons or nuclei at will requires an entirely different suite of capabilities.

In terms of theoretical frameworks, advancements in computational chemistry, quantum computing, and multiscale modeling are essential. Simulating the behavior of electrons and nuclei at picometer scales demands immense computational power and sophisticated algorithms. Quantum simulation platforms, using qubits to model molecular systems, could 1 day yield insights into how to selectively control electron orbitals or stabilize unusual nuclear states (Georgescu et al., 2014).

While nanotechnology emerged from decades of incremental progress in materials science, chemistry, and physics, picotechnology sits at a far more speculative frontier. Nanotechnology deals with stable arrangements of atoms and molecules, building nanoscale machines, drug delivery systems, and quantum dots by assembling known atomic constituents. By contrast, picotechnology's ambition is to restructure the fabric of matter at a scale where chemical bonds and atomic structures are defined by quantum and nuclear forces (Freitas Jr., 2005).

Femtotechnology, sometimes described in futurist literature, extends these ideas further down to manipulation at the femtometer scale, at which point the constituent parts of an atom's nucleus—protons, neutrons, and ultimately quarks—could be arranged at will. While femtotechnology rests firmly in the realm of speculation, it provides a conceptual boundary beyond which picotechnology may be seen as "less hypothetical." If femtotechnology deals with sub-nuclear arrangements, picotechnology "merely" seeks to sculpt electronic orbitals, influence isotopic compositions, and harness subtle quantum effects at the atomic scale.

Developing picotechnology is not just a matter of technical feasibility. It also raises deep theoretical and ethical questions. The fundamental limits imposed by quantum mechanics are not fully understood in terms of engineering practice. Is there a boundary beyond which matter cannot be stably configured, no matter how sophisticated the intervention? Do emergent quantum phenomena, such as entanglement and decoherence, impose practical ceilings on our ability to control subatomic structures?

From an ethical standpoint, picotechnology would grant humanity unprecedented power over the building blocks of the material world. Such capabilities might be used to create advanced weapons or surveillance tools at unimaginable scales. They might also exacerbate resource inequalities if only a handful of entities control the means of producing ultra-precise materials. On the other hand, picotechnology could alleviate resource constraints, allowing for the creation of abundant, tailored materials from common elements and potentially mitigating environmental challenges. As with all transformative technologies, careful consideration of governance, regulation, and international cooperation would be necessary (Bostrom & Ćirković, 2011).

12.2 Principles of Picotechnology in Biological Science

The convergence of cutting-edge technologies with biological sciences has consistently paved the way for transformative breakthroughs, from the advent of molecular biology to the rise of synthetic biology. As we stand on the precipice of the next technological revolution, the concept of *picotechnology*—the manipulation of matter at the picometer scale (10^{-12} m)—promises to redefine our understanding and control of biological systems. This exploration delves into the principles of picotechnology within biological sciences, examining its theoretical foundations, potential applications, challenges, and the interdisciplinary efforts required to realize its promise.

12.2.1 Defining Picotechnology in Biological Context

Picotechnology, as an extension of nanotechnology, operates at a scale three orders of magnitude smaller, targeting the manipulation of matter at the picometer level. In biological sciences, this implies controlling and engineering matter at the scale of atomic and subatomic particles within biological molecules and cellular structures. While nanotechnology has enabled the construction and manipulation of nanoparticles, DNA origami, and nanoscale drug delivery systems, picotechnology envisions a more granular level of control—altering electron distributions, nuclear configurations, and even manipulating isotopic compositions within biomolecules (Drexler, 2013).

12.2.2 Theoretical Foundations

12.2.2.1 Quantum Mechanics and Biological Systems

At the heart of picotechnology lies quantum mechanics, which governs the behavior of particles at atomic and subatomic scales. Biological processes, though often described by classical chemistry and biophysics, are fundamentally rooted in quantum phenomena. Enzymatic catalysis, photosynthesis, and even avian magnetoreception have been shown to involve quantum coherence and tunneling (Engel et al., 2007; Houk & Shaik, 2005). Picotechnology seeks to harness and manipulate these quantum aspects with unprecedented precision, enabling direct control over the quantum states that underpin biological functions.

12.2.2.2 Quantum Biology

Quantum biology is an emerging field that investigates the role of quantum phenomena in biological systems. It explores how quantum coherence, entanglement, and tunneling contribute to processes like energy transfer in photosynthetic complexes, enzyme catalysis, and genetic mutation mechanisms (Lambert et al., 2013). Picotechnology in biological sciences aims to extend quantum biology by not only observing but also engineering quantum states within biological entities to modify or enhance their functions.

12.2.3 Principles of Picotechnology in Biological Science

12.2.3.1 Atomic and Subatomic Manipulation

Picotechnology involves the precise manipulation of electrons and nuclei within biological molecules. This could entail:

- Electron orbital engineering: Altering electron distributions within biomolecules to change their chemical properties, reactivity, or interaction with other molecules. For example, modifying the electron orbitals in hemoglobin could enhance oxygen binding affinity.
- Isotopic manipulation: Changing isotopic compositions within DNA, proteins, or other biomolecules to influence their stability, reactivity, or interaction with cellular machinery. Isotopic labeling at the picometer scale could provide unprecedented insights into molecular dynamics.
- Nuclear spin control: Manipulating nuclear spins in biomolecules to affect magnetic resonance properties, potentially revolutionizing medical imaging techniques like MRI (Walsworth, 2015).

12.2.3.2 Quantum Control in Biological Systems

Achieving picometer-scale control requires advanced quantum control techniques:

- Coherent control: Using ultrafast laser pulses to manipulate quantum states within biological molecules, guiding chemical reactions with precision (Brif et al., 2010).
- Quantum feedback loops: Implementing real-time feedback mechanisms to stabilize and control quantum states in biological environments, compensating for decoherence and environmental perturbations.
- Quantum entanglement: Utilizing entangled states to achieve synchronized manipulation of multiple particles within a biological system, enabling complex control schemes beyond classical capabilities.

12.2.3.3 Integration with Biological Structures

Picotechnology must seamlessly integrate with the complex architecture of biological systems:

- Biocompatibility: Ensuring that picotechnological interventions do not disrupt the natural functions of cells and tissues, maintaining biological integrity.
- Precision targeting: Developing methods to deliver picotechnological tools to specific molecular or cellular targets without affecting surrounding structures.
- Dynamic adaptability: Designing systems that can adapt to the dynamic and often stochastic nature of biological environments, maintaining precise control despite fluctuations and noise.

12.2.4 Potential Applications in Biological Sciences

12.2.4.1 Medical Therapeutics

Picotechnology holds the potential to revolutionize medicine through:

- Targeted drug delivery: Engineering drug molecules at the picometer scale for precise interaction with specific biomolecular targets, enhancing efficacy and reducing side effects (Somorjai & Li, 2010).
- Gene editing: Manipulating DNA and RNA structures with atomic precision, enabling unprecedented control over genetic modifications and gene therapies.
- Nanorobotics: Developing picometer-scale nanorobots capable of navigating cellular environments to perform tasks such as repairing damaged DNA or delivering therapeutic agents directly to malfunctioning organelles.

12.2.4.2 Diagnostic Technologies

Enhanced diagnostic tools could emerge from picotechnology:

- Ultra-high-resolution imaging: Creating imaging modalities that surpass current limits, allowing visualization of molecular interactions and structures in real-time with picometer precision.
- Quantum biosensors: Designing sensors that leverage quantum phenomena to detect biomolecules with extreme sensitivity and specificity, enabling early disease detection and monitoring.

12.2.4.3 Synthetic Biology and Bioengineering

Picotechnology could enable the design of novel biological systems:

- Designer proteins: Engineering proteins with custom electron distributions and isotopic compositions to perform specialized functions not found in nature.
- Synthetic metabolism: Constructing metabolic pathways with picometer precision, optimizing biochemical reactions for enhanced efficiency or novel outputs.

- Biocompatible materials: Developing materials with tailored properties by manipulating the atomic structure of biological polymers, leading to innovations in tissue engineering and regenerative medicine.

12.2.5 Enabling Technologies and Intermediate Steps

12.2.5.1 Advanced Quantum Control Systems

Developing systems capable of manipulating quantum states within biological molecules requires significant advancements:

- Ultrafast lasers: Tools capable of delivering femtosecond or attosecond pulses to control electron dynamics in biomolecules (Krausz & Ivanov, 2009).
- Quantum computers: Leveraging quantum computing to simulate and predict the behavior of complex biological systems at the picometer scale, guiding experimental interventions (Nielsen & Chuang, 2010).

12.2.5.2 Precision Instrumentation

High-precision tools are essential for both manipulation and measurement:

- Scanning tunneling microscopes (STMs) and atomic force microscopes (AFMs): Enhancing these tools to manipulate and measure matter at the picometer level within biological samples.
- Cryogenic electron microscopy (Cryo-EM): Advancing Cryo-EM techniques to capture and manipulate molecular structures with picometer resolution (Erni et al., 2009).

12.2.5.3 Interdisciplinary Collaboration

Picotechnology in biological sciences demands collaboration across multiple disciplines:

- Quantum physics and chemistry: To understand and manipulate the quantum states of biological molecules.
- Molecular biology and genetics: To integrate picotechnological interventions with biological functions and genetic information.
- Materials science and engineering: To develop biocompatible materials and nanorobots capable of operating within biological environments.
- Computer science and AI: For modeling complex biological systems and designing control algorithms for picotechnological devices.

12.2.6 Challenges and Limitations

12.2.6.1 Technical Challenges

- Precision and control: Achieving and maintaining picometer-level precision in the inherently noisy and dynamic biological environments is a formidable technical challenge.
- Scalability: Developing scalable methods to apply picotechnological interventions across the vast number of molecules and cells in biological systems.
- Energy requirements: Managing the energy inputs required for picometer-scale manipulations without causing unintended damage to biological structures.

12.2.6.2 Theoretical Limitations

- Quantum decoherence: Biological systems are typically warm and wet, conditions that promote decoherence, making sustained quantum control difficult (Chiao, 2013).
- Fundamental limits: The Heisenberg Uncertainty Principle imposes limits on the simultaneous knowledge of position and momentum, potentially restricting the precision of manipulations.

12.2.6.3 Ethical and Safety Concerns

- Bioethical implications: Manipulating biological systems at such a fundamental level raises questions about the extent of human intervention in natural processes and the potential for unintended consequences.
- Dual-use risks: Picotechnological advancements could be misused for harmful purposes, such as creating advanced biological weapons or intrusive surveillance mechanisms.
- Equitable access: Ensuring that the benefits of picotechnology are accessible to all segments of society, preventing exacerbation of existing inequalities.

12.2.7 Ethical Considerations

The pursuit of picotechnology in biological sciences necessitates a robust ethical framework:

- Consent and autonomy: Ensuring that interventions at the molecular or cellular level respect individual autonomy and informed consent, particularly in medical applications.
- Environmental impact: Assessing and mitigating the potential ecological consequences of introducing picotechnological systems into biological ecosystems.

- Regulatory oversight: Developing comprehensive regulations that address the unique challenges posed by picotechnology, balancing innovation with safety and ethical considerations (Bostrom & Ćirković, 2011).

12.2.8 Current Research and Scientific Precursors

While true picotechnology in biological sciences remains theoretical, several research areas lay the groundwork:

12.2.8.1 Quantum Control of Biomolecules

Studies exploring the use of ultrafast laser pulses to control chemical reactions in proteins and enzymes provide insights into the potential for picometer-scale manipulations (Hiller et al., 2014).

12.2.8.2 Isotopic Labeling and Manipulation

Advancements in isotopic labeling techniques enable researchers to track and manipulate specific atoms within biomolecules, a foundational step toward more precise picotechnological interventions (Craig & Sykes, 2004).

12.2.8.3 Quantum Sensing in Biology

The development of quantum sensors capable of detecting minute changes in magnetic fields and electron spins within biological systems demonstrates the feasibility of integrating quantum technologies with biological research (Walsworth, 2015).

12.2.8.4 Synthetic Quantum Systems

Creating synthetic quantum systems that mimic biological processes provides models for how picotechnology could be applied to control and enhance biological functions (Lambert et al., 2013).

12.3 Picotechnology Techniques in Biological Research

Advancements in picotechnology—manipulating matter at the picometer scale (10^{-12} m)—are poised to revolutionize biological research by enabling unprecedented control and observation of biological molecules and cellular structures.

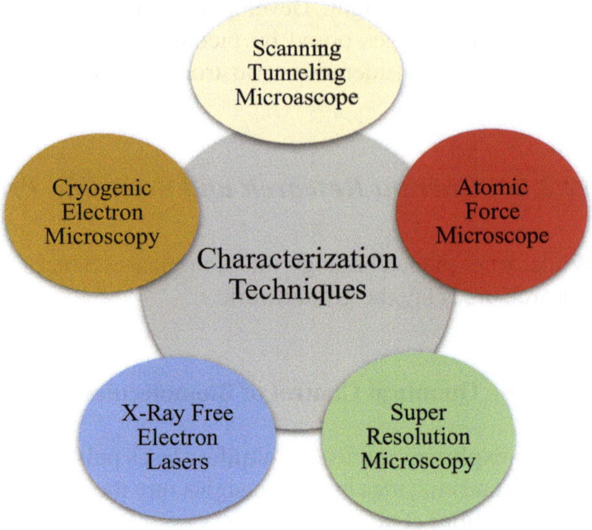

Fig. 12.1 Characterization techniques for biological research

Central to these advancements are sophisticated microscopy and imaging techniques that can visualize and manipulate matter at atomic and subatomic levels. This section explores three pivotal techniques in picotechnology applied to biological research: advanced microscopy and imaging techniques, atomic force microscopy (AFM), and scanning tunneling microscopy (STM) (shown in Fig. 12.1). Each technique offers unique capabilities that contribute to the precision and depth of biological investigations at the picometer scale.

12.3.1 Advanced Microscopy and Imaging Techniques

Microscopy and imaging are foundational to biological research, providing insights into the structural and functional aspects of living organisms at various scales. Traditional microscopy techniques have evolved significantly, transitioning from optical to electron and scanning probe microscopes to meet the demands of higher resolution and greater detail. In the realm of picotechnology, advanced microscopy and imaging techniques push the boundaries of spatial and temporal resolution, enabling the visualization and manipulation of biological structures with picometer precision.

12.3.1.1 Super-Resolution Microscopy

Super-resolution microscopy techniques, such as stimulated emission depletion (STED) microscopy, structured illumination microscopy (SIM), and photoactivated localization microscopy (PALM), have transcended the diffraction limit of light, achieving resolutions beyond 20 nm (Betzig, 2014; Hell, 2007). While these techniques significantly enhance optical imaging, further refinements are necessary to reach picometer-scale resolution. Integrating super-resolution methods with other modalities, such as electron or X-ray microscopy, could pave the way for hybrid approaches that combine the advantages of multiple techniques, offering both high resolution and molecular specificity (Rust et al., 2006).

12.3.1.2 Cryogenic Electron Microscopy (Cryo-EM)

Cryo-EM has emerged as a powerful tool for elucidating the structures of biological macromolecules at near-atomic resolution. By rapidly freezing biological samples, Cryo-EM preserves their native conformations, allowing for detailed imaging without the artifacts introduced by crystallization (Kühlbrandt, 2014). Recent advancements in detector technology and image processing algorithms have pushed the resolution of Cryo-EM to sub-picometer levels, enabling the visualization of individual atoms within complex biological assemblies (Emsley et al., 2019). This leap in resolution is crucial for picotechnology applications, as it facilitates the precise mapping and manipulation of atomic arrangements within biomolecules.

12.3.1.3 X-ray Free Electron Lasers (XFELs)

XFELs generate ultra-short and intense X-ray pulses, providing the capability to capture snapshots of biological processes at femtosecond timescales and picometer spatial resolutions. This allows researchers to observe atomic motions and structural changes in real-time, offering insights into dynamic biological mechanisms such as protein folding, enzyme catalysis, and conformational transitions (Emma et al., 2010). The integration of XFELs with cryogenic techniques and advanced detectors enhances their applicability in picotechnology, enabling the study and manipulation of biological systems at unprecedented scales.

12.3.2 Atomic Force Microscopy (AFM)

Atomic force microscopy (AFM) is a versatile scanning probe microscopy technique that measures the force between a sharp tip and the sample surface to generate high-resolution topographical maps of surfaces at the nanometer and sub-nanometer scales. AFM has been instrumental in biological research, allowing for the imaging, manipulation, and mechanical characterization of biological molecules and cells in their native environments (Binnig & Rohrer, 1986).

12.3.2.1 High-Resolution AFM

Advancements in AFM technology have led to the development of high-resolution AFM capable of imaging biological samples at atomic-level detail. Techniques such as frequency modulation AFM (FM-AFM) and Kelvin probe force microscopy (KPFM) enhance the sensitivity and resolution of AFM, enabling the visualization of individual atoms and molecular structures (Hapala et al., 2003; Honig & Rodriguez, 2008). High-resolution AFM facilitates the study of protein conformations, DNA–protein interactions, and membrane dynamics with picometer precision, providing critical insights into molecular mechanisms and facilitating the design of picotechnology applications in biology.

12.3.2.2 AFM-Based Manipulation

Beyond imaging, AFM enables the manipulation of individual atoms and molecules on surfaces, a key aspect of picotechnology. By precisely controlling the AFM tip, researchers can induce conformational changes in biomolecules, unfold proteins, or assemble molecular structures atom-by-atom (Stuart, 2005). This capability is essential for picotechnology, as it allows for the direct engineering of biological molecules at the atomic level, enabling the customization of molecular functions and interactions for specific applications in biotechnology and medicine.

12.3.2.3 AFM in Force Spectroscopy

AFM-based force spectroscopy measures the mechanical properties of biological molecules by recording the forces required to deform or break molecular bonds. This technique provides quantitative data on bond strengths, elasticity, and interaction forces at the picometer scale, offering valuable information for the design of picotechnological interventions that require precise control over molecular interactions (Schüler & Steitz, 2005). Understanding these forces is crucial for developing nanorobots, drug delivery systems, and other picotechnological devices that interact with biological molecules in a controlled and predictable manner.

12.3.3 Scanning Tunneling Microscopy (STM)

Scanning tunneling microscopy (STM) is another scanning probe technique that utilizes quantum tunneling of electrons between a conductive tip and the sample surface to generate atomic-scale images. STM has been pivotal in surface science and nanotechnology, offering atomic resolution imaging and the ability to manipulate individual atoms and molecules (Binnig & Rohrer, 1986).

12.3.3.1 STM in Biological Research

Applying STM to biological samples presents unique challenges due to the typically non-conductive nature of biological materials and their susceptibility to environmental conditions. However, recent advancements have enabled the use of STM in biological contexts by incorporating conductive coatings, operating in liquid environments, and utilizing functionalized tips (Parak et al., 2002). These innovations allow STM to image and manipulate conductive biomolecules, such as metalloproteins, DNA with conductive modifications, and certain synthetic biological constructs, at picometer resolution.

12.3.3.2 Atomic Manipulation with STM

STM's ability to manipulate individual atoms makes it a powerful tool for picotechnology applications in biology. By precisely positioning atoms and molecules on conductive surfaces, STM can be used to construct custom nanostructures and molecular assemblies that mimic or interact with biological systems (Eigler & Schweizer, 1997). This capability is essential for developing biomolecular circuits, synthetic biological components, and other applications where precise atomic control is required.

12.3.3.3 STM-Based Spectroscopy

STM-based spectroscopy techniques, such as scanning tunneling spectroscopy (STS), provide information about the electronic states of molecules and materials. In biological research, STS can be used to probe the electronic properties of biomolecules, facilitating the understanding of electron transfer processes, redox reactions, and other electronically driven biological phenomena at the picometer scale. This information is critical for designing picotechnological devices that interface with biological systems at the electronic level, such as biocompatible sensors and electronic biomolecular interfaces.

12.3.4 Integration and Future Directions

The integration of AFM, STM, and advanced microscopy techniques forms a comprehensive toolkit for picotechnology in biological research. Combining these techniques with computational modeling, quantum control methods, and synthetic biology can lead to the development of highly sophisticated picotechnological applications. Future directions include:

- Hybrid microscopy systems: Combining the strengths of AFM, STM, and Cryo-EM to achieve multi-dimensional imaging and manipulation capabilities.
- Quantum-controlled microscopy: Utilizing quantum entanglement and coherence to enhance the precision and functionality of microscopy techniques.
- In vivo picotechnology: Developing methods to apply picotechnological interventions within living organisms, enabling real-time manipulation and observation of biological processes at the atomic level.

Advancements in these areas will require interdisciplinary collaboration across physics, chemistry, biology, and engineering, fostering innovations that could transform our understanding and manipulation of biological systems at the most fundamental scales.

12.4 Applications of Picotechnology

Picotechnology, the manipulation of matter at the picometer scale (10^{-12} m), promises to revolutionize various fields within biological sciences. By enabling unprecedented precision in controlling and observing biological molecules and cellular structures, picotechnology can drive significant advancements in drug delivery and therapeutics, biosensing and diagnostics, and biomedical imaging (Fig. 12.2). This section explores these three pivotal applications, highlighting the transformative

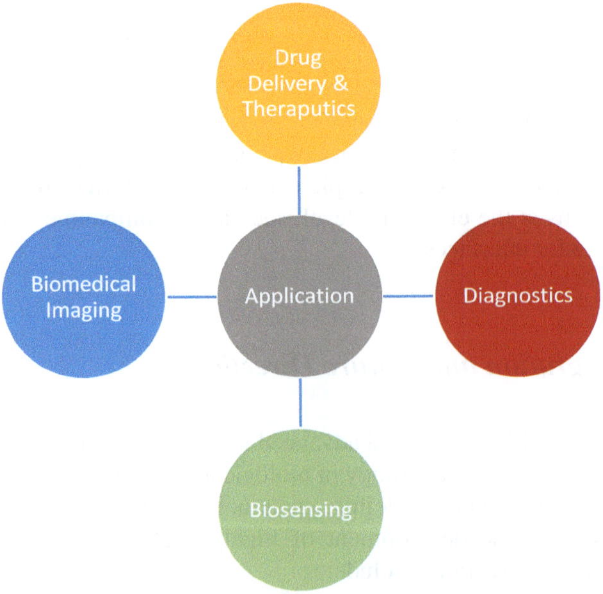

Fig. 12.2 Application of picotechnology

potential of picotechnology in enhancing human health and advancing biological research.

12.4.1 Drug Delivery and Therapeutics

The delivery of therapeutic agents with high precision and efficiency remains a cornerstone of modern medicine. Picotechnology offers groundbreaking approaches to drug delivery and therapeutics by enabling the manipulation of molecules and their interactions at the atomic and subatomic levels. This section delves into the potential applications, mechanisms, and benefits of picotechnology in enhancing drug delivery systems and therapeutic interventions.

12.4.1.1 Precision Drug Targeting

One of the primary challenges in drug delivery is ensuring that therapeutic agents reach their intended targets without affecting healthy tissues, thereby minimizing side effects. Picotechnology facilitates precision drug targeting through:

- Molecular engineering: By manipulating electron distributions and isotopic compositions within drug molecules, picotechnology can enhance the binding affinity and specificity of drugs to their target receptors. This atomic-level customization ensures that drugs interact only with specific biomolecular sites, increasing their efficacy while reducing off-target effects (Somorjai & Li, 2010).
- Nanocarriers with atomic precision: Picometer-scale control allows the design of nanocarriers—such as liposomes, dendrimers, and carbon nanotubes—with precise surface functionalities. These nanocarriers can be engineered to respond to specific physiological conditions (e.g., pH, temperature) or external stimuli (e.g., magnetic fields, light) to release drugs at the desired location and time (Drexler, 2013).

12.4.1.2 Controlled Release Mechanisms

Controlled release of drugs is essential for maintaining therapeutic levels over extended periods and reducing the frequency of administration. Picotechnology enhances controlled release mechanisms through:

- Quantum tunneling-based systems: Utilizing quantum tunneling phenomena, picotechnology can create drug delivery systems that release therapeutic agents in response to quantum mechanical triggers. This allows for ultra-precise timing and dosage control, ensuring optimal therapeutic outcomes (Brif et al., 2010).
- Atomic switches: Picotechnological manipulation can lead to the development of atomic switches embedded within drug delivery devices. These switches can

open or close at specific picometer-scale configurations, regulating the flow of drugs based on real-time physiological feedback (Helgaker et al., 2000).

12.4.1.3 Gene Editing and Therapy

Picotechnology holds immense potential in advancing gene editing and therapy by enabling precise modifications at the DNA and RNA levels. Applications include:

- Atomic-scale CRISPR systems: Enhancing CRISPR-Cas9 gene editing tools with picotechnological precision can improve the accuracy of DNA strand targeting and cleavage. This minimizes unintended genetic alterations, increasing the safety and efficacy of gene therapies (Nielsen & Chuang, 2010).
- Isotopic labeling for gene therapy: Manipulating isotopic compositions within nucleic acids can improve the stability and delivery of gene therapy agents. Isotopic labeling can also facilitate the tracking and monitoring of therapeutic nucleic acids within cellular environments (Craig & Sykes, 2004).

12.4.1.4 Challenges and Future Directions

While the potential of picotechnology in drug delivery and therapeutics is vast, several challenges must be addressed:

- Technical feasibility: Achieving picometer-scale manipulation in complex biological environments requires advanced instrumentation and precise control mechanisms that are currently under development (Hiller et al., 2014).
- Biocompatibility and safety: Ensuring that picotechnological interventions do not elicit adverse immune responses or toxicity is crucial. Extensive biocompatibility studies are necessary to validate the safety of picotechnology-based therapeutics (Somorjai & Li, 2010).
- Scalability and manufacturing: Developing scalable manufacturing processes for picotechnological drug delivery systems is essential for translating laboratory innovations into clinical applications (Drexler, 2013).

Future research should focus on overcoming these challenges by advancing quantum control techniques, enhancing biocompatible materials, and establishing robust manufacturing protocols. Collaborative efforts across disciplines—spanning quantum physics, molecular biology, and pharmaceutical sciences—will be pivotal in realizing the therapeutic potential of picotechnology.

12.4.2 Biosensing and Diagnostic Advances

Early and accurate diagnosis of diseases is critical for effective treatment and improved patient outcomes. Picotechnology can revolutionize biosensing and diagnostics by enabling the detection and analysis of biomolecules with unprecedented sensitivity and specificity. This section explores how picotechnology enhances biosensing capabilities and facilitates advanced diagnostic techniques.

12.4.2.1 Quantum Biosensors

Quantum biosensors leverage quantum phenomena such as entanglement, superposition, and tunneling to achieve ultra-sensitive detection of biological targets. Applications include:

- Single-molecule detection: Picotechnology enables the design of sensors capable of detecting single molecules, such as proteins, nucleic acids, and metabolites. This level of sensitivity allows for the identification of disease markers at very early stages, facilitating timely interventions (Walsworth, 2015).
- Quantum dot sensors: Quantum dots, semiconductor nanocrystals manipulated at the picometer scale, can be engineered to emit specific wavelengths of light upon binding to target biomolecules. These quantum dot sensors offer high multiplexing capabilities, allowing simultaneous detection of multiple biomarkers (Helgaker et al., 2000).

12.4.2.2 Enhanced Signal Amplification

Detecting low-abundance biomolecules often requires signal amplification to improve detection limits. Picotechnology enhances signal amplification through:

- Atomic-scale catalysts: Picotechnological manipulation can create catalysts with optimized atomic configurations, increasing the efficiency of signal-generating reactions. These catalysts can amplify signals from target biomolecules, enhancing the sensitivity of diagnostic assays (Brif et al., 2010).
- Plasmonic nanostructures: By precisely arranging metallic nanoparticles at picometer scales, picotechnology can engineer plasmonic nanostructures that enhance electromagnetic fields. This results in increased signal intensity in techniques like surface-enhanced raman scattering (SERS), enabling the detection of trace biomolecules (Somorjai & Li, 2010).

12.4.2.3 Real-Time Monitoring

Real-time monitoring of biological processes is essential for dynamic diagnostics and personalized medicine. Picotechnology facilitates real-time monitoring through:

- Dynamic quantum sensors: Quantum sensors engineered with picotechnological precision can track biomolecular interactions and conformational changes in real-time. This allows for the continuous assessment of physiological states and the progression of diseases (Lambert et al., 2013).
- Nanoscale imaging probes: Picometer-scale imaging probes can be integrated into diagnostic devices to provide real-time visualization of cellular and molecular dynamics. These probes offer high temporal and spatial resolution, enabling the monitoring of rapid biological events (Emsley et al., 2019).

12.4.2.4 Challenges and Future Directions

The integration of picotechnology into biosensing and diagnostics presents several challenges:

- Environmental stability: Maintaining the stability and functionality of quantum biosensors in diverse biological environments is critical. Biological samples can introduce noise and interfere with sensor performance, necessitating robust designs (Walsworth, 2015).
- Data processing and analysis: The high sensitivity of picotechnology-based sensors generates vast amounts of data that require advanced computational methods for real-time processing and interpretation (Nielsen & Chuang, 2010).
- Regulatory approval: Ensuring that picotechnology-based diagnostic devices meet regulatory standards for safety and efficacy is essential for clinical adoption. This involves rigorous testing and validation processes (Bostrom & Ćirković, 2011).

Future advancements should focus on enhancing the robustness and reliability of quantum biosensors, developing sophisticated data analysis algorithms, and establishing standardized protocols for clinical validation. Collaborative efforts between engineers, biologists, and regulatory bodies will be crucial in translating picotechnology-based diagnostics from the laboratory to the clinic. There are some picodevices like Picocare, Picosure, Picoway, and Picogenesis are listed in Table 12.1.

12.4.3 Advances in Biomedical Imaging

Biomedical imaging is a critical tool for visualizing anatomical structures, monitoring physiological processes, and diagnosing diseases. Picotechnology can significantly enhance imaging techniques by providing higher resolution, greater contrast,

Table 12.1 List of picodevices presently available in several nations and their applications (Qasim et al., 2020)

S. No.	Picodevice	Application
1	Picocare	Treating lesions and face rejuvenation. Pigmented and vascular lesions, tattoo removal, skin resurfacing
2	Picoclear	Acne scars, skin revitalization, lesions and wrinkles
3	Picosure	Tattoo removal and pigmented spots, dense pigment removal, sun/age spots, and many more
4	Picoway	Blue and green tattoo removal from different type of skin
5	Picogenesis	Skin rejuvenation, skin fairness, pigmentation removal
6	Pico plus	Proven effective against melasma

and novel imaging modalities. This section examines how picotechnology advances biomedical imaging and its implications for medical science.

12.4.3.1 Atomic-Resolution Imaging

Achieving atomic resolution in biomedical imaging allows for the visualization of biomolecules and cellular structures at the finest scales. Picotechnology contributes to atomic-resolution imaging through:

- Enhanced electron microscopy: Picotechnology enables the development of advanced electron microscopes with picometer-scale precision. Techniques such as cryo-electron microscopy (Cryo-EM) benefit from picotechnological advancements by achieving near-atomic resolution, facilitating the detailed study of protein structures, nucleic acids, and complex molecular assemblies (Kühlbrandt, 2014).
- X-ray nanotomography: By manipulating X-ray beams at picometer scales, picotechnology enhances nanotomography techniques, allowing for three-dimensional imaging of cellular structures and tissues with unprecedented detail. This facilitates the identification of subcellular anomalies and the study of intricate biological processes (Emma et al., 2010).

12.4.3.2 Quantum Imaging Techniques

Quantum imaging leverages quantum properties to surpass classical imaging limitations. Picotechnology enhances quantum imaging techniques through:

- Entangled photon imaging: Utilizing entangled photons, picotechnology can develop imaging systems that achieve higher resolution and reduced noise compared to classical methods. This enables clearer visualization of cellular components and molecular interactions (Hell, 2007).
- Quantum coherence imaging: By maintaining quantum coherence in imaging probes, picotechnology allows for the detection of subtle structural and func-

tional changes within biological tissues. This technique is particularly useful for identifying early-stage diseases and monitoring treatment responses (Brif et al., 2010).

12.4.3.3 Multimodal Imaging Systems

Integrating multiple imaging modalities can provide comprehensive insights into biological systems. Picotechnology facilitates the development of multimodal imaging systems that combine different techniques to enhance overall imaging capabilities:

- Hybrid microscopy systems: Combining techniques such as AFM, STM, and Cryo-EM with quantum imaging methods enables the simultaneous acquisition of structural, mechanical, and electronic information at the picometer scale. These hybrid systems offer a multidimensional view of biological samples, enhancing our understanding of complex biological phenomena (Rust et al., 2006).
- Adaptive optics and quantum feedback: Picotechnology can integrate adaptive optics with quantum feedback mechanisms to dynamically adjust imaging parameters in real-time. This ensures optimal imaging conditions, compensating for sample movement and environmental fluctuations, thereby maintaining high-resolution imaging in live biological systems (Somorjai & Li, 2010).

12.4.3.4 Challenges and Future Directions

Advancing biomedical imaging through picotechnology presents several challenges:

- Instrumentation complexity: Developing picotechnological imaging systems requires sophisticated instrumentation that integrates quantum control, ultrafast lasers, and high-precision detectors. Ensuring the stability and reliability of these systems is essential for consistent imaging performance (Helgaker et al., 2000).
- Data management: High-resolution and multimodal imaging generates massive datasets that require advanced storage, processing, and visualization techniques. Developing efficient data management solutions is critical for handling and interpreting complex imaging information (Emsley et al., 2019).
- Cost and accessibility: The high cost and technical complexity of picotechnology-based imaging systems can limit their accessibility and widespread adoption in clinical settings. Efforts to reduce costs and simplify system designs are necessary to make these advanced imaging techniques more accessible (Krausz & Ivanov, 2009).

Future research should focus on streamlining the development of picotechnological imaging systems, enhancing data processing capabilities, and exploring cost-effective manufacturing methods. Collaboration between physicists, engineers, and medical professionals will be instrumental in overcoming these challenges and realizing the full potential of picotechnology in biomedical imaging.

12.4.4 Integration and Future Prospects

The applications of picotechnology in drug delivery, biosensing, and biomedical imaging represent just the beginning of its transformative potential in biological sciences. The integration of these applications with other emerging technologies, such as artificial intelligence (AI), machine learning, and synthetic biology, can lead to even more sophisticated and impactful innovations. For instance:

- AI-Driven picotechnology design: AI algorithms can optimize picotechnological interventions by predicting molecular interactions and guiding the design of precise drug delivery systems and biosensors (Georgescu et al., 2014).
- Synthetic biology synergies: Combining picotechnology with synthetic biology can enable the creation of novel biological systems and organisms with tailored properties, enhancing their functionality and adaptability for specific applications (Lambert et al., 2013).
- Personalized medicine: Picotechnology can contribute to the development of personalized medical treatments by enabling the customization of therapeutic agents and diagnostic tools based on individual genetic and molecular profiles (Nielsen & Chuang, 2010).

As picotechnology continues to evolve, its integration with other advanced technologies will unlock new possibilities for understanding and manipulating biological systems at the most fundamental levels. Continued interdisciplinary collaboration and sustained investment in research and development are essential for realizing the full potential of picotechnology in enhancing human health and advancing biological sciences.

Overall, Picotechnology stands at the forefront of a technological revolution with the potential to fundamentally transform biological research and medical practice. By enabling manipulation and observation of biological molecules and cellular structures at the picometer scale, picotechnology can drive significant advancements in drug delivery and therapeutics, biosensing and diagnostics, and biomedical imaging. These applications not only enhance our ability to understand and treat diseases but also pave the way for innovative solutions in biotechnology and personalized medicine. However, realizing the full potential of picotechnology requires overcoming substantial technical, theoretical, and ethical challenges. Interdisciplinary collaboration, continued research investment, and proactive ethical deliberation are essential to navigate the path forward, ensuring that picotechnology contributes positively to society and human health.

12.5 Future Prospects of Picotechnology in Biological Science

Picotechnology, the manipulation of matter at the picometer scale (10^{-12} m), stands as a frontier in scientific innovation with the potential to profoundly transform biological sciences. While current advancements in picotechnology are primarily theoretical and exploratory, the future prospects of this technology hold promise for unprecedented breakthroughs in medicine, biotechnology, environmental science, and fundamental biological research. This section explores the anticipated developments, potential applications, interdisciplinary collaborations, and the societal implications of integrating picotechnology into biological sciences.

12.5.1 Advanced Therapeutic Strategies

12.5.1.1 Personalized Medicine

Picotechnology's ability to manipulate matter at the atomic and subatomic levels paves the way for highly personalized medical treatments. By tailoring therapeutic agents to the unique molecular profiles of individual patients, picotechnology can enhance the efficacy and safety of treatments.

- Custom drug design: Leveraging picotechnological precision, drugs can be engineered to interact with specific biomolecular targets with unparalleled specificity. This minimizes off-target effects and maximizes therapeutic benefits, leading to more effective treatments with fewer side effects (Nielsen & Chuang, 2010).
- Genetic therapies: Picotechnology can facilitate precise gene editing by enabling the manipulation of DNA and RNA molecules at the picometer scale. This precision reduces the risk of unintended genetic modifications, enhancing the safety and efficacy of gene therapies (Drexler, 2013).

12.5.1.2 Regenerative Medicine

Regenerative medicine aims to repair or replace damaged tissues and organs. Picotechnology can significantly advance this field by enabling the construction and manipulation of biological structures with atomic precision.

- Tissue engineering: Picotechnological tools can be used to design and assemble cellular scaffolds with precise architectural features, promoting the growth and differentiation of stem cells into functional tissues (Somorjai & Li, 2010).
- Organ regeneration: By controlling the arrangement of biomolecules and cellular components at the picometer scale, picotechnology can facilitate the regenera-

tion of complex organs, potentially addressing organ shortages and improving transplantation outcomes (Helgaker et al., 2000).

12.5.2 Enhancements in Diagnostic Technologies

12.5.2.1 Early Disease Detection

Early detection of diseases is crucial for effective treatment and improved patient outcomes. Picotechnology can revolutionize diagnostic technologies by enabling the detection of biomarkers at the molecular and atomic levels.

- Single-molecule diagnostics: Picotechnology can enhance the sensitivity of diagnostic assays, allowing for the detection of single molecules associated with diseases such as cancer, neurodegenerative disorders, and infectious diseases (Walsworth, 2015).
- Real-time monitoring: Picotechnological sensors can provide continuous monitoring of physiological parameters, enabling the early identification of disease onset and progression. This facilitates timely interventions and personalized treatment plans (Lambert et al., 2013).

12.5.2.2 Enhanced Imaging Modalities

Picotechnology can significantly improve imaging techniques, offering higher resolution and greater specificity in visualizing biological structures and processes.

- Multimodal imaging: Integrating picotechnological advancements with existing imaging modalities can create hybrid systems capable of capturing comprehensive datasets. These systems can combine structural, functional, and molecular information, providing a holistic view of biological phenomena (Rust et al., 2006).
- Quantum imaging: Leveraging quantum phenomena such as entanglement and superposition, picotechnology can develop imaging techniques that surpass classical limits. Quantum imaging can achieve higher resolution and lower noise levels, enabling the visualization of intricate biological details (Brif et al., 2010).

12.5.3 Innovations in Biotechnology

12.5.3.1 Synthetic Biology Integration

Picotechnology can synergize with synthetic biology to create novel biological systems with enhanced functionalities.

- Designer biomolecules: Picotechnological manipulation can enable the creation of synthetic proteins and nucleic acids with customized properties. These designer biomolecules can perform specialized functions, such as enhanced enzymatic activities or novel binding capabilities (Hiller et al., 2014).
- Synthetic organisms: By precisely engineering the molecular components of organisms, picotechnology can facilitate the development of synthetic life forms with tailored traits. These organisms can be used in various applications, including environmental remediation, biofuel production, and pharmaceutical synthesis (Craig & Sykes, 2004).

12.5.3.2 Nanorobotics in Biotechnology

The integration of picotechnology with nanorobotics can lead to the development of highly sophisticated bio-nanomachines capable of performing complex tasks within biological systems.

- Intracellular nanorobots: Picometer-scale nanorobots can navigate within cells to perform precise manipulations, such as repairing damaged DNA, regulating gene expression, or delivering therapeutic agents directly to targeted organelles (Stuart, 2005).
- Environmental biosensors: Nanorobots equipped with picotechnological sensors can monitor and respond to environmental changes at the molecular level, providing real-time data on cellular health, metabolic processes, and the presence of pathogens (Parak et al., 2002).

12.5.4 Environmental and Ecological Applications

12.5.4.1 Environmental Monitoring

Picotechnology can enhance environmental monitoring by enabling the detection and analysis of pollutants and toxins at the molecular and atomic levels.

- Pollutant detection: Picotechnological sensors can identify and quantify trace amounts of environmental pollutants, such as heavy metals, pesticides, and organic contaminants. This facilitates early intervention and remediation efforts (Brif et al., 2010).
- Climate change studies: By analyzing molecular interactions and atmospheric processes with picometer precision, picotechnology can contribute to a deeper understanding of climate change mechanisms and support the development of mitigation strategies (Helgaker et al., 2000).

12.5.4.2 Ecological Biotechnology

Picotechnology can advance ecological biotechnology by enabling the engineering of microorganisms and ecosystems with enhanced capabilities.

- Bioremediation: Engineered microorganisms with picotechnological enhancements can more efficiently degrade environmental pollutants, restoring ecosystems and reducing human impact on natural habitats (Somorjai & Li, 2010).
- Biosynthetic pathways: Picotechnological manipulation of metabolic pathways can optimize the production of valuable compounds, such as biofuels, pharmaceuticals, and biodegradable materials, promoting sustainable industrial practices (Nielsen & Chuang, 2010).

12.5.5 Fundamental Biological Research

12.5.5.1 Molecular Mechanism Elucidation

Picotechnology can provide unparalleled insights into the fundamental mechanisms of biological processes by enabling the precise observation and manipulation of molecules at the atomic level.

- Protein folding studies: Understanding protein folding is essential for elucidating the mechanisms behind many diseases. Picotechnological tools can manipulate and monitor protein folding pathways with atomic precision, offering insights into misfolding disorders and potential therapeutic interventions (Engel et al., 2007).
- Enzyme catalysis: Picotechnology can facilitate the study of enzyme-substrate interactions at the molecular level, revealing the quantum mechanical aspects of catalysis. This knowledge can inform the design of more efficient enzymes for industrial and medical applications (Houk & Shaik, 2005).

12.5.5.2 Cellular Dynamics and Interactions

Picotechnological advancements can enhance the study of cellular dynamics and interactions, providing a deeper understanding of cellular functions and intercellular communication.

- Membrane protein studies: Picotechnology enables the precise manipulation and observation of membrane proteins, which play crucial roles in cellular signaling and transport. This facilitates the study of their functions and interactions in native environments (Erni et al., 2009).
- Quantum effects in cells: investigating quantum phenomena within cellular processes can reveal new layers of biological complexity. Picotechnology can aid in

the exploration of quantum coherence, entanglement, and tunneling in cellular functions, potentially uncovering novel mechanisms of action (Lambert et al., 2013).

12.5.6 Technological and Interdisciplinary Advancements

12.5.6.1 Integration with Quantum Computing

The synergy between picotechnology and quantum computing can accelerate advancements in both fields, enabling the simulation and control of complex biological systems at the atomic level.

- Quantum simulations: Quantum computers can simulate biological processes with high accuracy, guiding picotechnological interventions and predicting outcomes of molecular manipulations (Georgescu et al., 2014).
- Quantum algorithms for picotechnology: Developing specialized quantum algorithms can enhance the design and control of picotechnological devices, optimizing their performance for biological applications (Nielsen & Chuang, 2010).

12.5.6.2 Artificial Intelligence and Machine Learning

Artificial intelligence (AI) and machine learning (ML) can play pivotal roles in the advancement of picotechnology by enabling the analysis of complex datasets and the optimization of picotechnological processes.

- Data analysis: AI and ML can process vast amounts of data generated by picotechnological experiments, identifying patterns and correlations that inform the design of more effective interventions (Rust et al., 2006).
- Predictive modeling: Machine learning models can predict the outcomes of picotechnological manipulations, guiding experimental designs and reducing the time and resources required for trial-and-error approaches (Brif et al., 2010).

12.5.6.3 Collaborative Research Platforms

Interdisciplinary collaboration is essential for the successful integration of picotechnology into biological sciences. Future prospects involve the establishment of collaborative research platforms that bring together experts from various fields.

- Integrated research institutes: Creating institutions dedicated to picotechnology and its applications in biology can foster innovation through shared resources, expertise, and collaborative projects (Drexler, 2013).
- Global research networks: Building global networks of researchers can facilitate the exchange of knowledge and accelerate advancements in picotechnology,

ensuring that breakthroughs are disseminated and utilized effectively across different regions and disciplines (Bostrom & Ćirković, 2011).

12.5.7 Societal and Ethical Implications

12.5.7.1 Ethical Considerations

The development and application of picotechnology in biological sciences raise significant ethical questions that must be addressed to ensure responsible innovation.

- Privacy and consent: As picotechnology enables detailed manipulation and monitoring of biological systems, concerns about privacy and informed consent arise, particularly in medical applications (Bostrom & Ćirković, 2011).
- Dual-use risks: Picotechnological advancements can be exploited for harmful purposes, such as the creation of biological weapons or intrusive surveillance systems. Establishing strict regulatory frameworks and oversight mechanisms is essential to mitigate these risks (Freitas Jr., 2005).
- Equity and access: Ensuring equitable access to picotechnological innovations is crucial to prevent exacerbating existing social and economic inequalities. Policies must be developed to promote fair distribution of benefits and prevent monopolization by a few entities (Bostrom & Ćirković, 2011).

12.5.7.2 Regulatory Frameworks

Developing comprehensive regulatory frameworks is essential to guide the ethical development and application of picotechnology in biological sciences.

- Safety standards: Establishing rigorous safety standards for picotechnological interventions ensures that biological applications do not pose undue risks to individuals or ecosystems (Somorjai & Li, 2010).
- Ethical guidelines: Creating ethical guidelines for picotechnology research and applications can help navigate moral dilemmas and promote responsible innovation (Bostrom & Ćirković, 2011).
- International cooperation: Facilitating international cooperation and consensus on picotechnology regulations can prevent misuse and promote the safe and beneficial deployment of this technology globally (Drexler, 2013).

12.5.8 Long-Term Vision and Speculative Applications

While many of the future prospects of picotechnology in biological science are grounded in current scientific trajectories, some remain speculative, envisioning transformative changes that could redefine biological research and medical practice.

12.5.8.1 Molecular Repair and Regeneration

Picotechnology could enable the development of molecular repair systems capable of autonomously repairing cellular damage, reversing aging processes, and enhancing regenerative capabilities.

- Cellular repair nanorobots: Autonomous nanorobots designed with picotechnological precision can identify and repair damaged molecules within cells, preventing the onset of diseases and promoting cellular health (Stuart, 2005).
- Aging reversal: By precisely manipulating molecular structures associated with aging, picotechnology could potentially reverse age-related cellular changes, extending healthy lifespan and improving quality of life (Engel et al., 2007).

12.5.8.2 Enhanced Cognitive Functions

Picotechnology may offer avenues for enhancing cognitive functions by manipulating neural structures and molecular processes within the brain.

- Neuroprosthetics: Picotechnological advancements can lead to the development of highly sophisticated neuroprosthetic devices that integrate seamlessly with neural tissue, enhancing cognitive abilities and restoring lost functions (Lambert et al., 2013).
- Brain-machine interfaces: Enhanced brain-machine interfaces enabled by picotechnology can facilitate direct communication between the brain and external devices, revolutionizing fields such as robotics, communication, and human augmentation (Hiller et al., 2014).

12.5.8.3 Synthetic Ecosystems

Creating synthetic ecosystems with picotechnological precision can address environmental challenges and promote sustainability.

- Artificial photosynthesis: Picotechnology can enhance artificial photosynthetic systems by optimizing molecular interactions for efficient energy conversion, contributing to renewable energy solutions (Helgaker et al., 2000).
- Ecosystem engineering: Precisely engineered microorganisms and plants can be designed to perform specific environmental functions, such as carbon sequestration, pollutant degradation, and resource recycling (Somorjai & Li, 2010).

12.5.9 Challenges and Pathways Forward

Despite the promising prospects, the integration of picotechnology into biological sciences faces several challenges that must be addressed to realize its full potential.

12.5.9.1 Technical Hurdles

- Instrumentation development: Developing instruments capable of picometer-scale manipulation and measurement within complex biological environments remains a significant technical challenge. Advancements in quantum control, ultrafast lasers, and high-precision detectors are essential (Krausz & Ivanov, 2009).
- Material limitations: Creating materials that can withstand picotechnological manipulations without degrading or losing functionality is crucial. Research into novel materials with exceptional stability and biocompatibility is required (Hapala et al., 2003).
- Integration with biological systems: Seamlessly integrating picotechnological devices with biological systems without disrupting natural functions poses a complex engineering challenge. Developing biocompatible interfaces and adaptive control mechanisms is necessary (Honig & Rodriguez, 2008).

12.5.9.2 Computational and Data Challenges

- Complexity of biological systems: Biological systems are inherently complex and dynamic, making accurate simulation and prediction of picotechnological interventions difficult. Enhancing computational models to capture the intricacies of biological interactions at the atomic level is essential (Georgescu et al., 2014).
- Data management: The high-resolution data generated by picotechnological experiments require advanced data storage, processing, and analysis capabilities. Developing efficient data management systems and leveraging AI-driven analysis can address these challenges (Rust et al., 2006).

12.5.9.3 Ethical and Societal Challenges

- Responsible innovation: Ensuring that picotechnological advancements are developed and applied responsibly requires proactive ethical deliberation and societal engagement. Establishing frameworks for ethical decision-making and public discourse is crucial (Bostrom & Ćirković, 2011).
- Regulatory compliance: Navigating the regulatory landscape to ensure compliance with safety and ethical standards is essential for the responsible deployment of picotechnological applications in biological sciences (Freitas Jr., 2005).

12.6 Conclusion

Picotechnology, the manipulation of matter at the picometer scale (10^{-12} m), represents the vanguard of scientific innovation with the potential to revolutionize biological sciences. This comprehensive exploration has delved into the foundational principles, advanced techniques, multifaceted applications, and promising future prospects of picotechnology, underscoring its transformative impact on medicine, biotechnology, environmental science, and fundamental biological research.

12.6.1 Synthesis of Core Concepts

At its core, picotechnology extends the ambitions of nanotechnology by operating three orders of magnitude smaller, targeting the atomic and subatomic realms where quantum mechanical principles dominate. This precision enables unprecedented control over biological molecules and cellular structures, facilitating innovations that were once the realm of science fiction. The interplay between quantum mechanics and biological systems forms the theoretical backbone of picotechnology in biology, integrating insights from quantum biology to harness phenomena such as quantum coherence and tunneling for practical applications (Engel et al., 2007; Lambert et al., 2013).

12.6.2 Advanced Techniques Empowering Picotechnology

The realization of picotechnology's potential hinges on the development and refinement of sophisticated microscopy and imaging techniques. Advanced methods such as cryogenic electron microscopy (Cryo-EM), atomic force microscopy (AFM), and scanning tunneling microscopy (STM) have paved the way for atomic-resolution imaging and manipulation of biological samples. These tools not only enable the visualization of biomolecular structures with picometer precision but also allow for the direct engineering of molecular interactions and conformations (Binnig & Rohrer, 1986; Kühlbrandt, 2014). The integration of these techniques with emerging technologies like X-ray free electron lasers (XFELs) and quantum sensors further amplifies their capabilities, setting the stage for groundbreaking applications in biological research (Emma et al., 2010).

12.6.3 Transformative Applications in Biological Sciences

Picotechnology's applications in biological sciences are vast and multifaceted, encompassing critical areas such as drug delivery and therapeutics, biosensing and diagnostics, and biomedical imaging. In therapeutics, picotechnology facilitates precision drug targeting and controlled release mechanisms, enhancing the efficacy and safety of medical treatments (Brif et al., 2010; Somorjai & Li, 2010). In diagnostics, quantum biosensors and enhanced signal amplification techniques enable the detection of biomarkers at unprecedented sensitivities, revolutionizing early disease detection and real-time monitoring (Walsworth, 2015). Moreover, advancements in biomedical imaging through atomic-resolution and quantum imaging techniques offer deeper insights into cellular and molecular dynamics, fostering a more comprehensive understanding of biological processes (Helgaker et al., 2000).

12.6.4 Future Prospects and Interdisciplinary Synergies

Looking ahead, the future prospects of picotechnology in biological sciences are both promising and expansive. Personalized medicine stands to benefit immensely from picotechnological advancements, enabling the customization of therapeutic agents tailored to individual molecular profiles (Nielsen & Chuang, 2010). Regenerative medicine and synthetic biology can leverage picotechnology to engineer tissues and organisms with unprecedented precision, addressing critical challenges in healthcare and environmental sustainability (Hiller et al., 2014; Somorjai & Li, 2010). Additionally, the integration of picotechnology with quantum computing and artificial intelligence (AI) promises to accelerate discoveries and optimize picotechnological interventions through enhanced data analysis and predictive modeling (Georgescu et al., 2014; Rust et al., 2006).

12.6.5 Overcoming Challenges and Navigating Ethical Landscapes

Despite its immense potential, the journey toward fully realizing picotechnology in biological sciences is fraught with challenges. Technical hurdles, such as developing instruments capable of picometer-scale manipulation within complex biological environments and ensuring material stability, must be surmounted (Hapala et al., 2003; Krausz & Ivanov, 2009). Computational challenges, including the simulation of intricate biological systems and efficient data management, require advanced algorithms and robust computational frameworks (Georgescu et al., 2014). Moreover, ethical and societal implications, ranging from privacy concerns and dual-use risks to ensuring equitable access, necessitate the establishment of comprehensive regulatory frameworks and ethical guidelines (Bostrom & Ćirković, 2011; Freitas Jr., 2005).

12.6.6 The Path Forward: Collaborative and Responsible Innovation

The successful integration of picotechnology into biological sciences demands a concerted effort across multiple disciplines. Interdisciplinary collaboration between quantum physicists, molecular biologists, materials scientists, and engineers is essential to address the multifaceted challenges and unlock the full potential of picotechnology (Drexler, 2013). Establishing collaborative research institutes and global research networks can facilitate knowledge exchange and accelerate technological advancements (Bostrom & Ćirković, 2011). Furthermore, fostering a culture of responsible innovation through ethical deliberation and proactive regulatory measures will ensure that picotechnological advancements are harnessed for the greater good, minimizing risks and maximizing societal benefits (Somorjai & Li, 2010).

12.6.7 Concluding Reflections

Picotechnology stands at the cusp of transforming biological sciences, offering tools and methodologies that redefine our capacity to understand and manipulate life at its most fundamental levels. The convergence of advanced microscopy techniques, quantum control methods, and interdisciplinary collaborations heralds a new era of biological research and medical innovation. As we navigate the complexities of picotechnological advancements, it is imperative to balance scientific ambition with ethical responsibility, ensuring that the pursuit of knowledge and technological prowess contributes positively to human health, environmental sustainability, and the broader fabric of society.

In essence, picotechnology embodies the next frontier in the manipulation of matter, poised to unlock unprecedented capabilities in biological sciences. By harnessing the principles of quantum mechanics and leveraging cutting-edge technologies, picotechnology promises to usher in transformative changes that enhance our understanding of life and empower us to engineer biological systems with atomic precision. The journey ahead is challenging yet exhilarating, filled with opportunities to redefine the boundaries of science and medicine in the quest for a healthier, more sustainable future.

References

Atkins, P., & de Paula, J. (2011). *Physical chemistry*. Oxford University Press.

Attwood, D., & Sakdinawat, A. (2017). *X-rays and extreme ultraviolet radiation: Principles and applications*. Cambridge University Press.

Betzig, E. (2014). Development of super-resolution fluorescence microscopy. *Journal of Cell Biology, 206*(1), 11–18. https://doi.org/10.1083/jcb.201401025

Binnig, G., & Rohrer, H. (1985). Scanning tunneling microscopy. *Surface Science, 152*, 17–26.

Binnig, G., & Rohrer, H. (1986). Scanning tunneling microscopy: From discovery to quantum engineering. *Advances in Physics, 35*(1), 1–70. https://doi.org/10.1080/00018738600101599

Bostrom, N., & Ćirković, M. M. (Eds.). (2011). *Global catastrophic risks*. Oxford University Press.

Brif, C., Chakrabarti, R., & Rabitz, H. (2010). Control of quantum phenomena: Past, present and future. *New Journal of Physics, 12*, 075008. https://doi.org/10.1088/1367-2630/12/7/075008

Chiao, R. Y. (2013). Biological quantum information: An emerging frontier. *Journal of Integrative Neuroscience, 12*(2), 117–131. https://doi.org/10.1162/NECO_a_00277
Craig, A. D., & Sykes, B. D. (2004). *Mass spectrometry: A textbook*. Wiley.
Dowling, J. P., & Milburn, G. J. (2003). Quantum technology: The second quantum revolution. *Philosophical Transactions of the Royal Society A: Mathematical, Physical and Engineering Sciences, 361*(1809), 1655–1674.
Drexler, K. E. (2013). *Radical abundance: How a revolution in nanotechnology will change civilization*. PublicAffairs.
Eigler, D. M., & Schweizer, K. S. (1997). Positioning single atoms with a scanning tunneling microscope. *Nature, 387*(663–664), 708–709. https://doi.org/10.1038/387708a0
Emma, P., et al. (2010). X-ray free-electron lasers. *Nature Photonics, 4*(8), 641–648. https://doi.org/10.1038/nphoton.2010.161
Emsley, L., Wang, J., & Roy-Chowdhury, S. (2019). Single-particle cryo-EM at atomic resolution. *Nature Methods, 16*(3), 197–206. https://doi.org/10.1038/s41592-019-0376-1
Engel, G. S., Calhoun, T. R., Read, E. L., Ahn, T., Mančal, T., Aspuru-Guzik, A., & Fleming, G. R. (2007). Evidence for wavelike energy transfer through quantum coherence in photosynthetic systems. *Nature, 446*(7137), 782–786. https://doi.org/10.1038/nature05678
Erni, R., Rossell, M. D., Kisielowski, C., & Dahmen, U. (2009). Atomic-resolution imaging with a sub-50-pm electron probe. *Physical Review Letters, 102*(9), 096101.
Freitas, R. A., Jr. (2005). Theoretical nanomedicine: Molecular mechanical devices that operate at the picometer scale. In T. Vo-Dinh (Ed.), *Nanotechnology in biology and medicine: Methods, devices, and applications* (pp. 177–225). CRC Press.
Georgescu, I. M., Ashhab, S., & Nori, F. (2014). Quantum simulation. *Reviews of Modern Physics, 86*(1), 153–185. https://doi.org/10.1103/RevModPhys.86.153
Hapala, P., Nenadic, I., Losonczi, J., & Sha, W. T. (2003). High resolution magnetic force microscopy. *Physical Review Letters, 91*(2), 026103. https://doi.org/10.1103/PhysRevLett.91.026103
Helgaker, T., Jørgensen, P., & Olsen, J. (2000). *Molecular electronic-structure theory*. Wiley.
Hell, S. W. (2007). Far-field optical nanoscopy. *Science, 316*(5828), 1153–1158. https://doi.org/10.1126/science.1130039
Hiller, S. L., Kaupp, M., & Schulte-Herbruggen, O. (2014). Photoinduced electron transfer in photosystem II: From fundamental aspects to applications in artificial photosynthesis. *Photosynthesis Research, 122*(2), 243–261. https://doi.org/10.1007/s11120-014-0017-4
Honig, J. A., & Rodriguez, J. M. (2008). Kelvin probe force microscopy: A tool for nanoscopic surface potential measurements. *Advanced Materials, 20*(7), 1348–1360. https://doi.org/10.1002/adma.200701430
Houk, N. P., & Shaik, S. (2005). The potential energy surface of enzymatic reactions: The role of quantum mechanical effects. *Chemical Reviews, 105*(10), 433–450.
Krausz, F., & Ivanov, M. (2009). Attosecond physics. *Reviews of Modern Physics, 81*(1), 163–234. https://doi.org/10.1103/RevModPhys.81.163
Kühlbrandt, W. (2014). The resolution revolution. *Science, 343*(6178), 1443–1444. https://doi.org/10.1126/science.1257998
Lambert, N., Nori, F., & Girvin, S. M. (2013). Quantum biology. *Nature Physics, 9*(1), 10–18. https://doi.org/10.1038/nphys2544
Merkle, R. C. (1997). Femtotechnology: Nuclear control of matter. *Nanotechnology, 8*(1), 47–54.
Nielsen, M. A., & Chuang, I. L. (2010). *Quantum computation and quantum information: 10th anniversary edition*. Cambridge University Press.
Parak, F., et al. (2002). Scanning tunneling microscopy of biological samples. *Nature Reviews Molecular Cell Biology, 3*(9), 737–746. https://doi.org/10.1038/nrm908
Qasim, M. Z., Hammad, H. M., Abbas, F., Saeed, S., Bakhat, H. F., Nasim, W., & Fahad, S. (2020). The potential applications of picotechnology in biomedical and environmental sciences. *Environmental Science and Pollution Research, 27*, 133–142.
Roco, M. C., & Bainbridge, W. S. (Eds.). (2001). *Societal implications of nanoscience and nanotechnology*. Kluwer Academic Publishers.

Rust, M. J., Bates, M., & Zhuang, X. (2006). Sub-diffraction-limit imaging by stochastic optical reconstruction microscopy (STORM). *Nature Methods, 3*(10), 793–796. https://doi.org/10.1038/nmeth918
Sakurai, J. J., & Napolitano, J. (2017). *Modern quantum mechanics*. Cambridge University Press.
Schüler, D., & Steitz, S. (2005). AFM-based force spectroscopy. *Methods in Enzymology, 398*, 368–390. https://doi.org/10.1016/S0076-6879(05)98028-7
Somorjai, G. A., & Li, Y. (2010). *Introduction to surface chemistry and catalysis*. Wiley.
Stuart, S. J. (2005). The power of atomic force microscopy for probing biomolecular interactions. *Trends in Biotechnology, 23*(8), 402–410. https://doi.org/10.1016/j.tibtech.2005.06.001
Walsworth, R. L. (2015). Quantum sensing for biomolecular studies. *Frontiers in Physics, 3*, 77. https://doi.org/10.3389/fphy.2015.00077

Chapter 13
Case Studies in Sustainable Engineering with Nanomaterials

K. Gayathri, P. Saranraj, B. Lokeshwari, Hasanain A. J. Gharban, and U. Subhalakshmi

13.1 Introduction

Nanotechnology is becoming essential to sustainable engineering, providing creative solutions for sectors including environmental remediation, energy generation, healthcare, and agriculture. The applications of nanomaterials are more efficient, selective, and eco-friendly than conventional technologies (Banerjee & Yu, 2018). Carbon nanotubes, graphene oxide, and metal oxide nanoparticles have been used in water filtration and purification. Graphene oxide membranes can filter salts, viruses, and bacteria and have many potential applications. They can efficiently degrade organic effluents in an aqueous medium upon UV irradiation; photocatalytic TiO_2 nanoparticles are easily recoverable and used for wastewater treatment. The use of nanomaterials in the future generation of water treatment technology (Chaturvedi et al., 2021) allows the use of non-scalable energy- and chemicals-demanding water purification techniques.

Nanomaterials are also improving the efficacy of energy technology, especially when it comes to using renewable energy sources like solar and wind power. Light-absorbing nanomaterials are engineered to outperform traditional solar cells by enhancing efficiency and enabling new functionalities. The ability to absorb light with a size-tenable feature is one of the unique optical characteristics of quantum dots, which are semiconductor particles on the nanometer scale that enable them to catch a more significant portion of the Sun's Spectrum (Cioffi & Rai, 2012). Carbon

K. Gayathri · P. Saranraj (✉) · B. Lokeshwari · U. Subhalakshmi
PG and Research Department of Microbiology, Sacred Heart College (Autonomous), Tirupattur, Tamil Nadu, India

H. A. J. Gharban
Department of Internal and Preventive Veterinary Medicine, College of Veterinary Medicine, University of Wasit, Kut, Iraq

S. Kulkarni (ed.), *Next-Generation Nanomaterials for Sustainable Engineering*, Lecture Notes in Nanoscale Science and Technology 43,
https://doi.org/10.1007/978-3-032-05719-8_13

nanotubes and graphene, two nanomaterials utilized in solar panel construction, have been shown to improve conductivity and flexibility, resulting in lighter, more durable, and more efficient solar cells. Solar energy is becoming one of the most practical renewable sources (Markovic et al., 2019).

The primary use of sustainable engineering in environmental remediation involves using nanomaterials to remove heavy metals, organic pollutants, chemicals, pesticides, and other hazardous elements from soil, water, and air that endanger human health and the environment. According to Gogotsi (2020), nanomaterials offer quick ways to eliminate or neutralize these contaminants. It is well known that zero-valent iron (ZVI) nanoparticles can reduce harmful metals like arsenic and chromium to less dangerous forms through chemical interactions. Metal oxide nanoparticles, such as ZnO and CuO, are utilized to treat wastewater in an environmentally beneficial manner by photocatalytically degrading organic contaminants in water (Hersam, 2020). These nanomaterial-based remediation techniques improve environmental sustainability by utilizing less energy and fewer toxic chemicals.

Nanomaterials are used in agriculture to increase crop yields and reduce the environmental impact of farming operations. Nanof ertilizers based on nanotechnology are improving the sustainability of agriculture. High nutrient discharge rates from conventional fertilizers can damage soils and contaminate water supplies (Kah et al., 2019). Nano fertilizers deliver nutrients, unlike synthetic fertilizers that are readily washed away. Various nanoparticles, such as nitrogen or urea, have been encapsulated in biodegradable polymers to provide slow-release fertilizers that lessen environmental effects and nutrient losses (Khan et al., 2019). Nanomaterials are advancing sustainable medical practices by improving diagnostics, facilitating tissue regeneration, and offering targeted and effective drug delivery systems (DDS).

Conventional chemotherapy can have serious adverse effects and targets both cancer and healthy cells. By altering nanoparticles, chemotherapy can be sent straight to tumor cells, protecting good tissue. Polymeric nanoparticles, liposomes, and dendrimers have all been developed to deliver medications to the site of illness and release them under regulated conditions (Kumar et al., 2020). Biomarkers at extremely low concentrations enable early diagnosis and individualized treatment; nanomaterials such as gold nanoparticles and quantum dots are increasingly being utilized in diagnostics (Lin & Lee, 2018). The use of nanoparticles in high surface area ratio may have adverse effects involving biomagnification and toxicity in organisms, the intrinsic properties of these nanomaterials, and their toxicity and environmental impact (Madhavan & Roopan, 2020). Nanomaterials have significant uses in energy generation, environmental remediation, agriculture, health, and water purification.

13.2 Nanotechnology and Agricultural Sustainable Development

Sustainable agriculture can benefit significantly from using nanotechnology to monitor water and pesticide quality and control nutrients (Gruère, 2012). The chemical composition, shape, surface structure, surface charge, and degree of particle agglomeration or disaggregation are additional features of NPs that could influence toxicity (Ion et al., 2010). The major applications include the nanotubes, fullerenes, biosensors, controlled delivery systems, and nanofiltration (Sabir et al., 2014). It is mainly used for food processing, packaging, biomass, and agricultural waste (Floros et al., 2010). Heavy metal trace detection is made by nanodetection technologies such as biosensors, electrochemical sensors, optical sensors, and devices. In addition to directly catalyzing the breakdown of waste and harmful chemicals, nanomaterials also aid in increasing the effectiveness of microbes in these processes (Dixit et al., 2015). Nanotechnology plays a pivotal role in advancing agricultural sustainability, as demonstrated in Fig. 13.1. Key applications include the detection of plant pathogens, nano-enabled delivery systems for agrochemicals, and nanosensors for mycotoxin detection. Additionally, nanoparticles exhibit potent antimicrobial and antifungal properties against phytopathogens, offering innovative solutions for crop protection.

13.2.1 Nanofertilizers

Nano pesticides and fertilizers link nanotechnology and sustainable agriculture, boost crop productivity, decrease pollution, and improve nutrient efficiency (Dimkpa, 2014). Nano fertilizers boost the bioavailability of nutrients, improving plant resilience to pests and diseases and ultimately reducing the need for chemical pesticides. Although they have been utilized in fertilizers, nanomaterials such as nanoz inc, nanoi ron, and nanos ilica have mainly been used to enhance plant metabolism and stress tolerance (Sadeghzadeh, 2013). Nano fertilizers have emerged as a viable alternative that meets the needs of environmental-friendly agriculture, as soil deterioration and climate change pose significant dangers to the agricultural sector. Their function in precision farming is essential for preserving soil fertility and maximizing resource utilization for future agricultural sustainability.

13.2.2 Nano Pesticides

Nanopesticides are an innovative strategy for integrated pest management, merging the concepts of nanotechnology with sustainable development. Applying chemical pesticides has led to environmental contamination and insect resistance,

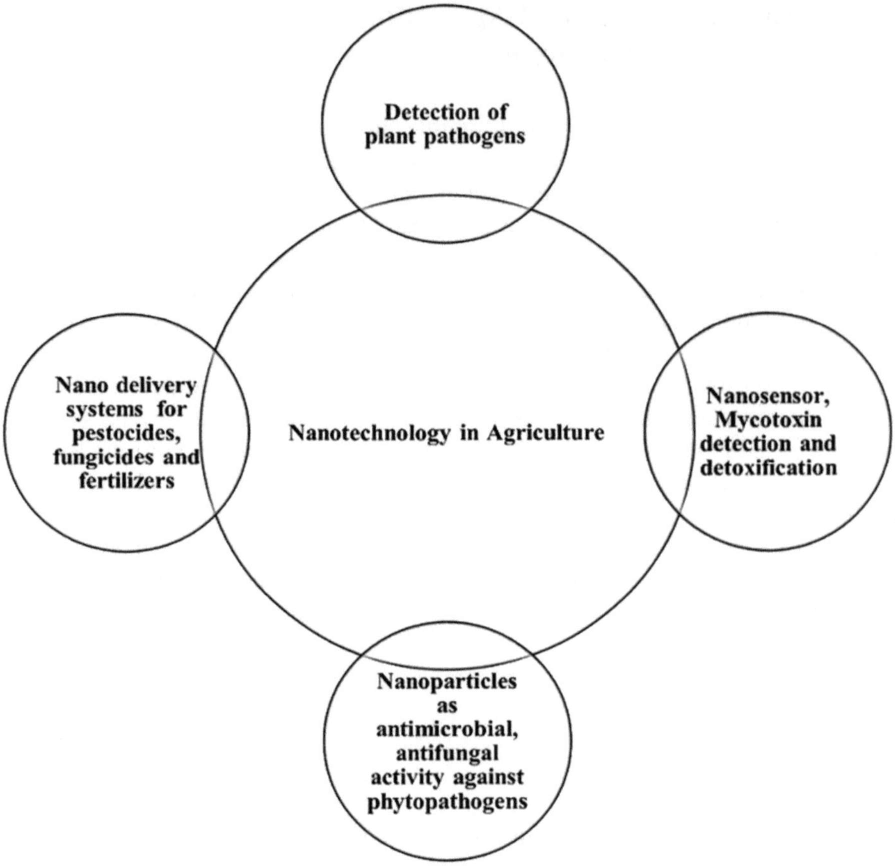

Fig. 13.1 The role of nanotechnology in agricultural sustainable development

jeopardizing human health and the ecosystem (De Oliveira et al., 2014). Nanoinsecticides are engineered at the nanoscale, facilitating controlled release, enhanced bioavailability, and improved adhesion to plant surfaces for effective pest targeting. Nanopesticides provide enhanced solubility and durability, preventing the rapid degradation of active constituents. Pesticide formulations Utilizing Nanoparticles in conventional chemical pesticides have demonstrated significant potential in enhancing pest resistance (Gouin, 2004). NPs may be vital in controlling insect pests and pathogens affecting the host. Developing secure and efficient pesticide delivery methods would improve global food production and reduce negative environmental impacts on ecosystems (Kandasamy & Prema, 2015).

13.2.3 Nanobiosensors

Nanobiosensors enhance agriculture by utilizing nanotechnology for sustainable growth through precise, real-time assessment of soil health, crop vitality, and environmental conditions (Fogel & Limson, 2016). Traditional agricultural monitoring systems frequently necessitate labor-intensive and time-consuming procedures. Nano biosensors have facilitated analysis optimally and effectively to enhance production. No chemical fertilizers, pesticides, or water will help for the more sustainable and less polluting agricultural system. This approach directly goes with precision agriculture and optimizes soil moisture and nutrient levels, thus reducing environmental pollution. Nanomaterials like gold nanoparticles, carbon nanotubes, and quantum dots are used to develop biosensors with higher sensitivity, selectivity, and stability for detecting specific agricultural parameters (Khota et al., 2012).

13.2.4 Ecotoxicological Implications of the Nanoparticles

Many types of nano biosensors, like enzyme-based, genetic, immune biosensors, etc., are integrated with sustainable development and nanotechnology in more smart agriculture and food-based industries that give accurate time testing on crop health, soil health, quality, and environmental conditions. It is a nanoscale sensor that monitors the biochemical changes that occur in mineral deficiency, plant stress, or soil or air contamination. In sustainable and ecological-friendly agricultural systems, nano biosensors help farmers save specific amounts of water, chemical fertilizers, and pesticides (Nuruzzaman et al., 2016). The soil can be mixed with moisture and nutrient levels to lower contaminations, which helps in precision farming. Nanomaterials with superior physical and chemical properties, including gold nanoparticles, carbon nanotubes, and quantum dots, are effective in enhancing the stability, sensitivity, and selectivity of biosensors for monitoring specific agrochemical properties. Precision agriculture systems improve productivity and decrease effects on environmental sustainability for a resilient future in agriculture (Vidotti et al., 2011). In nanoscale technology, synthetic NPs are increasingly released in the environment (Hoffmann et al., 2007).

Nanoparticle toxicity, widely applied in industries, has been utilized to the potential adverse effects of various nanomaterials on human health and plants (Yang & Watts, 2005). Metal nanoparticles have antibacterial (Aziz et al., 2016), anti-fungicidal, and anticandidal properties. In oxidative stress, cells upregulate the protective enzymatic or genetic pathways (Kovochich et al., 2005). The endpoint of toxicity and ecotoxicity is the intracellular production of reactive species (ROS) (Venkatachalam et al., 2017). The molecular scale of nanoparticular bioaccumulation and biotransformation in animals and some cell types as well as the cellular–cellular interaction with nanoparticles. Some materials have ecotoxicological factors and complexities of natural systems (Cox et al., 2017). Nanoparticles form in food webs and bioaccumulate in food chains moving in and through plant cellular systems, translocating and accumulating in each generation of aerial parts and concentrating their cycling across trophic levels. Their transpiration, respiration, photosynthesis, and nutrient translocation are essential physiological functions. The toxicity of the nanoparticles is highly associated with their surface properties and ecotoxicological effects on microbes, plants, and humans, as well as environmental risks (Rana & Kalaichelvan, 2013).

13.3 Case Studies in Sustainable Development

Numerous case studies worldwide show us the principles of sustainable development and the right balance between environmental integrity, social equity, and economic viability.

13.3.1 Solar Energy: Quantum Dot Solar Cells

Quantum dot (QD) solar cells are a novel solar energy converter that are more efficient and more cost-effective over silicon-based photovoltaics. The semiconductor materials known as quantum dots (QDs) are usually in the nanometer range (2–10 nm). The quantum dots, due to their unique characteristics such as size and increased absorption of light, could be utilized in solar cells (Shweta et al., 2016). The energy levels of nanometre-sized crystals of semi-conducting material, called quantum dots, allow the material to absorb a broader range of the light spectrum. This increases the energy harvesting efficiency in multi-junction solar cell types with several layers of quantum dots being customized to absorb different portions of the sun's spectrum. Quantum dot solar cells (QDSCs) use light-sensitive semiconductors called quantum dots as the active material. Quantum dots tend to degrade and become unstable due to environmental factors like oxygen, humidity, and UV light. It is also used to produce quantum dot solar cells in a scale required for commercial production (Porter et al., 2007).

13.3.2 Water Purification: Nanomaterial-Based Filters

Nanomaterial-based filters use nanomaterials with dimensions in the nanoscale range, typically between 1 and 100 nm that exhibit unique chemical, physical, and structural characteristics. The physical and chemical features, like a high ratio of surface area to volume, have high reactivity, and molecular interactions with pollutants and nanomaterials can effectively remove water pollutants like heavy metal ions, organic pollutants, and bacteria and viruses (Hussein et al., 2005).

Polymer-based nanomaterials have carbon-based nanomaterials like graphene and carbon nanotubes for water filtration. Graphene oxide membranes can be used to create nanometric holes, meaning that large amounts of water molecules can pass through them to filter more significant pollutants. These membranes were desalination and wastewater treatment solutions due to their capacity to efficiently reject salts, heavy metals and other hazardous substances (Ion et al., 2010). Iron oxide nanoparticles can adsorb heavy toxic elements like arsenic, lead, and mercury, thus lowering their concentration in drinking water. Nanomaterial-based filters trap specific contaminants, even in low concentrations, like activated carbon or sand filters (Gouin, 2004). The better efficiency and selectivity of removal of nanomaterial-based filters make them useful to small-scale industries as well as treatment plants in the design of compact and portable filtering units. The nanomaterials are toxic to the body and could persist in our environment or allow bioaccumulation. Nanomaterial-based filters should help their long-term safety in the environment and human health (Jampilek et al., 2015).

13.3.3 Healthcare: Nanomedicine and Drug Delivery

The medical application of nanotechnology includes drug delivery and nanomedicine that aims to deliver drugs and therapeutic agents with unnecessary side effects. Nanomedicine consists of materials at the nanoscale range from 1 to 100 nm at the molecular level. Nanoparticles' unique features, including high surface area, vary with size in drug delivery processes. The significance of nanomedicine is the delivery of medication to a specific part of the body at a required dosage with minimum adverse and maximum therapeutic effects. Nanoparticles can deliver proteins, nucleic acids, and small molecules to the body using nanocarriers that can act independently or with other agents to allow for controlled release (Grillo et al., 2016). This targeted delivery has enormous benefits for cancer treatment, as nanoparticles can be designed for tumor tissue through the enhanced permeation and retention (EPR) effect, and tumor blood vessels have a higher permeability than normal tissue. Functionalized nanoparticles with ligands or antibodies increase delivery methods so that cancer cells are targeted without harming healthy tissues/cells.

Nanomedicine can help with drug instability or poor solubility, and the drugs can be encapsulated into nanoparticles for targeted delivery at the right dosage to the target area (Du et al., 2017). The long-term therapeutic effects of drugs released from nanocarriers improve patients, enabling reduced dosing frequency. The most widely used forms of nHPs belong to liposomes, dendrimers, polymeric nanoparticles, and solid lipid nanoparticles; these have specific properties. Liposomes form the spherical vesicles that can deliver hydrophilic and hydrophobic drugs as well as linear polymers (Ezhilarasi et al., 2012). The potential toxicity of nanoparticles accumulates gradually in organs or tissues. The central nervous system (CNS) is the blood–brain barrier (BBB), a specialized endothelial barrier that protects the CNS from peripheral circulation of fabricating nanoparticles that can help deliver medication molecules across the blood–brain structure or bind more potently to receptors on the surface of brain endothelial cells (Feynman, 1996).

13.3.4 Agriculture: Nanofertilizers and Pest Management

Nanotechnology in agriculture may offer solutions to various issues, including fertilizer application and insect control regulation. Nano fertilizers are nanoscale compounds anticipated to enhance nutrient bioavailability and mitigate environmental impacts by their release, absorption, and activity in crops. Conventional fertilizers exhibit significant inefficiency in application, as a substantial portion is lost to the environment by leaching, runoff, and volatilization, contaminating ecosystems, mainly aquatic systems. Nano fertilizers are released gradually by the crop's requirements, resulting in no wastage (Floros et al., 2010).

In addition to reducing fertilizer overuse, these fertilizers can sequester essential components like potassium, phosphorus, and nitrogen into minute nanoparticles that dissolve gradually, giving plants a steady supply of nutrients. Nano fertilizers are a promising alternative because they control the amount released and help plants absorb nutrients even in unfavorable environmental circumstances (Vartholomeos et al., 2011). Nutrient particles can improve soil characteristics, such as water retention and nutrient storage capacity. Nanotechnologies offer novel possibilities for pest management by mitigating pests and diseases without the environmental contamination and health risks associated with chemical pesticides. They are utilized in agriculture as bionanomaterials. Insecticides, including nanoinsecticides, nano capsules, or nano emulsions generated in nanoforms, can be effectively produced for molecular-level targeting, thereby reducing exposure of non-target species to the pesticides. Specific formulations may facilitate the gradual release and non-repetitive application of pesticides in the environment (Venkatachalam et al., 2017). The nanoparticle also infiltrates the protective exoskeleton of insects or the cell membranes of pathogens, ensuring the efficient utilization of pest control agents. This integration has already produced benefits, such as using nanoparticles to encapsulate insecticides, so keeping them from degradation, boosting their efficacy, or reducing their toxicity to beneficial insects like pollinators.

Several nanoparticles have demonstrated the ability to suppress pathogenic bacterial growth and disrupt insect reproductive systems, thereby contributing to developing an environmentally sustainable pest management strategy. Nano-based pest management strategies can be included in precision agricultural frameworks (Vidotti et al., 2011). Control agents are administered exclusively to minimize their environmental impact. These possible benefits are associated with the inherent toxicity of nanomaterials that present serious hazards to the environment and human health. The long-term consequences of nanoparticles on soil health, water quality, and non-target species, even though these technologies have the potential to several agricultural challenges.

13.4 Applications of Nanomaterials in Sustainable Development

The wide applications of these nanomaterials in agriculture, water purification, and energy are capable of reducing energy waste and environmental damage caused by sustainable development. Nanostructured semiconductors can also be utilized in solar cell fabrication. Nanotechnology also plays a vital role in improving the energy storage devices of nanostructured materials used in advanced batteries and supercapacitors to enhance the charge capacity, stability, and cycling performance (Sayes et al., 2004). It can also improve renewable energy system performance to the extent that efficient energy storage strategies and variable renewables in the water purification systems are highly reactive for using nanomaterials.

Liquid nanofertilizers are nutrients stored in microunits that allow for the gradual release of the nutrient, lesser input waste, and better absorption due to the smaller amount of the nutrient in the environment (Singh et al., 2017). Nanomaterials have been utilized to formulate pesticides that are more effective, environmentally friendly, and less toxic. The encapsulation of nanoparticles can stabilize the active compound and enhance the efficiency, minimizing the negative impact on non-target animals and other ecosystems. Nanomaterials are driving innovations in sustainable development, as showcased in Fig. 13.2. Their applications span agricultural sustainability (e.g., nanopesticides, nanofertilizers, and crop protection) and

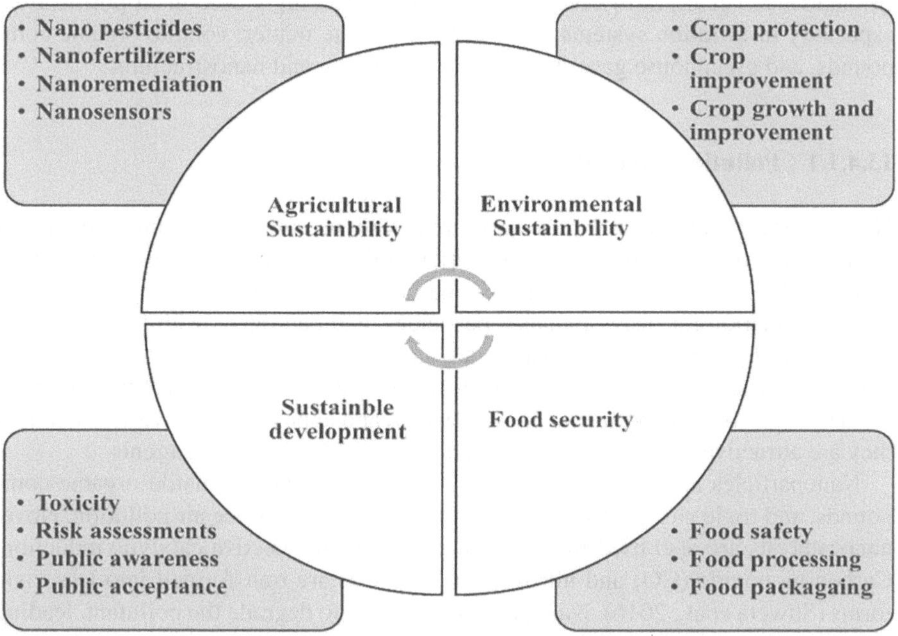

Fig. 13.2 Applications of nanomaterials in sustainable development

environmental sustainability (e.g., nanoremediation and nanosensors). Beyond production, nanomaterials enhance *food security* through improved safety, processing, and packaging. However, challenges such as toxicity risks and public acceptance must be addressed to ensure their safe and scalable adoption.

13.4.1 Environmental Applications

Nanomaterials have a high surface area, increased reactivity, and molecular interactions, making them capable of conventional materials for their environmental applications. Nanomaterials are used to enhance their sustainability in pollution control. Metal oxide nanoparticles, graphene, and carbon nanotubes degrade environmental pollutants (Llop et al., 2014). Nanoparticles are being utilized to repair oil spills since they can absorb hydrocarbons from the ocean more efficiently than solid materials. The high surface areas allow large volumes of oil to be trapped, which could facilitate oil extraction and lower environmental impacts from oil spills.

Nanomaterials could eliminate harmful contaminants from water, including heavy metals. Nanomaterials involved in environmental remediation may be poly zinc oxide (ZnO) and titanium dioxide (TiO_2) polymeric nanoparticles with photocatalytic properties that enable degradation of toxic substances under light exposure. These nanoparticles' ability to transform organic pollutants like pesticides and pharma into less harmful secondary metabolites improves water quality (Memarizadeh et al., 2014). Nanomaterials are increasingly used in air purification, especially in filtration systems, to absorb particulate matter, volatile organic compounds, and greenhouse gases using activated carbon and nanostructures.

13.4.1.1 Pollution Control

Nanomaterials make them highly versatile in various applications due to their vast surface area, high reactivity, and interaction at the molecular level with hazardous compounds, and they are being used to adsorb and decompose toxins. Decontamination for heavy metals, pesticides, and pathogens with functionalized nanomaterials such as carbon nanotubes, graphene oxide, and metal oxide nanostructures (Sekhon, 2014). ZnO and TiO_2 are effective photocatalysts for decomposing hazardous organic pollutants in wastewater, and exposed to ultraviolet radiation, they are attractive environmentally friendly wastewater treatment agents.

Nanoparticles are also used to adsorb particulate matter, volatile organic compounds, and toxic gasses in filtration systems, aiming to reduce air pollution. Novel nanocatalysts are also used in industrial processes of selective catalytic reduction. Carbon monoxide (CO) and nitrogen oxide (NOx) are transformed into less toxic forms (Shweta et al., 2016). Nanoparticles are used to degrade the pollutant, leading to soil recovery. Soil remediation techniques can be used, and nanoparticles help reduce the pollutants. Due to their tunability, which allows for the synthesis of such materials with targeted functions for the selective removal or degradation of contaminants, nanomaterials serve as pollution detection and remediation. The potential toxicity of the nanoparticles could have long-term effects on the environment, and they use nanoparticles to purify air, water, and soil for sustainable development for the clean-up of pollutants with a low environmental impact.

13.4.1.2 Water Purification

Nanomaterials are mainly used in water purification, which is enhanced due to their unique properties, such as large surface area, reactivity, and ability to target several contaminants at the molecular level. Metal oxide nanoparticles include zinc oxide (ZnO) and titanium dioxide (TiO_2), which can have potential applications for cleaning contaminated wastewater due to their photocatalytic properties. They also allow them to degrade organic pollutants upon exposure to light. Nanomaterials hold great potential for developing water purification technologies through effective, efficient and sustainable removal and treatment of water contaminants (Marzbani et al., 2015).

The superior porosity and selective permeability of carbon nanotubes, graphene, its derivatives and other nanostructures have been used in water filtration. Graphene oxide membranes are highly suitable for desalination. Nanomaterial-based filters exhibit higher stability and more effective filtration than traditional filters. They are known to have superior processes to conventional filters, leading to filtration replacement and lower maintenance costs. With their large surface area and tunability, nanoparticles can be used in the filtration systems to optimize the removal of impurities from the system and improve the purification process. These nanoparticles can potentially revolutionize water treatment for assessing potential toxicity and the environmental impact of these nanomaterials on humans and non-human animals to ensure their safe application (Sertova, 2015).

13.4.1.3 Remediation

Novel nanomaterials contribute functionalities or nanostructures that can be beneficial for environmental factors. The surface area, high reactivity, and unique molecular interactions can decompose various pollutants rapidly and effectively. Nanoparticles have been applied in soil and water treatment by utilizing new zero-valent iron (ZVI) and associated nanoparticles to eliminate toxic metals such as arsenic, lead, and mercury through adsorption, reduction, and precipitation. They are used almost instantaneously to bind and neutralize various toxic chemicals. Therefore, these nanoparticles are very effective for cleaning contaminated soil and water. When exposed to light, $TiO_{2,}$ and other metal oxide nanoparticles act as photocatalysts and dissolve particulate matter and volatile organic compounds (VOCs) by reducing air pollution (Mukhopadhyay, 2014). This technology is compatible with industrial or urban areas where it is installed to purify air. In addition, their adsorption and exchange capacities contribute to soil and water decontamination, with a wide range of carbonaceous and carbon-based nanomaterials, including activated carbon and carbon nanotubes, being used for that purpose (Bhattacharyya et al., 2016). The purpose of nanomaterials in the remediation process is to be safe and regulated to prevent adverse effects. Nanomaterials offer an innovative class of adaptable, efficient, and sustainable techniques for targeted repair of damaged ecosystems as part of the next generation of environmental remediation solutions.

13.4.2 *Healthcare and Medicine*

The characteristics of the nanomaterials, including the size, surface area, and reactivity, can improve the interactions of molecules with biological systems that facilitate the specificity, efficiency, and progression of medical therapy. The nanosized carriers are nanoparticles, nano shells, dendrimers, and liposomes (Llop et al., 2014). They have a wide variety of uses, especially in drug delivery, as they are an essential part of this process. The therapeutic agents in liposomes, dendrimers, and micelles decrease toxicity and enhance efficacy. Nanoparticles have been utilized for sustainable disease diagnosis, treatment, and prevention development.

Nanomaterials may specifically target malignant cells, and chemotherapeutic drugs can be targeted locally to minimize harm to surrounding tissues. The nanomaterials enable the personalized detection of cancer biomarkers and disease markers at low abundances and have also been used in diagnostic imaging and sensors (Kah & Hofmann, 2014). Target cells or tissues can be assessed using nanoparticle-based diagnostic techniques, making identifying diseases like cancer and cardiovascular conditions easier. Nanomaterials are used in regenerative medicine structures and biological scaffolds that regulate cellular growth for tissue engineering, which involves the repair of bodily tissues. Nanoparticles are essential for cost-effective vaccine delivery, diagnostics, and medical equipment solutions. Nanomaterials used their potential toxicity and environmental impact must be evaluated, particularly concerning their effects on environmental interfaces.

13.4.2.1 Drug Delivery Systems

New solutions are designed to address numerous diseases with enhanced precision and efficacy, promoting sustainability and minimizing the necessity for complex treatment protocols. Large surface area, adjustable characteristics, and diminutive size characterize nanomaterials like nanoparticles, liposomes, and dendrimers. They can be engineered for drug encapsulation and controlled release at the target site, enhancing therapeutic efficacy and minimizing systemic side effects. Targeted drug administration, exemplified by site-specific distribution, will be highly beneficial in treating diseases such as cancer when conventional therapies indiscriminately damage healthy tissue (Grillo et al., 2016). Ligands or antibodies on the nanoparticle surface can target therapeutic drugs to tumor cells, resulting in enhanced drug delivery to disease areas and enabling the use of lower drug doses.

Nanoparticles can enhance the bioavailability of weakly water-soluble medicines, ensuring adequate concentration-time profiles at the target site for bioactive molecules. Using nanomaterials diminishes drug delivery's environmental impact by improving drug carriers and decreasing active content, waste materials, and hazardous chemical applications. Nanomaterials exhibit significant potential for drug delivery applications; nevertheless, issues related to biocompatibility, toxicity, and environmental effects must be addressed before their utilization. The application of nanomaterials holds significant promise for advancing optimal selective and effective medication delivery systems in the medical field.

13.4.2.2 Diagnostics

Nanoparticle-based diagnostic innovations offer a cost-effective, sensitive, and ecologically sustainable approach for illness diagnosis. Consequently, nanomaterials with elevated surface area, optical properties, and reactivity facilitate the detection of diseases at the molecular level, including in the early stages of the disease. In diagnostic applications, the predominant types of nanoparticles are quantum dots, gold nanoparticles, and magnetic nanoparticles; these can be functionalized to bind with target disease biomarkers (proteins, DNA) to detect cancer, infectious diseases, or cardiovascular diseases (Patra & Baek, 2017). Consequently, gold nanoparticles are extensively utilized in biosensors and imaging systems to increase signals for identifying low-concentration biomarkers with great sensitivity. Moreover, nanomaterials enhance the advancement of point-of-care (POC) diagnostic devices for bedside diagnosis, which is typically more rapid and precise than conventional methods and has markedly increased testing efficiency, especially in peripheral regions or remote healthcare facilities that necessitate immediate diagnosis. Nanoparticle-based diagnostic solutions reduce the expenses and duration linked to laboratory testing, enhancing the affordability and accessibility of healthcare. The intrinsic optical and electrical characteristics of nanoparticles influence the released therapeutic medicines, biocompatibility, toxicity, and environmental impact of the nanomaterials. These nanomaterials are used for diagnostics, potentially leading to sustainable disease detection due to their higher sensitivity and affordable, dependable, quick diagnostics.

13.4.2.3 Personalized Medicine

Nanomaterials are specifically designed and engineered at the nanoscale to enhance biodistribution, increase anticancer efficacy, minimize toxicity, and facilitate efficient targeted drug delivery, thereby possessing transformative potential to meet unmet medical needs in personalized medicine. Nanoscale properties of nanomaterials, including nanoparticles, liposomes, and quantum dots, enable precise binding and interaction at the molecular and cellular levels, allowing for targeted delivery of drugs and diagnostic agents to disease sites, thereby minimizing systemic side effects and enhancing therapeutic efficacy. In cancer treatment, nanoparticles can preferentially transport chemotherapeutic drugs into neoplastic cells, sparing adjacent healthy tissues and enhancing therapeutic efficacy. This is crucial in personalized medicine in terms of an individual's genetics and their responses to treatments.

Nanomaterial-based diagnostic methods enable the development of novel instruments for the molecular-level detection of disease heterogeneity (Rana & Kalaichelvan, 2013). They may serve as viable candidates for creating functionalized agents with disease-specific targeting biomarkers to detect the presence of diseases such as cancer and cardiovascular conditions at minimal concentrations in the patient's body, thereby alerting clinicians to take timely actions through treatment protocols designed to optimize patient outcomes. Nanomaterials are used in the construction of specialized nanocarrier systems in drug delivery, which allow the drugs to be released in a precise, controlled, and sustained manner, allowing the concentration of the drug to remain in the target range for long periods.

13.4.3 Food and Agriculture

Applications for nanomaterials in food and agriculture are crucial because they improve sustainability, boost agricultural yields, improve food safety, and lessen environmental effects. Nanomaterials are also used in fertilizers, insecticides, and herbicides to improve agricultural nutrient delivery and pest management effectiveness. Nano fertilizers, for instance, release nutrients such as nitrogen, phosphorous, and potassium in a regulated manner, thereby minimizing nutrient runoff into groundwater while enhancing crop nutrient uptake. This also mitigates the excessive application of chemical fertilizers, which depletes soil fertility and contaminates aquatic ecosystems. Nanoparticles can be tailored for pest control to enhance pesticide delivery and target pests at the molecular level, diminishing the need for more potent pesticides (Sertova, 2015).

All advancement their vast surface area enables the encapsulation of active compounds, safeguarding them from the microenvironment by regulating release and reducing environmental effects without compromising efficacy. Nanotechnology is utilized in food packaging, as nanocomposites and nanoparticles improve the strength of packaging materials, especially their barrier qualities and the shelf life of products. This mitigates food waste by extending the shelf life of perishable goods and diminishes the necessity for preservatives. Nano sensors in food safety and quality control can identify pollutants, including pathogens, heavy metals, or poisons in food products, thereby ensuring safety from production to consumption. Nanomaterials possess significant potential for application in the development of functional foods to prevent and treat malnutrition by enhancing the bioavailability of nutrients and enabling their absorption in the body. Solutions produced from nanoparticles can promote sustainability by improving agricultural output, food safety, and environmental health.

13.4.3.1 Food Safety

The small size, large surface area, and increased reactivity of nanomaterials like nanosensors, nanoparticles, and nanocomposites allow for various highly sensitive methods for the detection of different foodborne pathogens, toxins, and dangerous substances, even at very low concentrations. The food safety and quality supply chain can be obtained by nanosensors in food packaging materials to monitor temperature, humidity, and pollutants. Health organizations can detect foodborne risks, improving consumer safety and reducing contamination (Ezhilarasi et al., 2012). Nanoparticles are used in packaging to strengthen food product barriers, increase shelf life and reduce the need for preservatives. Nanocomposite materials that contain nanoparticles such as clay, silver, or carbon-based compounds with antibacterial properties that restrict the growth of microorganisms, food safety. Nanoparticles are also used to detect pesticides, heavy metals, and other toxins in food, ensuring that products are free of harmful substances.

13.4.3.2 Crop Improvement

Nanomaterials are revolutionizing crop enhancement and providing sustainable ways to boost agricultural output, lessen environmental impacts, and solve global food security. Nano fertilizers significantly apply nanomaterials to boost crop productivity since they provide essential nutrients and minerals to plants more efficiently than conventional fertilizers. These fertilizers impart slow-release properties by encapsulating nutrients in nanoparticles, providing crops with a steady supply of nutrients over an extended period. This also reduces leaching, minimizing environmental nutrient loss, reducing water pollution and soil degradation, and better crop growth and yields. Nanotechnology serves as an adjunct in pest management by developing tailored nanopesticides that specifically address pest species, hence minimizing the application of broad-spectrum chemical pesticides (Daroczi et al., 2006). This nano composition will enhance pesticide efficacy, mitigate environmental risks, preserve beneficial insects, and promote humane agricultural practices. Nanotechnology enables the treatment of plants to enhance their resilience to diseases. Nanoparticles administered medication directly to plant tissues and elicited enhanced immune responses. Furthermore, the ability of nanoparticles to absorb and retain moisture and fertilizers can be employed to improve soil quality, hence facilitating enhanced water and nutrient uptake by plants, even under adverse conditions such as arid or nutrient-deficient soils. Nanotechnology is utilized to develop sensors that monitor soil conditions, crop health, and environmental stressors in real time, thereby equipping farmers with information to enhance crop management.

The dynamic imaging and manipulation of nanomaterials offer a promising pathway to enhance agricultural sustainability.

13.5 Regulatory and Ethical Concerns

Many ethical and regulatory issues are using the nanomaterials and sustainable engineering. Nanomaterials with small sizes and more significant surface areas enable them to interact with biological systems, and the production of many kinds of nanomaterials has caused their potential toxicity. These compounds have long-term impacts on human health and the environment. Bioaccumulation of nanoparticles in organisms is a potential mechanism underlying their health risks (Tripathi et al., 2017). A key challenge is the increasing prevalence of nanomaterials in consumer products amidst low public awareness of their function. Nanomaterials have great potential to advance sustainability engineering mechanisms that improve the interventions to assist and benefit.

13.6 Challenges and Limitations

The superior potential of nanoparticles to sustainable development exists, but challenges and barriers limit their wide and safe use. The potential toxicity and environmental concerns of their small size and distinct physicochemical properties as compared to bulk materials, nanoparticles mediate different interactions with biological systems, leading to concerns regarding bioaccumulation and chronic toxicity in humans and the environment (Prasad et al., 2014). There are no standardized testing methods available for nanoparticles in the risk assessment. Nanomaterial synthesis is an energy-consuming process that requires sophisticated technology that may restrict its wide application in sustainable development.

13.7 Conclusion

Sustainable engineering is experiencing a shift with the incorporation of nanomaterials, providing creative solutions to urgent environmental and industrial issues. Various case studies illustrate how nanomaterials like carbon nanotubes, graphene, and nanocomposites enhance energy efficiency, cut down waste, and facilitate cleaner production methods. Innovations such as self-cleaning coatings that reduce water usage and lightweight nanocomposites that boost transportation fuel efficiency highlight the crucial role of nanotechnology in sustainability. However, despite their potential, challenges persist regarding scalability, cost, and long-term environmental effects. Continued research, along with responsible manufacturing and regulatory backing, can enable nanomaterials to contribute to a more sustainable future. By integrating advanced science with eco-friendly engineering, nanomaterials are emerging as vital components for a circular economy, lower carbon footprints, and more efficient resource use. As this field advances, sustainable engineering through nanomaterials will increasingly address global environmental issues while promoting technological advancement. The pathway to sustainability is intricate, yet nanomaterials offer an effective means to create a greener, more resilient world.

References

Aziz, N., Pandey, R., Barman, I., & Prasad, R. (2016). Leveraging the attributes of *Mucor hiemalis* - Derived silver nanoparticles for a synergistic broad-spectrum antimicrobial platform. *Frontiers in Microbiology, 7*, 1984–1998.

Banerjee, I., & Yu, X. (2018). *Nanotechnology in modern animal biotechnology: Concepts and applications*. Springer.

Bhattacharyya, A., Duraisamy, P., Govindarajan, M., Buhroo, A. A., & Prasad, R. (2016). Nanobiofungicides: Emerging trend in insect pest control. *Advanced Fungal Nanobiotechnology, 45*, 307–319.

Chaturvedi, P., Rajput, M., & Arya, S. K. (2021). Nanotechnology for sustainable agriculture: Recent trends and prospects. *Sustainable Chemistry and Pharmacy, 21*, 100400.

Cioffi, N., & Rai, M. (2012). *Nano-antimicrobials: Progress and prospects*. Springer.

Cox, A., Venkatachalam, P., Sahi, S., & Sharma, N. (2017). Reprint of: Silver and titanium dioxide nanoparticle toxicity in plants: A review of current research. *Plant Physiology and Biochemistry, 110*, 33–49.

Daroczi, B., Kari, G., McAleer, M. F., Wolf, J. C., Rodeck, U., & Dicker, A. P. (2006). *In vivo* radioprotection by the fullerene nanoparticle DF-1 as assessed in a zebra fish model. *Clinical Cancer Research, 12*, 7086–7091.

De Oliveira, J. L., Campos, E. V., Bakshi, M., Abhilash, P. C., & Fraceto, L. F. (2014). Application of nanotechnology for encapsulating botanical insecticides for sustainable agriculture: Prospects and promises. *Biotechnology Advances, 32*, 1550–1561.

Dimkpa, C. O. (2014). Can nanotechnology deliver the promised benefits without negatively impacting soil microbial life? *Journal of Basic Microbiology, 54*, 889–904.

Dixit, R., Wasiullah, M. D., Pandiyan, K., Singh, U. B., & Sahu, A. (2015). Bioremediation of heavy metals from soil and aquatic environment: An overview of principles and criteria of fundamental processes. *Sustainability, 7*, 2189–2212.

Du, W., Tan, W., Peralta-Videa, J. R., Gardea-Torresdey, J. L., Ji, R., & Yin, Y. (2017). Interaction of metal oxide nanoparticles with higher terrestrial plants: Physiological and biochemical aspects. *Plant Physiology and Biochemistry, 110*, 210–225.

Ezhilarasi, P. N., Karthik, P., Chhanwal, N., & Anandharamakrishnan, C. (2012). Nanoencapsulation techniques for food bioactive components: A review. *Food and Bioprocess Technology, 6*, 628–647.

Feynman, R. P. (1996). *No ordinary genius: The illustrated richard feynman*. W.W. Norton and Company.

Floros, J. D., Newsome, R., Fisher, W., Barbosa Cánovas, G. V., Chen, H., & Dunne, C. P. (2010). Feeding the world today and tomorrow: The importance of food science and technology. *Comprehensive Reviews in Food Science and Food Safety, 9*, 572–599.

Fogel, R., & Limson, J. (2016). Developing biosensors in developing countries: South Africa as a case study. *Biosensors, 6*, 5–13.

Gogotsi, Y. (2020). *Nanomaterials handbook*. CRC Press.

Gouin, S. (2004). Microencapsulation: Industrial appraisal of existing technologies and trends. *Trends in Food Science and Technology, 15*, 330–347.

Grillo, R., Abhilash, P. C., & Fraceto, L. F. (2016). Nanotechnology applied to bio-encapsulation of pesticides. *Journal of Nanoscience and Nanotechnology, 16*, 1231–1234.

Gruère, G. P. (2012). Implications of nanotechnology growth in food and agriculture in OECD countries. *Food Policy, 37*, 191–198.

Hersam, M. C. (2020). *Nanotechnology applications for clean water: Solutions for improving water quality*. Elsevier.

Hoffmann, M., Holtze, E. M., & Wiesner, M. R. (2007). Reactive oxygen species generation on nanoparticulate material. *Environmental Nanotechnology: Applications and Impacts of Nanomaterials, 14*, 155–203.

Hussein, M. Z., Yahaya, A. H., Zainal, Z., & Kian, L. H. (2005). Nanocomposite-based controlled release formulation of an herbicide, 2,4-dichloro phenoxyacetate incapsulated in zinc-aluminium-layered double hydroxide. *Science and Technology of Advanced Materials, 6*, 956–962.

Ion, A. C., Ion, I., & Culetu, A. (2010). Carbon-based nanomaterials: Environmental applications. *University of Science and Technology Politehnica Bucharest, 38*, 129–132.

Jampilek, J., Zaruba, K., Oravec, M., Kunes, M., Babula, P., & Ulbrich, P. (2015). Preparation of silica nanoparticles loaded with nootropics and *in vivo* permeation through blood-brain barrier. *BioMed Research International, 15*, 812673.

Kah, M., & Hofmann, T. (2014). Nanopesticides research: Current trends and future priorities. *Environment International, 63*, 224–235.

Kah, M., Tufenkji, N., & White, J. C. (2019). Nano-enabled strategies to enhance crop nutrition and protection. *Nature Nanotechnology, 14*(6), 532–540.

Kandasamy, S., & Prema, R. S. (2015). Methods of synthesis of nano particles and its applications. *Journal of Chemical and Pharmaceutical Research, 7*, 278–285.

Khan, M. I., Lee, I. K., & Kim, Y. (2019). Nanotechnology: An emerging tool for contaminant removal and environmental sustainability. *Environmental Science & Technology, 53*(15), 8945–8966.

Khota, L. R., Sankarana, S., Majaa, J. M., Ehsania, R., & Schuster, E. W. (2012). Applications of nanomaterials in agricultural production and crop protection: A review. *Crop Protection, 35*, 64–70.

Kovochich, M., Xia, T., Xu, J., Yeh, J. I., & Nel, A. E. (2005). Principles and procedures to assess nanoparticles. *Environmental Science & Technology, 39*, 1250–1256.

Kumar, R., Kumar, G., Bansal, V., & Yadav, M. (2020). Nanotechnology for a sustainable development: A review. *Journal of Cleaner Production, 263*, 121522.

Lin, S., & Lee, T. C. (2018). *Nanotechnology for agriculture: Crop production & protection*. Springer.

Llop, J., Estrela Lopis, I., Ziolo, R. F., González, A., Fleddermann, J., & Dorn, M. (2014). Uptake, biological fate, and toxicity of metal oxide nanoparticles. *Particle Systems Characterization, 31*, 24–35.

Madhavan, J., & Roopan, S. M. (2020). Sustainable nanotechnology: Advancement and challenges. *Current Opinion in Green and Sustainable Chemistry, 26*, 100395.

Markovic, Z., Diamanti, E. K., & Hoek, E. M. V. (2019). Nano-enabled renewable energy for sustainable development. *Sustainable Energy Technologies and Assessments, 31*, 83–93.

Marzbani, P., Afrouzi, Y. M., & Omidvar, A. (2015). The effect of nano-zinc oxide on particleboard decay resistance. *Maderas. Ciencia y Tecnología, 17*, 63–68.

Memarizadeh, N., Ghadamyari, M., Adeli, M., & Talebi, K. (2014). Preparation, characterization and efficiency of nanoencapsulated imidacloprid under laboratory conditions. *Ecotoxicology and Environmental Safety, 107*, 77–83.

Mukhopadhyay, S. S. (2014). Nanotechnology in agriculture: Prospects and constraints. *Nanotechnology, Science and Applications, 7*, 63–71.

Nuruzzaman, M., Rahman, M. M., Liu, Y., & Naidu, R. (2016). Nanoencapsulation, nano-guard for pesticides: A new window for safe application. *Journal of Agricultural and Food Chemistry, 64*, 1447–1483.
Patra, J. K., & Baek, K. H. (2017). Antibacterial activity and synergistic antibacterial potential of biosynthesised silver nanoparticles against foodborne pathogenic bacteria and its anticandidal and antioxidant effects. *Frontiers in Microbiology, 8*, 167–178.
Porter, A. E., Gass, M., Muller, K., Skepper, J. N., Midgley, P. A., & Welland, M. (2007). Direct imaging of single-walled carbon nanotubes in cells. *Nature Nanotechnology, 2*, 713–717.
Prasad, R., Kumar, V., & Prasad, K. S. (2014). Nanotechnology in sustainable agriculture: Present concerns and future aspects. *African Journal of Biotechnology, 13*, 705–713.
Rana, S., & Kalaichelvan, P. T. (2013). Ecotoxicity of nanoparticles. *ISRN Toxicology, 13*, 574648.
Sabir, S., Arshad, M., & Chaudhari, S. K. (2014). Zinc oxide nanoparticles for revolutionising agriculture: Synthesis and applications. *Scientific World Journal, 14*, 925494.
Sadeghzadeh, B. (2013). A review of zinc nutrition and plant breeding. *Journal of Soil Science and Plant Nutrition, 13*, 905–927.
Sayes, C. M., Fortner, J. D., Guo, W., Lyon, D., Boyd, A. M., Ausman, K. D., & Tao, Y. J. (2004). The differential cytotoxicity of water-soluble fullerenes. *Nano Letters, 4*, 1881–1887.
Sekhon, B. S. (2014). Nanotechnology in Agri-food production: An overview. *Nanotechnology, Science and Applications, 7*, 31–53.
Sertova, N. M. (2015). The application of nanotechnology in the detection of mycotoxins in the agricultural sector. *Journal of Central European Agriculture, 16*, 117–130.
Shweta, S., Tripathi, D. K., Singh, S., Singh, S., Dubey, N. K., & Chauhan, D. K. (2016). Impact of nanoparticles on photosynthesis: Challenges and opportunities. *Materials Focus, 5*, 405–411.
Singh, S., Vishwakarma, K., Singh, S., Sharma, S., Dubey, N. K., & Singh, V. K. (2017). Understanding the plant and nanoparticle interface at transcriptomic and proteomic level: A concentric overview. *Plant Gene, 11*, 265–272.
Tripathi, D. K., Singh, S., Singh, V. P., Prasad, S. M., Dubey, N. K., & Chauhan, D. K. (2017). Silicon nanoparticles more effectively alleviated UV-B stress than silicon in wheat (*Triticum aestivum*) seedlings. *Plant Physiology and Biochemistry, 110*, 70–81.
Vartholomeos, P., Fruchard, M., Ferreira, A., & Mavroidis, C. (2011). MRI-guided nanorobotic systems for therapeutic and diagnostic applications. *Annual Review of Biomedical Engineering, 13*, 157–184.
Venkatachalam, P., Jayaraj, M., Manikandan, R., Geetha, N., Rene, E. R., & Sharma, N. C. (2017). Zinc oxide nanoparticles (ZnONPs) alleviate heavy metal induced toxicity in *Leucaena leucocephala* seedlings: A physiochemical analysis. *Plant Physiology and Biochemistry, 110*, 59–69.
Vidotti, M., Carvalhal, R. F., Mendes, R. K., Ferreira, D. C. M., & Kubota, L. T. (2011). Biosensors based on gold nanostructures. *Journal of the Brazilian Chemical Society, 22*, 3–20.
Yang, L., & Watts, D. J. (2005). Particle surface characteristics may play an important role in phytotoxicity of alumina nanoparticles. *Toxicology Letters, 158*, 122–132.

Chapter 14
Future Trends in Nanotechnology

Jincemon Cyriac, A. R. Niranjana, Gouri G. Nair, K. Sreelatha, Tom Cherian, Jijo Francis, and V. N. Sheeja

14.1 Introduction to Emerging Nanotechnology Trends

Nanotechnology is a field of research that opens doors to endless opportunities to any and every field of science and technology. The historic talk of the great physicist Richard Feynman at the annual meeting of American Physics Society in 1959, "There's Plenty of Room at the Bottom—An Invitation to Enter a New Field of Physics" is a testament to the potential of the field of nanotechnology (Islam & Vashist, 2014).

14.1.1 Overview of Nanotechnology Evolution

The beginning of the nanoparticle research was initiated by the U.S. National Science Foundation (NSF). A unifying definition for Nanotechnology was formulated in 1999. This was followed by the organization of the National Nanotechnology

J. Cyriac (✉)
Department of Physics, Christ College (Autonomous), Irinjalakuda, Kerala, India

A. R. Niranjana · G. G. Nair · V. N. Sheeja
Department of Physics, Marian College Kuttikkanam (Autonomous), Kuttikkanam, Kerala, India

K. Sreelatha
Post Graduate Department of Physics, Sree Narayana College, Chengannur, Kerala, India

T. Cherian
Department of Chemistry, Christ College (Autonomous), Irinjalakuda, Kerala, India

J. Francis
Department of Zoology, Christ College (Autonomous), Irinjalakuda, Kerala, India

S. Kulkarni (ed.), *Next-Generation Nanomaterials for Sustainable Engineering*, Lecture Notes in Nanoscale Science and Technology 43,
https://doi.org/10.1007/978-3-032-05719-8_14

Initiative (NNI)—a research initiative entirely dedicated to nanotechnology research, in the 2000 (in the USA), which inspired similar initiatives worldwide. The development and foreseeable future of nanotechnology is generally categorized into four generations. The First Foundational Phase (2000–2010) was entirely focused on synthesis of nanoscale materials for future applications along with the development of tools for nanoscale measurements and synthesis. The Second Foundational Phase (2010–2020) was focused on application—based research effort made toward nanotechnology integration and mass production resulting in a transition from passive nanostructures to active nanostructures. This era projected nanotechnology to the title of a global-purpose technology with applications spanning healthcare, medicine, energy storage, space exploration, and many more. The third phase (2020–2030) marked by research to develop self-assembled nanostructures. This current era also focuses on identifying the economic, environmental, and societal impacts of nanotechnology and addressing these concerns. The fourth generation of nanotechnology, projected for 2030 and beyond, is anticipated to emphasize the integration of computation and artificial intelligence. This advancement aims to enable automation within the field, enhancing efficiency and innovation (Islam & Vashist, 2014; Roco, 2011).

14.1.2 Importance of Forecasting Future Trends

Nanotechnology is an evolving field with active research in electronics, energy, medicine, and materials science. Tracking progress and predicting future developments is challenging due to its steady and systematic growth. Therefore, formulating a strategic policy for investment decisions is crucial. Additionally, it is essential to anticipate potential challenges and address ethical concerns to ensure responsible advancement in the field.

Forecasting also assists time relevant inventions and discoveries. It bridges the gap between research and real-world applications (Forloni, 2012; Islam & Vashist, 2014).

14.1.3 Current State of Nanotechnology

Nanotechnology, an interdisciplinary domain blurring the boundaries between various branches of science, has grown significantly over the last two decades. Its contributions include advanced nanomaterials, biomedical application, enhanced energy storage devices etc., with each benefaction not only intriguing researchers but also increasing its industrial and economic value. Nanotechnology is also being explored for solutions to water purification, clean energy, and climate change mitigation. With the spotlight focused on nanotechnology, addressing the toxicity of nanoparticles, ethical implications and safety regulations is crucial. The transition of nanotechnology from laboratory to homes requiring mass production is also a concern due to scalability and cost (Forloni, 2012; Roco, 2011).

14.2 Advanced Nanofabrication Techniques

14.2.1 Atomic Layer Deposition and Precision Engineering

Atomic layer deposition (ALD) is a bottom-up vapor-deposition technique of the synthesis of materials. The characteristic feature of ALD is that it is a self-assembly technique that processes through self-limiting reaction. Low growth rate of ALD was considered as a down side in its application in material fabrication. However, the growing need of nanofilms for the fabrication of scaled down devices has brought ALD back on spot light. The advantages of ALD include thickness control at the atomic scale, production of highly conformal films, low temperature growth, and wide-area uniformity. The main steps of synthesis by ALD include:

1. Exposure of substrate to first precursor for chemical adsorption onto substrate surface.
2. Removal of byproducts and excess precursor by flow of inert gas.
3. Exposure to second substrate aiding the formation of monolayer of the precursor.
4. Removal of byproducts and excess precursor by flow of inert gas.

The cycle repeats till desired film thickness is attained. Among the several advantages of ALD the excellent conformality and atomic scale thickness control are of particular interest to precision engineering techniques and studies are done to enhance these very features of ALD (Kim et al., 2009; Zhang et al., 2022).

14.2.2 High-Throughput Synthesis Methods

High-throughput screening (HTS) was developed primarily to assist drug discovery involving automated testing of large chemical libraries to identify compounds with biological activities. HTS is an excellent addition to nanofabrication enabling accurate testing of new materials and optimizing nanofabrication processes. It is a cost and time efficient technique that accelerates the identification of novel nanomaterials with desired properties and their optimization. Advanced computational techniques and data analytics are often coupled with HTS for more reliable results in material selection and analysis. HTS is transforming nanofabrication and analysis by enabling rapid screening and reducing material and operational costs (Attene-Ramos et al., 2014; Mayr & Bojanic, 2009).

14.2.3 Hybrid Fabrication Approaches

A different approach to nanofabrication is by combining the bottom-up and top-town methods. A practical employment of the hybrid technique resulted in the fabrication of plasmonic nano-platforms with sub-10 nm gaps between the nanostructures. This gap range between the nanostructures was found to enhance the surface-enhanced Raman scattering (SERS) signals. The enhanced sensitivity of the material to SERS provides hope for development of further techniques that bridges the gap between nanoscale fabrication and practical large-area sensing applications (Dhawan et al., 2011).

14.3 Nanoscale Materials for Next-Generation Applications

14.3.1 Two-Dimensional Materials

Two-dimensional nanomaterial is ultrathin (one dimension in nanoscale) materials that range from monoatomic to a few layers of atoms in thickness. These materials possess unique properties different from their respective bulk materials which can be attributed primarily to the large aspect ratio (length or width/thickness) of the two-dimensional nanomaterials. Few of the novel characteristics of 2D nanomaterials include its anisotropic nature, shape, surface charge, and tunable functionalities (Baig, 2023; Chimene et al., 2015).

Graphene is an allotrope of carbon with hexagonal cells forming a honeycomb structure. The unique properties of grapheme encompass superconductivity, flexibility, and mechanical strength. Graphene and graphene-based nanomaterials find applications in flexible transparent electronics, energy-storage devices, biomedical applications, separation membranes, printed electronics, etc. An important application of graphene two-dimensional nanomaterial (graphene membrane) is its efficiency in water desalination. This is achieved by modeling graphene membranes with precise nanometer-sized pores which permits the passage of water molecules but restricts larger entities hence acting as filter. A similar principle is employed in the use of graphene membrane for gas separation (Baig, 2023; Nguyen & Nguyen, 2016).

MXenes are two dimensional materials composed of transition metal carbides, nitrides, or carbonitrides. The notable properties of MXenes include high conductivity, stability, energy storage, and tunability. The high metallic conductivity and hydrophobic nature make it an excellent candidate in the fields of supercapacitors, batteries, and flexible electronics (Perera et al., 2023).

Table 14.1 Carbon nanostructures and applications (Kasálková et al., 2021; Liu et al., 2018)

Carbon nanostructure	Type of nanomaterial	Field of application
Carbon-based quantum dots (QD)	0D	Bio-imaging, drug delivery
Fullerenes	0D	Biosensors, reinforced composites, photonics
Diamond like carbon	2D	Biomedical implants, wear-resistant surface
Carbon nanotubes (CNT)	1D	Biosensors, reinforced composites, photonics
Graphene, graphene composites	2D	Tissue engineering, energy storage, structural materials

14.3.2 Carbon Nanostructures and Beyond

Carbon being a unique and versatile element, carbon nanomaterials, and nanostructures occupy a distinctive place in the field of nanomaterials. This can be credited to the different possible hybridization states of carbon (sp, sp^2, sp^3) endowing the allotropes formed with distinct physical and chemical properties. Some of the common carbon nanostructures and their applications are given in Table 14.1 (Kasálková et al., 2021; Liu et al., 2018).

14.3.3 Perovskites and Advanced Functional Materials

Perovskites are a class of highly functional materials of the ABX_3 crystal structure. Their notable tunability finds them applications in quantum devices, optoelectronics, solar energy, and many more. Functional perovskite materials are sought after due to their wide spectrum of potential applications attributed to their ferroelectric, ferromagnetic, multiferroic, and conducting properties (Li et al., 2023).

The advanced growth techniques offered by nanotechnology has contributed to optimization of perovskites. The nanotechnology enabled development of epitaxial perovskite films with exceptional ferroelectric and multiferroic properties are used in fabrication of memory devices, sensors and actuators. Nanotechnology has also made possible synthesis of superlattices with enhanced tunability by multilayer arrangement of different perovskites. The development of self-assembled nanostructures enables large scale ferroelectric arrays. It also facilitates integration of ferroelectric and magnetic perovskites for spintronics applications (IonelaVrejoiu et al., 2008).

14.4 Artificial Intelligence and Machine Learning in Nanotechnology

14.4.1 Data-Driven Design of Nanomaterials

The tremendous development of data science and machine learning in the past decades has led way to its incorporation in research fields. The viability of data driven research in the field of nanotechnology is currently limited due to inadequate number of datasets available. However, given the enormous scope of nanotechnology efforts are being made to expand the database for the potential automation of this field. In the context of data driven research, discovery is defined as observing a previously unknown phenomena or material and design as devising an object based on particular vision of its properties. Since designing requires a prerequisite about necessary features it can be rightfully said that discovery precedes designs. To put simply we cannot design a method for synthesis of a material with no available data on the nature of the material itself. Data driven design helps in enhancement of available (reported data) design so as to align with our specific requirements (Pollice et al., 2021; Yang et al., 2022).

An example of data driven research is a study of the efficiency of desalination membranes using Density Functional Theory-Machine Learning (DFT-ML) framework. The process begins with selection of membranes with desirable salt ion retention, service life, water flux and material properties by particle swarm algorithms. The DFT-ML results show that efficient desalination membranes can be signaled out by energy densities, adsorption energy and mechanical property parameters (Meng et al., 2023).

14.4.2 Role of AI in Predictive Modeling

Artificial intelligence (AI) and nanotechnology are two fields of research with immense revolutionary potential. The careful intersection of these disciplines holds great scope for development and enhancement of applications of nanotechnology. The use of AI in nanotechnology can be used for predictive modeling and material analysis which could speed up the rate of discoveries in nano science, hence paving way for transformative advancements in diverse domains (Naik & Jagtap, 2024).

14.4.3 Automating Nanotechnology Research

With reliable developments in AI and ML complete automation of the field of nanotechnology research is possible. This would also accelerate the rates of work done in the field. AI-integrated nanotechnology could mean material property production and AI optimized nano sensors for biological monitoring. A major hurdle, however, in the potential automation of nanotechnology include availability of limited datasets, data privacy, algorithmic bias, and societal impacts like employment loss from automation (Saif Al-Deen et al., 2023).

14.5 Nanotechnology in Healthcare and Medicine

14.5.1 Targeted Drug Delivery Systems

The conventional use of macro-sized drug delivery mechanisms poses several challenges including poor solubility and absorption and issues with target specific delivery in addition to side effects of the drugs. Hence switching to an alternative drug delivery mechanism is imperative. A solution to this issue involves integration of nanotechnology in the picture. Emergence of nanomedicine which are drug delivery systems in the nanonometer size that contain dispersed, encapsulated, adsorbed, or conjugated drug and imaging agents is a feasible alternative. Nanomedicine facilitates the administration of drugs without the use of needles or capillaries. The two primary techniques for drug delivery in nanomedicine are active targeting and passive targeting. The passive targeting mechanism exploits the physiology of diseased tissues of having a leaky microvascular. The nanoparticle introduced diffuses into the leaking sites of the target (diseased) tissue by utilizing the EPR effect (Enhanced Permeability and Retention) of the damaged tissue. Active targeting uses ligand functionalized nanoparticles that selectively bind to overexposed epitopes and receptors for drug delivery. The active targeting technique is more target specific than the passive technique and involves complex synthesis (nanoparticle modification required). Passive targeting though comparatively less efficient is more than sufficient for various general disease treatments (Antonio Di Stefano et al., 2023; Koo et al., 2005).

14.5.2 Nano-Immunotherapy and Cancer Treatment

Cancer is a severe disease threatening human health. Apart from the gravity of the disease itself, the intractable problems posed by the traditional treatment methods is a major concern. The emergence of nano-immunotherapy as a cancer treatment method is a relief to the concerns caused by the existing standard methods of treatment by radiation, surgery and/or chemotherapy. Currently most of the nano-immunotherapy research is based on the T-cell (cancer immunity cell) mediated cellular immunity. Cancer results from uncontrolled growth of malignant tumor cells. It persists in the body due to no response of the immune system to combat these cells. Unlike the conventional treatment methods which focus on the removal of the malignant tumor which pose an inevitable threat of recurrence of tumor

Table 14.2 Nanomaterial-based biosensors (Barbosa et al., 2021; Nagraik et al., 2021)

Biosensor type	Nanomaterial used	Detection
Electrochemical	AuNP, CNT	Measure electrical signals from biomolecular interaction
Optical	QD, AgNP, plasmonic NP	By fluorescence, raman spectroscopy or surface Plasmon resonance
Piezoelectric	Nanowires (Si, ZnO), CNT	Measure mechanical vibration due to molecular binding

nano-immunotherapy focuses on the root cause of the problem. Cancer-immunotherapy boosts the inherent capability of the immune system to identify tumor cells and generate anti-tumor responses providing a new avenue to combat cancer (Rana et al., 2023; ShiyiZuo et al., 2021).

14.5.3 Biosensors and Diagnostic Devices

Early and accurate detection of diseases is essential for timely medical attention to combat them efficiently. Biosensors are a group of biotechnologically synthesized analytical devices used for disease diagnostics. Biosensors, composed of a bio-receptor, a transducer and a signal reading device detects biological molecules and convert their presence into measurable signal. The quantification process of traditional biomarkers is time-consuming and expensive require high-end technological application. The integration of nanostructured materials to modify the traditional biosensors aids in ameliorating the properties of biosensors and expanding their limit of detection. Table 14.2 tabulates nanomaterial-based biosensors.

14.6 Sustainable Nanotechnology

With the increase in the global human population, the adequate supply of resources has become limited. Pollution-free technologies must be developed to guarantee clean energy solutions and environmental restoration. With major advantages for the environment and human health, nanotechnology is essential to the development of greener and cleaner solutions. Its uses are being investigated to improve the effectiveness of traditional environmental cleanup technologies while managing, mitigating, and remediating pollutants in the air, water, and soil.

By lowering energy consumption during synthesis and manufacturing, encouraging product recyclability, and creating environmentally friendly materials, green nanotechnology focuses on sustainability. Even though nanotechnology has a lot of potential to solve sustainability issues, it is important to evaluate and reduce any possible dangers related to nanomaterials, especially those that could affect the environment and human health (Khan, 2020).

14.7 Green Synthesis of Nanomaterials

Green synthesis of nanomaterials provides a sustainable substitute for conventional synthesis techniques by using biological systems (Sivaraj et al., 2020). This method reduces more extensive, effectiveness. Green synthesis guarantees an eco-friendly, economical, safe, and clean approach clean approach of creating nanomaterials by using low-energy techniques and organically available starting ingredients. Microorganisms (including bacteria, yeast, fungi, algae, and some plants) act as/ supply biological substrates for the synthesis of nanomaterials. Certain active molecules and precursors, such as metal salts, influence nanoparticulate final morphology and size (Sivaraj et al., 2020). Moreover, green-derived nanomaterials also showed useful properties such as antimicrobial activity, natural reducing potentials, and detoxication. The active molecules in the biological substrates are responsible for these properties, having only been discovered recently and not stated in the last review of nanomaterials by Saratale et al. (Saratale et al., 2018). The biological components responsible for nanomaterial synthesis include specific enzymes (Birla et al., 2009), amino acid groups, proteins (Ganesh Babu & Gunasekaran, 2009), or either unique chemical structure (Pugazhendhi et al., 2018), which facilitate nanoparticle formation, reduction, and stabilization. Being biocatalysts, enzymes catalyze the redox reaction for precipitation of metal ions to NP form. The stabilization mechanism of nanoparticles is the presence of amino acid groups and proteins that serve as capping agents, preventing aggregation, and maintaining a uniform dispersion.

14.7.1 Nanotechnology for Renewable Energy Solutions

Greenhouse gases released from the combustion and processing of fossil-based fuels result in disastrous ramifications to modern society, and thereby, the development of alternative renewable energy sources is an inevitable tendency. However, various obstructions remain in new energy systems. The research and development of nanotechnology (such as the innovation of nanomaterials and nanoreactors) are considered one of the most encouraging approaches to improve the efficiency of clean energy systems, facilitating their implementation to satisfy the world's energy demand. In this chapter, renewable energy sources, such as solar energy, wind power, geothermal energy, and hydrogen energy, using nanotechnology.

14.7.2 Addressing Environmental and Societal Concerns

While nanoparticle synthesis offers significant benefits across various fields, it also raises critical environmental and societal concerns. Large-scale production and use of nanomaterials, many of which remain insufficiently studied, have led to unintended ecological consequences, such as the accumulation of inorganic nanoparticles in soil and water, toxicity to plant and animal life, and challenges in waste management. Engineered nanoparticles have the potential to disrupt natural ecosystems, making comprehensive risk assessments and effective disposal strategies essential to mitigate their impact.

Green nanotechnology and other sustainable synthesis techniques need to be given the greatest importance in order to address these worries. Energy-efficient procedures and bio-based, non-toxic precursors can lower environmental concerns while preserving scientific advancement. Furthermore, in order to create regulatory frameworks, encourage responsible innovation, and guarantee the moral and long-term integration of nanotechnology into society, interdisciplinary partnerships between scientists, legislators, and industry stakeholders are crucial.

14.8 Nanotechnology in Electronics and Quantum Computing

One technique that promises to address the shortcomings of traditional complementary metal-oxide-semiconductor (CMOS) technology for high density and high-performance applications is quantum computing. But because quantum aspects are involved, many novices find it hard to follow the field (Bhat et al., 2022). Ensuring resilience is one of the major issues in the era of nano-scopic computing. It is now considerably harder to provide high dependability in robust computer system design due to quantum physical, probabilistic, and even biological phenomena. The uncertainty in ensuring "zero-one" based computation and probabilistic effects may provide additional issues as devices are scaled down to single electron functioning. By using redundancy and replication, tiny electronics—which suggest billions of units on a single chip—may be able to lessen the uncertainty problem. But with the sheer scale, such device densities will make design and validation a nightmare (Shukla & Bahar, 2004). Instead of being limited to two states, quantum computers encode data using quantum bits, or qubits, which may be either 0 or 1 or both 0 and 1 simultaneously in a phenomenon known as superposition. At the atomic level, atoms, photons, ions, or electrons, together with the control devices that go along with them, may physically create quantum computers that function as memory and a computer processor. Many alternative physical realizations of qubits have been suggested and investigated at the level of basic research exploratory facilities.

Solid-state solutions have garnered a lot of attention lately because of their ability to scale to high numbers of qubits.

14.8.1 Miniaturization of Electronic Devices

Component shrinking has been the foundation for the long-term improvement in information processing technology performance. Examples are used to illustrate the history of miniaturization (Keyes, 1988). Electronics are now able to permeate society more deeply and extensively thankful to miniaturization. Examples of the downsizing of electronics that the general public first learned about include pocket calculators, electronic watches, tiny color television receivers, and similar devices. Miniaturized electronic systems already had a big influence on the commercial, industrial, and military sectors before they were created. Space exploration, communications, equipment and process management, and data handling and processing were all aided by miniaturization. Some people believe that the integrated circuit is the reason behind the somewhat late development of electronics miniaturization, but miniaturization based on size, weight, and power requirements was already underway long before the integrated circuit was created and even before transistors were commercially available. Smaller electronic systems were made possible by the screen printing of resistive and other passive components, the idea of electronic modules, and the remarkable success of valve (vacuum-tube) manufacturers in creating miniature and subminiature valves, some of which were smaller than a modern power transistor. The integrated circuit, which created tiny and then massive circuits on a single silicon chip, was the catalyst for the significant acceleration toward microelectronics. As a result, the systems were far more intricate and bigger than previously thought (Atherton, 1984).

14.8.2 Quantum Dots and Quantum Computing Advances

Over the past 20 years, researchers have become more interested in quantum dots (QDs) due to their intriguing physical and chemical characteristics. Quantum dots are fluorescent semiconductor nanoparticles made up of a core substance encased in a 2–10 nm diameter shell of another semiconductor material. Characteristics such as optical characteristics, absorbance, and photoluminescence are dependently reflected in the size of QDs (Karakoti et al., 2015). Its optical characteristics and quantum confinement are indicated by the moniker "quantum dots." They are a good fit for biological function and imaging because of this particular characteristic. They are very promising for real-time monitoring, tracking, imaging, and sensing. Because they have the same amount of electrons as atoms and may move in three dimensions with a narrow electronic energy level, QDs are sometimes referred to as artificial atoms (Ai et al., 2013). The use of quantum dots in different sectors determines which ones are chosen.

Driving fundamental research that will contribute to the advancement of humanity's understanding of the cosmos is made possible by quantum computing. Findings in this discipline have the potential to lead to revolutionary new knowledge and applications, as is the case with any fundamental scientific research. The development of a massive, error-corrected quantum computer is fraught with difficulties. Unprecedented control over quantum coherence will be necessary for successful quantum computation, pushing the envelope of what is feasible through the development of new methods and tools or the improvement of current ones. These developments may also benefit related technologies that also depend on quantum coherence management, such quantum sensing and quantum communication (Fan et al., 2020).

14.8.3 Flexible and Wearable Electronics

A new age of wearability and adaptability in personal electronics is emerging, opening up a plethora of possibilities in sports, entertainment, and healthcare. The traditional power source's large and inflexible design acts as a roadblock to the wearable smart devices, industry's quick development. Power supply with great flexibility, bendability, and stretchability have drawn a lot of attention in order to satisfy the demands of next-generation wearable electronics (National Academies of Sciences, Engineering, and Medicine, 2019). Exploring electronics for real-time data acquisition has become imperative due to the significant technical advancements in the era of intelligence and the integration of the physical environment and human life's information network. The investigation of wearable and flexible electronics of the future has demonstrated encouraging promise for the fourth industrial revolution, which will change the paradigm of personal computing and create a society that is highly networked (Fan et al., 2020). Wearables are known for the excellent portability, conformal contact characteristic, and user-friendly interfaces for a variety of applications (e.g., smart clothes, e-textiles, electronic skins, on body health monitors, and flexible displays), ushering in the Internet-of-Everything (IoE) wherein devices are mated with the human body intimately. Power supplies, which serve as the power source for wearables, should be extremely flexible and shape-adaptive in order to blend in with the intricate curves and soft tissue surfaces of people. Because of this particular, existing energy storage devices are difficult to use because of their flat, heavy, and inflexible structures, which make them unsuitable for wearable electronics that are affixed to the soft, curved, and elastic human body. Wearable and flexible power sources that can simultaneously function as stretchy, bending, and malleable materials are potential tools for fusing the human body with power sources. Generally, requirements for flexible and wearable power sources include the following aspects: (1) flexible configuration design, (2) development of flexible materials, (3) high energy density, and (4) other requirements such as safety issues and environmental adaptability.

14.9 Nanotechnology for Energy Applications

14.9.1 Nanomaterials for Solar Energy Conversion

The development of appliances and components smaller than 100 nm offers new ways to capture, store, and exchange energy, making nanotechnology an enabling technology that offers a wide range of resources to address energy-related issues. The enormous amount of energy that the sun emits each day is produced by nuclear fusion. More energy is emitted in a single second by the sun than by humans since the dawn of time. The technical potential of solar energy has been observed to be several times greater than the current global need for primary energy. The technology of sun harvesting using nanoparticles has been thoroughly examined in this paper. There has been discussion of the various forms of contemporary solar collection technology that successfully and efficiently employ nanomaterials. Among other devices, nanomaterials have been utilized to boost efficiency in solar photovoltaics, fuel cells, solar collectors, and photocatalysis. It has been discovered that employing nanomaterials can boost incident radiation by nine times while increasing the solar collector's efficiency by 10% when compared to a traditional flat plate solar collector (Lee et al., 2018).

To meet the need for sustainable energy worldwide, new projects aimed at more efficiently gathering incident solar radiation are required. Three generations of solar cell technologies have developed sequentially (Chen et al., 2018). A single crystalline semiconductor wafer serves as the foundation for first generation solar cells. Inorganic thin-film structures are used in the cell assembly of second-generation solar cells. Although they are less expensive to manufacture, their efficiency, less than 14% in amorphous thin-film solar cells is less than that of the first generation's single junction crystalline solar cells, which can achieve up to 27% efficiency. According to Shockley–Queisser thermodynamics, single junction cells should theoretically have an efficiency of up to roughly 33% (Abdin et al., 2013). Therefore, a new technology for solar cells is needed to get efficiencies above 33% at a cheaper cost of production. Third-generation solar cells are the first to make this achievement. The third generation of solar cells, which may produce affordable, highly efficient cells and may soon become a new technology, are currently the focus (Lee et al., 2018).

14.9.2 Energy Storage Devices: Batteries and Supercapacitors

In our contemporary world, energy storage devices are essential because they make it possible to use renewable energy sources efficiently, power portable electronics, and ease the switch to electric vehicles. The intermittent nature of renewable energy sources, such the sun and wind, makes it difficult to continuously deliver energy to meet demand, even while demand for these sources rises. Effective and durable energy storage solutions are needed for portable electronics and transportation electrification. Effective energy storage technologies that can store excess energy available during periods of low demand and use it during periods of high demand are therefore desperately needed (Werner, 2004). Energy storage techniques are essential for stabilizing power networks, maximizing the use of energy resources, and mitigating the intermittent and variable nature of renewable energy sources (Shockley & Queisser, 1961). There are many different energy storage techniques available today, each with special traits, benefits, and drawbacks. They fall into four general categories: chemical, mechanical, electrochemical, and thermal storage methods. Because of their adaptability, effectiveness, and ongoing development, batteries and supercapacitors have become important actors among the various energy storage technologies. Applications for batteries are numerous and range from large-scale grid storage to tiny handheld gadgets. They are a dependable and portable power source because they store electrical energy through reversible electrochemical reactions.

Batteries are electrochemical energy storage devices that use a sequence of electrochemical reactions to convert chemical energy into electrical energy. They are made up of a separator, an electrolyte, and positive and negative electrodes in one or more electrochemical cells. An electrochemical process takes place inside the cell when a load is attached to the battery, causing electrons to flow and electrical current to be produced (Lens & Uddandarao, 2023). An electrochemical cell, which consists of two electrodes submerged in an electrolyte, is the central component of a battery. Usually composed of distinct materials, the electrodes are referred to as the cathode and the anode. The positive electrode in this case is called the cathode, and the negative electrode is called the anode. The redox (reduction–oxidation) reactions that take place during operation are facilitated by these electrodes. In order to keep the electrodes apart and enable ion movement between them, the electrolyte serves as a medium for ion transport. The type of battery will determine whether it is a liquid, gel, or solid. The electrolyte completes the electrical circuit and transports the ions required for the redox processes. Batteries frequently have additional parts to improve performance and safety in addition to the electrodes and electrolyte. Separators, which keep the electrodes from making direct contact and creating short circuits, may be one of these. Usually made of porous membranes, separators let ion flow while avoiding physical contact between the electrodes. Additionally, current collectors-conductive elements that gather and disperse the electrical current produced-may be found in batteries. Typically, these collectors consist of metallic foils that are affixed to the electrodes, like copper or aluminum.

Supercapacitors, sometimes referred to as ultracapacitors or double-layer capacitors, are electrochemical energy storage devices that have a significantly lower energy density than batteries but a very high-power density. This gets them ready for the best uses, like hybrid cars, electric cars, or electronic gadgets that call for brief spikes in power. Two methods are used by supercapacitors to store energy. The predominant mechanism in the majority of supercapacitors is electric double-layer capacitance. It happens when the electrolyte's positive and negative charges are drawn to the corresponding positive and negative electrodes. As a result, a small layer of electrolyte separates the control layer from each electrode. Supercapacitors have a high surface area-to-volume ratio because of this layer's thinness, which is only a few nanometers (Aneke & Wang, 2016). This phenomenon can happen in supercapacitors that employ active redox materials. In this instance, energy can be stored in the form of chemical bonds by oxidizing or reducing the ions in the electrolyte at the electrodes (Funabashi, 2016). Supercapacitors are perfect for applications that need high power but little energy since they can be charged and discharged rapidly (Winter & Brodd, 2004).

Supercapacitors and batteries have several uses and are widely used in many different sectors. Mobile phones, tablets, and PCs are the primary devices that use these energy storage technologies. They are also essential for driving electric and hybrid cars. In medicine, cardiac pacemakers employ batteries. They supply the energy required for spacecraft and satellites in the aerospace industry. In renewable energy systems, batteries also store energy generated by wind turbines and solar panels. Ultracapacitors, also known as supercapacitors, are exceptional to conventional batteries in that they can accommodate high-power density demands and rapid energy release rates. They are therefore a necessary innovation in many different applications. When powering devices that require rapid energy bursts, such the regenerative braking systems found in electric automobiles, the capacity to charge and release energy considerably more quickly is essential. Furthermore, supercapacitors are essential parts of renewable energy systems, and they need fast storage options to improve energy management and power load balancing in the face of increasing volatility. Additionally, even in the face of severe temperatures, these devices provide dependable backup power sources that are essential for the long-term survival of emergency equipment or uninterruptible power supplies (UPS).

14.9.3 Hydrogen Production and Fuel Cells

Nowadays, many people consider hydrogen and fuel cells to be among the most important energy options for the twenty-first century. Reduced environmental effect, increased energy security (and diversity), and the development of new energy sectors will all be greatly aided by these technologies. Energy storage systems, distributed heat and power generating, and transportation can all make use of hydrogen and fuel cells. However, there are substantial scientific, technological, and socioeconomic obstacles to the adoption of hydrogen and fuel cells as future clean energy solutions in the shift from a carbon-based (fossil fuel) energy system to a hydrogen-based economy (Edwards et al., 2008). Traditionally, the chemical industry has produced hydrogen on a considerable scale through the reforming of light hydrocarbons, primarily natural gas. The quality of the raw materials, the conversion efficiency, and the security requirements for the integration of hydrogen production, purification, and use are just a few of the technical and scientific obstacles that the reforming technologies currently in use must overcome. The generation of hydrogen from biogas and other renewable resources contributes significantly to the reduction of greenhouse gas emissions. In this regard, combining biogas reforming procedures with hydrogen to activate fuel cells is a significant way to provide clean energy with increased energy efficiency (Alves et al., 2013).

14.10 Challenges and Ethical Considerations

14.10.1 Toxicological Impact of Nanomaterials

Toxicology traditionally addresses adverse poisoning effects of chemicals to humans, animals, and the environment. While nanotechnology and the production of nanoparticles are growing exponentially, research into the toxicological impact and possible hazard of nanoparticles to human health and the environment is still in its infancy (Elsaesser & Howard, 2025). Because of their small size and high surface reactivity, nanoparticles behave differently than bigger materials and are thus more prone to interact with biological systems in unanticipated ways. Because airborne nanoparticles may enter the body through the lungs, the gut, or even the skin and move to key organs like the liver, heart, and brain, this has sparked worries about uncontrolled exposure. Asbestos and air pollution serve as reminders of the need of early risk assessment. However, nanoparticles are complex to research because, unlike conventional chemicals, their effects rely on size, shape, charge, and how they disperse or cluster together in addition to composition. These subtle effects are challenging for standard toxicity tests to capture, and scientists are still working to understand how nanoparticles travel and interact within the body. Some can even imitate viruses by moving through cells utilizing the body's own transport systems, particularly those smaller than 50 nm. Nanotoxicology is essential to maximizing the advantages of nanotechnology while reducing its potential risks because there is still so much to learn (Elsaesser & Howard, 2012).

14.10.2 Regulations and Standards for Nanotechnology

In a number of fields, including environmental science, electronics, and medicine, nanotechnology—the manipulation of matter at the atomic and molecular level—has created new opportunities. However, strict laws and standards must be put in place to guarantee the safe and efficient use of nanomaterials due to their special characteristics, which include their size-dependent behaviors and possible interactions with biological systems. Standards have been created by regulatory organizations such as ASTM, international to address these issues. One thorough framework for teaching the workforce about the unique characteristics and safe handling of nanoparticles is ASTM's E3089-22 Standard Guide. This standard guarantees that those involved in the creation, manufacture, and research of nanotechnology are prepared to handle the difficulties posed by these innovative materials (ASTM, 2025).

Through the ISO/TS 80004 series, international organizations including the International Organization for Standardization (ISO) have attempted to develop standard language and guidelines. By providing a consistent vocabulary for nanotechnology, these standards aim to promote effective communication between scientists, authorities, and producers. ISO helps prevent misconceptions and encourage uniformity in regulatory initiatives across nations by standardized language. The Food and Drug Administration (FDA) (FDA, 2019) has adopted a science-based regulatory approach, where each nanotechnology product is assessed individually. This approach ensures that goods are assessed for safety and efficacy in a way that tackles their potential dangers by taking into consideration the unique features of nanoparticles. With the help of these guidelines and standards, nanotechnology may develop further while maintaining the safety and social benefits of its uses.

14.10.3 Addressing Socioeconomic Inequities

Rapid advancements in research and development related to nanomedical technology raise various ethical and legal challenges in areas relevant to disease detection, diagnosis, and treatment. With its advances in illness diagnosis and treatment, nanomedicine holds the potential to completely transform the healthcare industry. But technology also presents moral and legal dilemmas that, if not handled wisely, might exacerbate already-existing disparities. According to Wasti et al., some of the main worries include the privacy hazards associated with gathering a large quantity of personal health data, the unknown health dangers of nanomaterials, and the challenge of ensuring individuals fully comprehend and agree to these therapies. Nanomedicine has the potential to revolutionize the healthcare industry due to its advances in the diagnosis and treatment of illnesses. But technology also poses moral and legal challenges that with poor management could widen existing gaps (Wasti et al., 2023). A second major issue is access; without proper controls, only wealthier individuals may be able to access these life-altering drugs, leaving poorer behind. Furthermore, regulatory loopholes blur safety standards as nanomedical items become harder to track and define. In addition, low-income countries are at risk of falling behind the benefits of nanomedicine as most of the current governance is oriented toward high-income countries. In addition, the legislative systems are prevalent on high-income countries; thus, impoverished countries may be at the risk of lacking benefits from nanomedicine. We need equitable legislation, educational investments, and international collaboration to ensure that these innovations serve the common good. To make sure nanomedicine is capable of improving lives, regardless of race, religion, or income status, ethical and legal mechanisms need to develop in step with technology (Wasti et al., 2023).

14.11 Future Directions and Innovations in Nanotechnology

According to leading scientists, developments in nanotechnology will fundamentally alter our understanding of and approach to developing materials, technologies, and systems. Innovations based on nanotechnology are frequently discussed in terms of their revolutionary character, with many referring to them as such in the media and technical domains. It is anticipated that these developments would cause significant changes for companies, upending established practices and having a significant impact on society. Advances in scalable nanomanufacturing, bio-nanotechnology, and quantum materials are anticipated to influence the future of nanotechnology. Future studies will probably concentrate on using the special qualities of nanomaterials to create molecularly-level enhanced medical diagnostics and treatments, self-healing materials, and extremely efficient energy storage devices. It is projected that the combination of nanotechnology with artificial intelligence (AI) would speed up the search for new materials and improve nanoscale manufacturing techniques, leading to previously unheard-of levels of accuracy and productivity. Furthermore, advancements in green nanotechnology—such as water purification membranes, biodegradable nanomaterials, and nanoscale catalysts for carbon capture—will become more and more important as environmental concerns increase. In addition to gaining a competitive edge, businesses and sectors that effectively incorporate these developments into their innovation portfolios will also spur sector-wide transformations, therefore solidifying nanotechnology's position as a key component of next industrial revolutions (Shea, 2005).

14.11.1 The Role of Collaborative Research

Science policy makers cherish particular hopes in interdisciplinary nanoscale research, there is not a single report that does not emphasize the importance of interdisciplinarity, and no financing program that does not specifically address internal transdisciplinary techniques, demonstrating how much science policy makers value multidisciplinary nanoscale research (Meyer & Persson, 1998). However, the terms "nanoscience" and "nanotechnology," which are sometimes used interchangeably, are rather ambiguous. Collaborative research brings together a variety of resources, perspectives, and areas of expertise, it is essential to the advancement of scientific knowledge and technological innovation. In a time when multidisciplinary approaches are needed to solve complex global issues, cooperation encourages synergy between researchers from many disciplines, organizations, and even nations. It facilitates better idea sharing, speeds up issue solving, and encourages the economical use of resources and infrastructure. Furthermore, by guaranteeing that breakthroughs and discoveries are disseminated widely rather than being restricted to specialized research groups, cooperative efforts help democratize knowledge. In fields such as nanotechnology, engineering, and medicine, teamwork has emerged as a key factor in life-changing breakthroughs and advancements (Schummer, 2004).

14.11.2 Emerging Markets and Industries

Nanotechnology is the fastest growing technology in the world, This rapid pace of technological revolution can especially be seen in the developed world, where nano-scale markets have taken over rapidly in the past decade (Malik et al., 2023). The phrases "nanotechnology," "nanoscience," and "nanomaterials"are now widely used in both everyday life and study. Scientists, engineers, and technicians have recently shown a great deal of interest in nanotechnology research projects. Nanoscience is the study of the manipulation and characterization of matter on length scales typically between atomic and micron-size. Nanotechnology is a new field of science and engineering that uses techniques or procedures from nanoscience to produce nanoproducts.

Nanotechnology demonstrates the crucial characteristics of common materials and structures. The development of the scanning tunneling microscope (STM), carbon nanotubes (CNTs), and fullerenes (also known as buckyballs) paved the way for nanotechnology as they allowed for the performance of investigations at the atomic and molecular level. Engineering, technology, applied sciences, biomedicine, chemistry, pharmaceuticals, food and agriculture, and construction are just a few of the sectors that use this technology (Khan et al., 2025). However, it is becoming more and more challenging to discern between actual applications that are now on the market and promises for future products due to the enormous number of research initiatives related to nanotechnology. The fact that 'nano' is not a well-defined and protected word further complicates matters. As a result, it is also abused to promote non-nano items. The goal of this study is to provide a clearer picture of scientific advancements and the commercialization of nanotechnological applications (Forster et al., 2011).

14.11.3 Vision for Nanotechnology in the Next Decade

Over the next 10 years, nanotechnology has the potential to transform a number of sectors, with significant implications for computers, sustainable development, and medicine. Nanotechnology breakthroughs are predicted to transform health prevention, diagnosis, and treatment in the medical field. By delivering medications straight to the intended cells, engineered nanoparticles can reduce adverse effects and increase therapeutic efficacy. Furthermore, nanomaterials that promise to improve patient care are being created for application in regenerative medicine, including tissue engineering and wound healing. The healthcare industry is changing as a result of these advancements, which portend more accurate and successful treatments in the future (Patil et al., 2008). Nanotechnology provides solutions for clean energy, water purification, and environmental remediation, all of which are in line with the Sustainable Development Goals of the UN. More effective solar cells and sophisticated filtration devices that eliminate impurities from water sources are being developed using nanomaterials. By encouraging the prudent use of resources, these applications not only address urgent environmental issues but also help ensure a sustainable future (Pokrajac et al., 2021). Significant improvements are also anticipated as a result of the incorporation of nanotechnology into computers. Nanotechnology-driven approaches are being explored to overcome current computing limitations, potentially leading to breakthroughs in quantum computing and advanced data storage systems. These innovations could revolutionize data processing and storage, paving the way for new possibilities in information management and technology (National Science Foundation, 2016).

As advancements in nanotechnology research and development continue, multidisciplinary collaboration will be essential for translating these innovations into practical applications. Over the next decade, this collective effort will ensure that nanotechnology achieves its full potential and delivers significant societal benefits. Guided by the 1999 vision of nanotechnology, the National Nanotechnology Initiative (NNI) and other global R&D programs have laid the groundwork for future breakthroughs while fostering a comprehensive understanding of nanoscale research. As the field progresses, equipping the next generation of leaders with the necessary expertise to maximize nanotechnology's potential is crucial, promoting scientific exploration, critical thinking, and creativity through educational integration (Roco, 2011). By aligning these initiatives across various industries and academic disciplines, nanotechnology's transformative impact can be fully harnessed, paving the way for advancements in computing, sustainability, and healthcare (Tawiah et al., 2024).

References

Abdin, Z., et al. (2013). Solar energy harvesting with the application of nanotechnology. *Renewable and Sustainable Energy Reviews, 26*, 837–852. https://doi.org/10.1016/j.rser.2013.06.023

Ai, Z., Gao, Z., Zhang, L., He, W., & Yin, J. J. (2013). Core–shell structure dependent reactivity of Fe@Fe2O3 nanowires on aerobic degradation of 4-chlorophenol. *Environmental Science & Technology, 47*(10), 5344–5352. https://doi.org/10.1021/es4005202

Alves, H. J., Bley Junior, C., Niklevicz, R. R., Frigo, E. P., Frigo, M. S., & Coimbra-Araújo, C. H. (2013). Overview of hydrogen production technologies from biogas and the applications in fuel cells. *International Journal of Hydrogen Energy, 38*(13), 5215–5225. https://doi.org/10.1016/j.ijhydene.2013.02.057

Aneke, M., & Wang, M. (2016). Energy storage technologies and real life applications—A state of the art review. *Applied Energy, 179*, 350–377. https://doi.org/10.1016/j.apenergy.2016.06.097

Antonio Di Stefano, G., d'Annunzio, C.-P., & Via deiVestin. (2023). Nanotechnology in targeted drug delivery. *International Journal of Molecular Sciences, 24*, 8194. https://doi.org/10.3390/ijms24098194

ASTM. (2025). *Standard guide for nanotechnology workforce education in material properties and effects of size*. ASTM. Retrieved February 2, 2025, from https://www.astm.org/e3089-22.html

Atherton, W. A. (1984). Miniaturization of electronics. In W. A. Atherton (Ed.), *From compass to computer: A history of electrical and electronics engineering* (pp. 237–267). Macmillan Education. https://doi.org/10.1007/978-1-349-17365-5_10

Attene-Ramos, M. S., Austin, C. P., & Xia, M. (2014). High throughput screening. In *Encyclopedia of toxicology* (Vol. 2). Elsevier. https://doi.org/10.1016/B978-0-12-386454-3.00209-8

Baig, N. (2023). Two-dimensional nanomaterials: A critical review of recent progress, properties, applications, and future directions. *Composites Part A: Applied Science and Manufacturing, 165*, 107362.

Barbosa, A. I., Rebelo, R., Reis, R. L., Bhattacharya, M., & Correlo, V. M. (2021). Current nanotechnology advances in diagnostic biosensors. *Medical Devices & Sensors, 4*, e10156.

Bhat, H. A., Khanday, F. A., Kaushik, B. K., Bashir, F., & Shah, K. A. (2022). Quantum computing: Fundamentals, implementations and applications. *IEEE Open Journal of Nanotechnology, 3*, 61–77. https://doi.org/10.1109/OJNANO.2022.3178545

Birla, S. S., Tiwari, V. V., Gade, A. K., Ingle, A. P., Yadav, A. P., & Rai, M. K. (2009). Fabrication of silver nanoparticles by Phomaglomerata and its combined effect against *Escherichia coli, Pseudomonas aeruginosa* and *Staphylococcus aureus*. *Letters in Applied Microbiology, 48*(2), 173–179. https://doi.org/10.1111/j.1472-765X.2008.02510.x

Chen, B., et al. (2018). Three-dimensional ultraflexibletriboelectricnanogenerator made by 3D printing. *Nano Energy, 45*, 380–389. https://doi.org/10.1016/j.nanoen.2017.12.049

Chimene, D., Alge, D. L., & Gaharwar, A. K. (2015). Two-dimensional nanomaterials for biomedical applications: Emerging trends and future prospects. *Advanced Materials, 27*(45), 2422. https://doi.org/10.1002/adma.201502422

Dhawan, A., Yan, D., Batchelor, D., Wang, H.-N., & Vo-Dinh, T. (2011). Hybrid top-down and bottom-up fabrication approach for wafer-scale plasmonic nanoplatforms. *Small, 7*(6), 727–731. https://doi.org/10.1002/smll.201002186

Edwards, P. P., Kuznetsov, V. L., David, W. I. F., & Brandon, N. P. (2008). Hydrogen and fuel cells: Towards a sustainable energy future. *Energy Policy, 36*(12), 4356–4362. https://doi.org/10.1016/j.enpol.2008.09.036

Elsaesser, A., & Howard, C. V. (2012). Toxicology of nanoparticles. *Advanced Drug Delivery Reviews, 64*(2), 129–137. https://doi.org/10.1016/j.addr.2011.09.001

Elsaesser, A., & Howard, C. V. (2025). Nanoparticle toxicology - An overview. *Nanoparticle Toxicology*. Retrieved from https://www.sciencedirect.com/topics/pharmacology-toxicology-and-pharmaceutical-science/nanoparticle-toxicology

Fan, X., et al. (2020). Flexible and wearable power sources for next-generation wearable electronics. *Batteries & Supercaps, 3*(12), 1262–1274. https://doi.org/10.1002/batt.202000115

FDA. (2019). *FDA's approach to regulation of nanotechnology products*. FDA. Retrieved February 2, 2025, from https://www.fda.gov/science-research/nanotechnology-programs-fda/fdas-approach-regulation-nanotechnology-products

Forloni, G. (2012). Responsible nanotechnology development. *Journal of Nanoparticle Research, 14*, 1007. https://doi.org/10.1007/s11051-012-1007-1

Forster, S. P., Olveira, S., & Seeger, S. (2011). Nanotechnology in the market: Promises and realities. *International Journal of Nanotechnology, 8*(6–7), 592–613. https://doi.org/10.1504/IJNT.2011.040193

Funabashi, T. (2016). Chapter 1 - Introduction. In T. Funabashi (Ed.), *Integration of distributed energy resources in power systems* (pp. 1–14). Academic Press. https://doi.org/10.1016/B978-0-12-803212-1.00001-5

Ganesh Babu, M. M., & Gunasekaran, P. (2009). Production and structural characterization of crystalline silver nanoparticles from Bacillus cereus isolate. *Colloids and Surfaces. B, Biointerfaces, 74*(1), 191–195. https://doi.org/10.1016/j.colsurfb.2009.07.016

IonelaVrejoiu, M. A., Hesse, D., & Gösele, U. (2008). Functional perovskites—From epitaxial films to nanostructured arrays. *Advanced Functional Materials, 18*(24), 3892–3906. https://doi.org/10.1002/adfm.200800560

Islam, N., & Vashist, S. K. (2014). *Nanotechnology: Recent trends, emerging issues and future directions*. Nova Science Publishers.

Karakoti, A. S., Shukla, R., Shanker, R., & Singh, S. (2015). Surface functionalization of quantum dots for biological applications. *Advances in Colloid and Interface Science, 215*, 28–45. https://doi.org/10.1016/j.cis.2014.11.004

Kasálková, N. S., Slepicka, P., & Švorcík, V. (2021). Carbon nanostructures, nanolayers, and their composites. *Nanomaterials, 11*, 2368. https://doi.org/10.3390/nano11092368

Keyes, R. W. (1988). Miniaturization of electronics and its limits. *IBM Journal of Research and Development, 32*(1), 84–88. https://doi.org/10.1147/rd.321.0024

Khan, S. H. (2020). Green nanotechnology for the environment and sustainable development. In M. Naushad & E. Lichtfouse (Eds.), *Green materials for wastewater treatment* (pp. 13–46). Springer. https://doi.org/10.1007/978-3-030-17724-9_2

Khan, W. S., Asmatulu, E., & Asmatulu, R. (2025). Chapter 1: Nanotechnology emerging trends, markets and concerns. In R. Asmatulu, W. S. Khan, & E. Asmatulu (Eds.), *Nanotechnology safety* (2nd ed., pp. 1–21). Elsevier. https://doi.org/10.1016/B978-0-443-15904-6.00017-4

Kim, H., Lee, H.-B.-R., & Maeng, W.-J. (2009). *Applications of atomic layer deposition to nanofabrication and emerging nanodevice* (pp. 790–784). Pohang University of Science and Technology (POSTECH).

Koo, O. M., Rubinstein, I., & Onyuksel, H. (2005). Role of nanotechnology in targeted drug delivery and imaging: A concise review. *Nanomedicine: Nanotechnology, Biology, and Medicine, 1*, 193–212.

Lee, H. E., et al. (2018). Novel electronics for flexible and neuromorphic computing. *Advanced Functional Materials, 28*(32), 1801690. https://doi.org/10.1002/adfm.201801690

Lens, P. N. L., & Uddandarao, P. (2023). *Nanobiohybrids for advanced wastewater treatment and energy recovery*. IWA Publishing. https://doi.org/10.2166/9781789063592

Li, H., Guo, P., Wang, H., Guo, Y., Wei, Y., Cao, F., Li, L., & Wang, H. (2023). Recent advances on nanocrystals embedding for high performance perovskite solar cells. *Advanced Functional Materials, 33*(8), 2210885.

Liu, D., Ni, K., Ye, J., Xie, J., Zhu, Y., & Song, L. (2018). Tailoring the structure of carbon nanomaterials toward high-end energy applications. *Advanced Materials, 30*, 1802104.

Malik, S., Muhammad, K., & Waheed, Y. (2023). Nanotechnology: A revolution in modern industry. *Molecules, 28*(2), 2. https://doi.org/10.3390/molecules28020661

Mayr, L. M., & Bojanic, D. (2009). Novel trends in high-throughput screening. *Current Opinion in Pharmacology, 9*, 580–588.

Meng, K., Niu, Y., Zhao, X., Zhang, Y., Zhao, Y., Xiaohua, Y., Rong, J., & Hou, H. (2023). Data-driven design of high-performance graphene-based seawater desalination membranes. *ACS Applied Nano Materials, 6*(7), 5889–5900.

Meyer, M., & Persson, O. (1998). Nanotechnology-interdisciplinarity, patterns of collaboration and differences in application. *Scientometrics, 42*(2), 195–205. https://doi.org/10.1007/BF02458355

Nagraik, R., Sharma, A., Kumar, D., Mukherjee, S., FatihSen, A. P., & Kumar. (2021). Amalgamation of biosensors and nanotechnology in disease diagnosis: Mini-review. *Sensors International, 2*, 100089.

Naik, G. G., & Jagtap, V. A. (2024). Two heads are better than one: Unravelling the potential impact of artificial intelligence in nanotechnology. *Nano TransMed, 3*, 100041.

National Academies of Sciences, Engineering, and Medicine. (2019). *Computing, quantum computing: Progress and prospects*. National Academies Press.

National Science Foundation. (2016). A federal vision for future computing: A nanotechnology-inspired grand challenge.

Nguyen, B. H., & Nguyen, V. H. (2016). Promising applications of graphene and graphene-based nanostructure. *Advances in Natural Sciences: Nanoscience and Nanotechnology, 7*, 023002.

Patil, M., Mehta, D. S., & Guvva, S. (2008). Future impact of nanotechnology on medicine and dentistry. *Journal of Indian Society of Periodontology, 12*(2), 34–40. https://doi.org/10.4103/0972-124X.44088

Perera, A. A. P. R., Madhushani, K. A. U., Punchihewa, B. T., Kumar, A., & Gupta, R. K. (2023). MXene-based nanomaterials for multifunctional applications. *Materials, 16*, 1138. https://doi.org/10.3390/ma16031113

Pokrajac, L., Abbas, A., Chrzanowski, W., Dias, G. M., Eggleton, B. J., Maguire, S., Maine, E., Malloy, T., Nathwani, J., Nazar, L., & Sips, A. (2021). Nanotechnology for a sustainable future: Addressing global challenges with the international network4sustainable nanotechnology. *ACS Nano, 15*(12), 18608–18623.

Pollice, R., dos Passos, G., Gomes, M. A., Hickman, R. J., Krenn, M., Lavigne, C., Lindner-Daddario, M., Nigam, A. K., Ser, C. T., Yao, Z., & Aspuru-Guzik, A. (2021). Data-driven strategies for accelerated materials design. *Accounts of Chemical Research, 54*, 849–860.

Pugazhendhi, A., Prabakar, D., Jacob, J. M., Karuppusamy, I., & Saratale, R. G. (2018). Synthesis and characterization of silver nanoparticles using Gelidiumamansii and its antimicrobial property against various pathogenic bacteria. *Microbial Pathogenesis, 114*, 41–45. https://doi.org/10.1016/j.micpath.2017.11.013

Rana, I., Oh, J., JuwonBaig, J. H. M., Son, S., & Nam, J. (2023). Nanocarriers for cancer nano-immunotherapy. *Drug Delivery and Translational Research, 13*, 1936–1954.

Roco, M. C. (2011). The long view of nanotechnology development: The national nanotechnology initiative at 10 years. *Journal of Nanoparticle Research, 13*, 427–445. https://doi.org/10.1007/s11051-010-0192-z

Saif Al-Deen, H., Hassan, M. N., Almaliki, S., Hussein, Z. A., Albehadili, H. M., Banoon, S. R., Abboodi, A., & Al-Saady, M. (2023). Development of nanotechnology by artificial intelligence: A comprehensive review. *Journal of Nanostructures, 13*(4), 915–932. https://doi.org/10.22052/JNS.2023.04.002

Saratale, R. G., et al. (2018). A comprehensive review on green nanomaterials using biological systems: Recent perception and their future applications. *Colloids and Surfaces. B, Biointerfaces, 170*, 20–35. https://doi.org/10.1016/j.colsurfb.2018.05.045

Schummer, J. (2004). Multidisciplinarity, interdisciplinarity, and patterns of research collaboration in nanoscience and nanotechnology. *Scientometrics, 59*(3), 425–465. https://doi.org/10.1023/B:SCIE.0000018542.71314.38

Shea, C. M. (2005). Future management research directions in nanotechnology: A case study. *Journal of Engineering and Technology Management, 22*(3), 185–200. https://doi.org/10.1016/j.jengtecman.2005.06.002

ShiyiZuo, J. S., Zhang, J., He, Z., Sun, B., & Sun, J. (2021). Nano-immunotherapy for each stage of cancer cellular immunity: Which, why, and what? *Theranostics, 11*(15), 59953. https://doi.org/10.7150/thno.59953

Shockley, W., & Queisser, H. J. (1961). Detailed balance limit of efficiency of p-n junction solar cells. *Journal of Applied Physics, 32*(3), 510–519. https://doi.org/10.1063/1.1736034

Shukla, S. K., & Bahar, R. I. (2004). *Nano, quantum and molecular computing*. Kluwer Academic Publishers. https://doi.org/10.1007/b116438

Sivaraj, A., Kumar, V., Sunder, R., Parthasarathy, K., & Kasivelu, G. (2020). Commercial yeast extracts mediated green synthesis of silver chloride nanoparticles and their anti-mycobacterial activity. *Journal of Cluster Science, 31*(1), 287–291. https://doi.org/10.1007/s10876-019-01626-4

Tawiah, B., Seidu, R. K., Fobiri, G. K., & Chichi, C. A. (2024). Nanotechnology and education: Preparing the next generation of future leaders. In S. C. George & B. Tawiah (Eds.), *Nanotechnology in societal development* (pp. 247–270). Springer. https://doi.org/10.1007/978-981-97-6184-5_7

Wasti, S., Lee, I. H., Kim, S., Lee, J.-H., & Kim, H. (2023). Ethical and legal challenges in nanomedical innovations: A scoping review. *Frontiers in Genetics, 14.* https://doi.org/10.3389/fgene.2023.1163392

Werner, J. H. (2004). Second and third generation photovoltaics—Dreams and reality. In B. Kramer (Ed.), *Advances in solid state physics* (pp. 51–66). Springer. https://doi.org/10.1007/978-3-540-39970-4_5

Winter, M., & Brodd, R. J. (2004). What are batteries, fuel cells, and supercapacitors? *Chemical Reviews, 104*(10), 4245–4270. https://doi.org/10.1021/cr020730k

Yang, R. X., McCandler, C. A., Oxana, A. M., Rachel, S., Woods-Robinson, M. K., Horton Kristin, A., & Persson. (2022). Big data in a nano world: A review on computational, data-driven design of nanomaterials structures, properties, and synthesis. *ACS Nano, 16*(12), 19873–19891.

Zhang, J., Li, Y., Cao, K., & Chen, R. (2022). Advances in atomic layer deposition. *Nanomanufacturing and Metrology, 5*, 191–208. https://doi.org/10.1007/s41871-022-00136-8

Index

S. Kulkarni (ed.), *Next-Generation Nanomaterials for Sustainable Engineering*, Lecture Notes in Nanoscale Science and Technology 43,
https://doi.org/10.1007/978-3-032-05719-8